Recent Advancements in Space Flight

Recent Advancements in Space Flight

Edited by Philips Wright

STATES
ACADEMIC PRESS
www.statesacademicpress.com

States Academic Press,
109 South 5th Street,
Brooklyn, NY 11249, USA

ISBN: 978-1-63989-458-1

Cataloging-in-Publication Data

Recent advancements in space flight / edited by Philips Wright.
p. cm.
Includes bibliographical references and index.
ISBN 978-1-63989-458-1
1. Space flight. 2. Aeronautics--Flights. 3. Aerospace engineering.
4. Aeronautics. 5. Astronautics. I. Wright, Philips.
TL790 .R43 2022
629.4--dc23

For information on all States Academic Press publications visit our website at www.statesacademicpress.com

STATES
ACADEMIC PRESS

Contents

Preface

Space flight refers to flying spacecraft into or through outer space, either with or without humans on board. It is an application of astronautics. Uncrewed spaceflights such as space probes and satellites are operated either by telerobotic control or are fully autonomous. Space flights are used for space exploration, space observatories, intelligence satellites and other Earth observation satellites. It is also used for commercial activities like satellite telecommunications and space tourism. Various types of launch systems, like rocket launching, are used to achieve spaceflight. The ever growing need of advanced technology is the reason that has fueled the research in the field of space flight in recent times. This book is a valuable compilation of topics, ranging from the basic to the most complex advancements in this field. It aims to equip students and experts with the advanced topics and upcoming concepts in this area.

This book unites the global concepts and researches in an organized manner for a comprehensive understanding of the subject. It is a ripe text for all researchers, students, scientists or anyone else who is interested in acquiring a better knowledge of this dynamic field.

I extend my sincere thanks to the contributors for such eloquent research chapters. Finally, I thank my family for being a source of support and help.

Editor

1

On Six DOF Relative Orbital Motion of Satellites

Daniel Condurache

Abstract

In this chapter, we reveal a dual-tensor-based procedure to obtain exact expressions for the six degree of freedom (6-DOF) relative orbital law of motion in the specific case of two Keplerian confocal orbits. The result is achieved by pure analytical methods in the general case of any leader and deputy motion, without singularities or implying any secular terms. Orthogonal dual tensors play a very important role, with the representation of the solution being, to the authors' knowledge, the shortest approach for describing the complete onboard solution of the 6-DOF orbital motion problem. The solution does not depend on the local-vertical–local-horizontal (LVLH) properties involves that is true in any reference frame of the leader with the origin in its mass center. A representation theorem is provided for the full-body initial value problem. Furthermore, the representation theorems for rotation part and translation part of the relative motion are obtained.

Keywords: relative orbital motion, full body problem, dual algebra, Lie group, Lie algebra, closed form solution

1. Introduction

The relative motion between the leader and the deputy in the relative motion is a six-degrees-of-freedom (6-DOF) motion engendered by the joining of the relative translational motion with the rotational one. Recently, the modeling of the 6-DOF motion of spacecraft gained a special attention [1–5], similar to the controlling the relative pose of satellite formation that became a very important research subject [6–10]. The approach implies to consider the relative translational and rotational dynamics in the case of chief-deputy spacecraft formation to be modeled using vector and tensor formalism.

In this chapter we reveal a dual algebra tensor based procedure to obtain exact expressions for the six D.O.F relative orbital law of motion for the case of two Keplerian confocal orbits.

Orthogonal dual tensors play a very important role, the representation of the solution being, to the authors' knowledge, the shortest approach for describing the complete onboard solution of the six D.O.F relative orbital motion problem. Because the solution does not depend on the LVLH properties involves that is true in any reference frame of the Leader with the origin in its mass center. To obtain this solution, one has to know only the inertial motion of the Leader spacecraft and the initial conditions of the deputy satellite in the local-vertical-local-horizontal (LVLH) frame. For the full body initial value problem, a general representation theorem is given. More, the real and imaginary parts are split and representation theorems for the rotation and translation parts of the relative orbital motion are obtained. Regarding translation, we will prove that this problem is super-integrable by reducing it to the classic Kepler problem.

The chapter is structured as following. The second section is dedicated to the rigid body motion parameterization using orthogonal dual tensors, dual quaternions and other different vector parameterization. The Poisson-Darboux problem is extended in dual Lie algebra. In the third section, the state equations for a rigid body motion relative to an arbitrary non-inertial reference frame are determined. Using the obtained result, in the fourth section, the representation theorem and the complete solution for the case of onboard full-body relative orbital motion problem is given. The last section is designated to the conclusions and to the future works.

2. Rigid body motion parameterization using dual Lie algebra

The key notions that will be presented in this section are tensorial, vectorial and non-vectorial parameterizations that can be used to properly describe the rigid-body motion. We discuss the properties of proper orthogonal dual tensorial maps. The proper orthogonal tensorial maps are related with the skew-symmetric tensorial maps via the Darboux–Poisson equation. Orthogonal dual tensorial maps are a powerful instrument in the study of the rigid motion with respect to an inertial and noninertial reference frames. More on dual numbers, dual vectors and dual tensors can be found in [2, 16–23].

2.1. Isomorphism between Lie group of the rigid displacements $S\mathbb{E}_3$ and Lie group of the orthogonal dual tensors $\underline{S\mathbb{O}}_3$

Let the orthogonal dual tensor set be denoted by.

$$\underline{S\mathbb{O}}_3 = \{\underline{\boldsymbol{R}} \in \mathrm{L}(\underline{\mathbf{V}}_3, \underline{\mathbf{V}}_3) | \underline{\boldsymbol{R}}\underline{\boldsymbol{R}}^T = \underline{\boldsymbol{I}}, \det\underline{\boldsymbol{R}} = 1\} \tag{1}$$

where $\underline{S\mathbb{O}_3}$ is the set of special orthogonal dual tensors and $\underline{\boldsymbol{I}}$ is the unit orthogonal dual tensor.

The internal structure of any orthogonal dual tensor $\underline{\boldsymbol{R}} \in \underline{S\mathbb{O}}_3$ is illustrated in a series of results which were detailed in our previous work [17, 18, 23].

Theorem 1. (Structure Theorem). *For any* $\underline{\boldsymbol{R}} \in \underline{S\mathbb{O}}_3$ *a unique decomposition is viable*

$$\underline{R} = (I + \varepsilon\widetilde{\rho})Q \tag{2}$$

where $Q \in S\mathbb{O}_3$ and $\mathbf{\rho} \in V_3$ are called **structural invariants**, $\varepsilon^2 = 0$, $\varepsilon \neq 0$.

Taking into account the Lie group structure of $\underline{S\mathbb{O}}_3$ and the result presented in previous theorem, it can be concluded that any orthogonal dual tensor $\underline{R} \in \underline{S\mathbb{O}}_3$ can be used globally parameterize displacements of rigid bodies.

Theorem 2 (Representation Theorem). *For any orthogonal dual tensor* $\underline{\mathbf{R}}$ *defined as in Eq. (2), a dual number* $\underline{\alpha} = \alpha + \varepsilon d$ *and a dual unit vector* $\underline{\mathbf{u}} = \mathbf{u} + \varepsilon \mathbf{u}_0$ *can be computed to have the following Eq.* [17, 18]:

$$\underline{R}(\underline{\alpha}, \underline{\mathbf{u}}) = I + \sin\underline{\alpha}\widetilde{\underline{u}} + (1 - \cos\underline{\alpha})\widetilde{\underline{u}}^2 = \exp\left(\underline{\alpha}\widetilde{\underline{u}}\right) \tag{3}$$

The parameters $\underline{\alpha}$ and $\underline{\mathbf{u}}$ are called the **natural invariants** of $\underline{\mathbf{R}}$. The unit dual vector $\underline{\mathbf{u}}$ gives the Plücker representation of the Mozzi-Chalses axis [16, 24] while the dual angle $\underline{\alpha} = \alpha + \varepsilon d$ contains the rotation angle α and the translated distance d.

The Lie algebra of the Lie group $\underline{S\mathbb{O}}_3$ is the skew-symmetric dual tensor set denoted by $\underline{s\mathbb{o}}_3 = \left\{\widetilde{\underline{\alpha}} \in \mathbf{L}(\underline{\mathbf{V}}_3, \underline{\mathbf{V}}_3) \middle| \widetilde{\underline{\alpha}} = -\widetilde{\underline{\alpha}}^{\mathrm{T}}\right\}$, where the internal mapping is $\langle\widetilde{\underline{\alpha}}_1, \widetilde{\underline{\alpha}}_2\rangle = \widetilde{\widetilde{\underline{\alpha}}_1\,\underline{\alpha}_2}$.

The link between the Lie algebra $\underline{s\mathbb{o}}_3$, the Lie group $\underline{S\mathbb{O}}_3$, and the exponential map is given by the following.

Theorem 3. *The mapping is well defined and surjective.*

$$\begin{gathered}\exp : \underline{s\mathbb{o}}_3 \rightarrow \underline{S\mathbb{O}}_3, \\ \exp\left(\widetilde{\underline{\alpha}}\right) = e^{\widetilde{\underline{\alpha}}} = \sum_{k=0}^{\infty} \frac{\widetilde{\underline{\alpha}}^k}{k!}\end{gathered} \tag{4}$$

Any screw axis that embeds a rigid displacement is parameterized by a unit dual vector, whereas the screw parameters (angle of rotation around the screw and the translation along the screw axis) is structured as a dual angle. The computation of the screw axis is bound to the problem of finding the logarithm of an orthogonal dual tensor $\underline{R}$, that is a multifunction defined by the following equation:

$$\begin{gathered}\log : \underline{S\mathbb{O}}_3 \rightarrow \underline{s\mathbb{o}}_3, \\ \log\underline{R} = \left\{\widetilde{\underline{\psi}} \in \underline{s\mathbb{o}}_3 \middle| \exp\left(\widetilde{\underline{\psi}}\right) = \underline{R}\right\}\end{gathered} \tag{5}$$

and is the inverse of Eq. (4).

From **Theorem 2** and **Theorem 3**, for any orthogonal dual tensor $\underline{R}$, a dual vector $\underline{\boldsymbol{\psi}} = \underline{\alpha}\underline{\mathbf{u}} = \boldsymbol{\psi} + \varepsilon\boldsymbol{\psi}_0$ is computed, represents the **screw dual vector** or **Euler dual vector** (that includes the screw axis and screw parameters) and the form of $\underline{\boldsymbol{\psi}}$ implies that $\widetilde{\underline{\psi}} \in \log\underline{\mathbf{R}}$. The types of rigid displacements that is parameterized by the Euler dual vector $\underline{\boldsymbol{\psi}}$ as below:

i. roto-translation if $\boldsymbol{\psi} \neq \mathbf{0}$, $\boldsymbol{\psi}_0 \neq \mathbf{0}$ and $\boldsymbol{\psi} \cdot \boldsymbol{\psi}_0 \neq 0 \Leftrightarrow \left|\underline{\boldsymbol{\psi}}\right| \in \underline{\mathbb{R}}$ and $\left|\boldsymbol{\psi}\right| \notin \varepsilon\mathbb{R}$;

ii. pure translation if if $\boldsymbol{\psi} = \mathbf{0}$ and $\boldsymbol{\psi}_0 \neq \mathbf{0} \Leftrightarrow \left|\underline{\boldsymbol{\psi}}\right| \in \varepsilon\underline{\mathbb{R}}$;

iii. pure rotation if $\boldsymbol{\psi}\neq\mathbf{0}$ and $\boldsymbol{\psi}\cdot\boldsymbol{\psi}_0=0 \Leftrightarrow \left|\underline{\boldsymbol{\psi}}\right| \in \mathbb{R}$.

Also, $\|\boldsymbol{\psi}\| < 2\pi$, **Theorem 2** and **Theorem 3** can be used to uniquely recover the screw dual vector $\underline{\boldsymbol{\psi}}$, which is equivalent with computing $\log \underline{\boldsymbol{R}}$.

Theorem 4. *The natural invariants* $\underline{\alpha} = \alpha + \varepsilon d$, $\underline{\mathbf{u}} = \mathbf{u}+\varepsilon\mathbf{u}_0$ *can be used to directly recover the structural invariants* $\boldsymbol{Q}$ *and* $\boldsymbol{\rho}$ *from Eq. (2):*

$$\begin{aligned} \boldsymbol{Q} &= \boldsymbol{I} + \sin\alpha\tilde{\boldsymbol{u}} + (1 - \cos\alpha)\tilde{\boldsymbol{u}}^2 \\ \boldsymbol{\rho} &= d\mathbf{u} + \sin\alpha\mathbf{u}_0 + (1 - \cos\alpha)\mathbf{u}\times\mathbf{u}_0 \end{aligned} \tag{6}$$

To prove Eq. (6), we need to use Eqs. (2) and (3). If these equations are equal, then the structure of their dual parts leads to the result presented in Eq. (6).

Theorem 5. (Isomorphism Theorem): *The special Euclidean group* $(\mathbb{SE}_3,\cdot)$ *and* $\left(\underline{\mathbb{SO}}_3,\cdot\right)$ *are connected via the isomorphism of the Lie groups*

$$\begin{aligned} &\Phi : \mathbb{SE}_3 \to \underline{\mathbb{SO}}_3, \\ &\Phi(g) = (\boldsymbol{I} + \varepsilon\tilde{\boldsymbol{\rho}})\boldsymbol{Q} \end{aligned} \tag{7}$$

where $g = \begin{bmatrix} \boldsymbol{Q} & \boldsymbol{\rho} \\ 0 & 1 \end{bmatrix}$, $\Phi \in \mathbb{SO}_3$, $\boldsymbol{\rho} \in V_3$.

Proof. For any $g_1, g_2 \in \mathbb{SE}_3$, the map defined in Eq. (7) yields

$$\Phi(g_1\cdot g_2) = \Phi(g_1)\cdot\Phi(g_2) \tag{8}$$

Let $\underline{\boldsymbol{R}} \in \underline{\mathbb{SO}}_3$. Based on **Theorem 1**, which ensures a unique decomposition, we can conclude that the only choice for g, such that $\Phi(g) = \underline{\boldsymbol{R}}$ is $g = \begin{bmatrix} \boldsymbol{Q} & \boldsymbol{\rho} \\ 0 & 1 \end{bmatrix}$. This underlines that Φ is a bijection and keeps all the internal operations.

Remark 1: The inverse of Φ is

$$\Phi^{-1} : \underline{\mathbb{SO}}_3 \leftrightarrow \mathbb{SE}_3; \Phi^{-1}(\underline{\boldsymbol{R}}) = \begin{bmatrix} \boldsymbol{Q} & \boldsymbol{\rho} \\ 0 & 1 \end{bmatrix} \tag{9}$$

where $\boldsymbol{Q} = \mathrm{Re}(\underline{\boldsymbol{R}})$, $\boldsymbol{\rho} = vect\left(Du(\underline{\boldsymbol{R}})\cdot\boldsymbol{Q}^{\mathrm{T}}\right)$.

2.2. Dual tensor-based parameterizations of rigid-body motion

The Lie group $\underline{\mathbb{SO}}_3$ admits multiple parameterization and few of them will be discussed in this section.

2.2.1. The exponential parameterization (the Euler dual vector parameterization)

If $\underline{\boldsymbol{R}} = \underline{\boldsymbol{R}}(\underline{\alpha}, \underline{\mathbf{u}})$, then we can construct the **Euler dual vector** (screw dual vector) $\underline{\boldsymbol{\psi}} = \underline{\alpha}\underline{\mathbf{u}}$, $\underline{\boldsymbol{\psi}} \in \underline{\mathbf{V}}_3$ which combined with **Theorem 2** and **Theorem 3** lead to

$$\underline{\boldsymbol{R}} = \exp\left(\underline{\tilde{\boldsymbol{\psi}}}\right) = \underline{\boldsymbol{I}} + \operatorname{sinc}\left|\underline{\boldsymbol{\psi}}\right|\underline{\tilde{\boldsymbol{\psi}}} + \frac{1}{2}\operatorname{sinc}^2\frac{\left|\underline{\boldsymbol{\psi}}\right|}{2}\underline{\tilde{\boldsymbol{\psi}}}^2 \tag{10}$$

where

$$\operatorname{sinc}(|\underline{\mathrm{x}}|) = \begin{cases} \dfrac{\sin|\underline{\mathrm{x}}|}{|\underline{\mathrm{x}}|}, & |\underline{\mathrm{x}}| \notin \varepsilon\mathbb{R} \\ 1, & |\underline{\mathrm{x}}| \in \varepsilon\mathbb{R} \end{cases} \tag{11}$$

2.2.2. Dual quaternion parameterization

One of the most important non-vectorial parameterizations for the orthogonal dual tensor $\underline{\mathbb{SO}}_3$ is given by the dual quaternions [20, 21]. A dual quaternion can be defined as an associated pair of a dual scalar quantity and a free dual vector:

$$\underline{\hat{\mathbf{q}}} = \left(\underline{q}, \underline{\mathbf{q}}\right), \underline{q} \in \underline{\mathbb{R}}, \underline{\mathbf{q}} \in \underline{\mathbf{V}}_3 \tag{12}$$

The set of dual quaternions will be denoted $\underline{\mathbf{Q}}$ and is organized as a $\underline{\mathbb{R}}$-module of rank 4, if dual quaternion addition and multiplication with dual numbers are considered.

The product of two dual quaternions $\underline{\hat{\mathbf{q}}}_1 = \left(\underline{q}_1, \underline{\mathbf{q}}_1\right)$ and $\underline{\hat{\mathbf{q}}}_2 = \left(\underline{q}_2, \underline{\mathbf{q}}_2\right)$ is defined by

$$\underline{\hat{\mathbf{q}}}_1\underline{\hat{\mathbf{q}}}_2 = \left(\underline{q}_1 \cdot \underline{q}_2 - \underline{\mathbf{q}}_1 \cdot \underline{\mathbf{q}}_2, \underline{q}_1\underline{\mathbf{q}}_2 + \underline{q}_2\underline{\mathbf{q}}_1 + \underline{\mathbf{q}}_1 \times \underline{\mathbf{q}}_2\right) \tag{13}$$

From the above properties, results that the $\underline{\mathbb{R}}$-module $\underline{\mathbf{Q}}$ becomes an associative, non-commutative linear dual algebra of rank 4 over the ring of dual numbers. For any dual quaternion defined by Eq. (12), the conjugate denoted by $\underline{\hat{\mathbf{q}}}^* = \left(\underline{q}, -\underline{\mathbf{q}}\right)$ and the norm denoted by $\left|\underline{\hat{\mathbf{q}}}\right|^2 = \underline{\hat{\mathbf{q}}}\underline{\hat{\mathbf{q}}}^*$ can be computed. For $\left|\underline{\hat{\mathbf{q}}}\right| = 1$, any dual quaternion is called unit dual quaternion. Regarded solely as a free $\underline{\mathbb{R}}$-module, $\underline{\mathbf{Q}}$ contains two remarkable sub-modules: $\underline{\mathbf{Q}}_{\underline{\mathbb{R}}}$ and $\underline{\mathbf{Q}}_{\underline{V}_3}$. The first one composed from pairs $\left(\underline{q}, \underline{\mathbf{0}}\right), \underline{q} \in \underline{\mathbb{R}}$, isomorphic with $\underline{\mathbb{R}}$, and the second one, containing the pairs $\left(\underline{0}, \underline{\mathbf{q}}\right), \underline{\mathbf{q}} \in \underline{V}_3$, isomorphic with $\underline{V}_3$. Also, any dual quaternion can be written as $\underline{\hat{\mathbf{q}}} = \underline{q} + \underline{\mathbf{q}}$, where $\underline{q} = \left(\underline{q}, \underline{0}\right)$ and $\underline{\mathbf{q}} = \left(\underline{0}, \underline{\mathbf{q}}\right)$, or $\underline{\hat{\mathbf{q}}} = \hat{\mathbf{q}} + \varepsilon\hat{\mathbf{q}}_0$, where $\hat{\mathbf{q}}, \hat{\mathbf{q}}_0$ are real quaternions. The scalar and the vector parts of a dual unit quaternion are also known as **dual Euler parameters** [19].

Let denote with $\mathbb{U}$ the set of unit quaternions and with $\underline{\mathbb{U}}$ the set of unit dual quaternions. For any $\underline{\hat{\mathbf{q}}} \in \underline{\mathbb{U}}$, the following equation is valid [17, 20]:

$$\underline{\widehat{\mathbf{q}}} = \left(1 + \varepsilon \frac{1}{2}\widehat{\boldsymbol{\rho}}\right)\widehat{\mathbf{q}} \tag{14}$$

where $\boldsymbol{\rho} \in \mathbf{V}_3$ and $\widehat{\mathbf{q}} \in \mathbb{U}$. This representation is the quaternionic counterpart to Eq. (2). Also a dual number $\underline{\alpha}$ and a unit dual vector $\underline{\mathbf{u}}$ exist so that:

$$\underline{\widehat{\mathbf{q}}} = \cos\frac{\underline{\alpha}}{2} + \underline{\mathbf{u}}\sin\frac{\underline{\alpha}}{2} = \exp\left(\frac{1}{2}\underline{\alpha}\,\underline{\mathbf{u}}\right). \tag{15}$$

Remark 2: The mapping $\exp : \underline{V}_3 \to \underline{\mathbb{U}}, \underline{\widehat{\mathbf{q}}} = \exp\frac{1}{2}\underline{\boldsymbol{\Psi}}$, is well defined and surjective.

Remark 3: The dual unit quaternions set $\underline{\mathbb{U}}$, by the multiplication of dual quaternions, is a Lie group with $\underline{\mathbf{V}}_3$ being it's associated Lie algebra (with the cross product between dual vectors as the internal operation).

Using the internal structure of any element from $\underline{S\mathbb{O}_3}$ the following theorem is valid:

Theorem 6. *The Lie groups $\underline{\mathbb{U}}$ and $\underline{S\mathbb{O}_3}$ are linked by a surjective homomorphism*

$$\Delta : \underline{\mathbb{U}} \to \underline{S\mathbb{O}_3}, \Delta\left(\underline{q} + \underline{\mathbf{q}}\right) = \underline{\boldsymbol{I}} + 2\underline{q}\underline{\widetilde{q}} + 2\underline{\widetilde{q}}^2 \tag{16}$$

Proof. Taking into account that any $\underline{\widehat{\mathbf{q}}} \in \underline{\mathbb{U}}$ can be decomposed as in Eq. (15), results that $\Delta\left(\underline{\widehat{\mathbf{q}}}\right) = \exp\left(\underline{\alpha}\underline{\widetilde{u}}\right) \in \underline{S\mathbb{O}_3}$. This shows that relation Eq. (16) is well defined and surjective. Using direct calculus, we can also acknowledge that $\Delta\left(\underline{\widehat{\mathbf{q}}}_2\underline{\widehat{\mathbf{q}}}_1\right) = \Delta\left(\underline{\widehat{\mathbf{q}}}_2\right)\Delta\left(\underline{\widehat{\mathbf{q}}}_1\right)$.

An important property of the previous homomorphism is that for $\underline{\widehat{\mathbf{q}}}$ and $-\underline{\widehat{\mathbf{q}}}$ we can associate the same orthogonal dual tensor, which shows that Eq. (16) is not injective and $\underline{\mathbb{U}}$ is a double cover of $\underline{S\mathbb{O}_3}$.

2.2.3. N-order modified fractional Cayley transform for dual vectors

Next, we present a series of results that are the core of our research. These results are obtained after using a set of Cayley transforms that are different than the ones already reported in literature [17, 25–27].

Theorem 7. *The fractional order Cayley map $f : \underline{V}_3 \to \underline{\mathbb{U}}$*

$$cay_{\frac{n}{2}}(\underline{\mathbf{v}}) = f(\underline{\mathbf{v}}) = (1 + \underline{\mathbf{v}})^{\frac{n}{2}}(1 - \underline{\mathbf{v}})^{-\frac{n}{2}}, n \in \mathbb{N}^* \tag{17}$$

is well defined and surjective.

Proof. Using direct calculus results that $f(\underline{\mathbf{v}})f^*(\underline{\mathbf{v}}) = 1$ and $|f(\underline{\mathbf{v}})| = 1$. The surjectivity is proved by the following theorem.

Theorem 8. *The inverse of the previous fractional order Cayley map, is a multifunction with n branches* $f^{-1}: \underline{\mathbb{U}} \rightarrow \underline{V}_3$ given by

$$\underline{\mathbf{v}} = \frac{\sqrt[n]{\widehat{\underline{\mathbf{q}}}^2} - 1}{\sqrt[n]{\widehat{\underline{\mathbf{q}}}^2} + 1}. \tag{18}$$

Remark 4: *If* $|\underline{\mathbf{v}}| \in \mathbb{R}$ *then* $cay_{\frac{n}{2}}(\underline{\mathbf{v}})$ *is the parameterization of a pure rotation about an axis which does not necessarily pass through the origin of reference system. Meanwhile, if* $|\underline{\mathbf{v}}| \in \varepsilon\mathbb{R}$ *the mapping* $cay_{\frac{n}{2}}(\underline{\mathbf{v}})$ *is the parameterization of a pure translation. Otherwise,* $cay_{\frac{n}{2}}(\underline{\mathbf{v}})$ *is the parameterization of roto-translation.*

Taking into account that a dual number $\underline{\alpha}$ and a dual vector $\underline{\mathbf{u}}$ exist in order to have

$$\widehat{\underline{\mathbf{q}}} = \cos\frac{\underline{\alpha}}{2} + \underline{\mathbf{u}}\sin\frac{\underline{\alpha}}{2}, \tag{19}$$

from Eq. (18), results that:

$$\underline{\mathbf{v}} = \tan\frac{\underline{\alpha} + 2k\pi}{2n}\underline{\mathbf{u}}, k = \{0, 1, ..., n-1\}. \tag{20}$$

The previous equation contains both the principal parameterization $\underline{\mathbf{v}}_0 = \tan\frac{\underline{\alpha}}{2n}\underline{\mathbf{u}}$, which is the **higher order Rodrigues dual vector,** while for $k = \{1, ..., n-1\}$ the dual vectors $\underline{\mathbf{v}}_k = \tan\frac{\underline{\alpha}+2k\pi}{2n}\underline{\mathbf{u}}$ are the **shadow parameterization** [25] that can be used to describe the same pose. Based on $|\underline{\mathbf{v}}_0| = \tan\frac{\underline{\alpha}}{2n}$ and $|\underline{\mathbf{v}}_k| = \tan\frac{\underline{\alpha}+2k\pi}{2n}$, results that $|\underline{\mathbf{v}}_k| = \frac{|\underline{\mathbf{v}}_0| + \tan\frac{k\pi}{n}}{1 - |\underline{\mathbf{v}}_0|\tan\frac{k\pi}{n}}$.

If $\mathrm{Re}(|\underline{\mathbf{v}}_0|) \rightarrow \infty$ then $\mathrm{Re}(|\underline{\mathbf{v}}_k|) \rightarrow -\cot\frac{k\pi}{n}$, which allows the avoidance of any singularity of type $\mathrm{Re}(\frac{\underline{\alpha}}{2n}) = \frac{\pi}{2} + \pi\mathbb{Z}$.

Theorem 9. *If* $\underline{\mathbf{v}} \in \underline{\mathbf{V}}_3$ *is the parameterization of a displacement obtained from Eq. (20), then*

$$\pm\widehat{\underline{\mathbf{q}}} = \frac{1}{\sqrt{\left(1 + |\underline{\mathbf{v}}|^2\right)^n}}\left[p_n(|\underline{\mathbf{v}}|) + q_n(|\underline{\mathbf{v}}|)\underline{\mathbf{v}}\right] \tag{21}$$

where

$$p_n(X) = \sum_{k=0}^{[n/2]} (-1)^k \binom{2k}{n} X^{2k} \tag{22}$$

$$q_n(X) = \sum_{k=0}^{[(n-1)/2]} (-1)^k \binom{2k+1}{n} X^{2k} \tag{23}$$

In Eqs. (22) and (23), $[.]$ represents the floor of a number and $\binom{k}{n}$ are binomial coefficients.

Remark 5. The structure of the polynomials $p_n(X)$ and $q_n(X)$, given by Eqs. (22) and (23), can be used to obtain the following iterative expressions:

$$\begin{aligned} p_{n+1}(X) &= p_n(X) - X^2 q_n(X) \\ q_{n+1}(X) &= q_n(X) + q_n(X) \\ p_1(X) &= 1, q_1(X) = 1. \end{aligned} \tag{24}$$

In order to evaluate the usefulness of the iterative expressions, we provide the second to third order polynomials and the resulting dual quaternions and dual orthogonal tensors:

$$\begin{aligned} &p_1(X) = 1; q_1(X) = 1; \underline{\mathbf{v}} = \tan\frac{\alpha}{2}\underline{\mathbf{u}}; \\ &\pm\underline{\hat{\mathbf{q}}} = \frac{1}{\sqrt{1+|\underline{\mathbf{v}}|^2}}[1+\underline{\mathbf{v}}]; \\ &\underline{\mathbf{R}} = \underline{\mathbf{I}} + \frac{2}{1+|\underline{\mathbf{v}}|^2}\left[\underline{\tilde{v}} + \underline{\tilde{v}}^2\right]; \end{aligned} \tag{25}$$

$$\begin{aligned} &p_2(X) = 1 - X^2; q_2 = 2; \underline{\mathbf{v}} = \tan\frac{\alpha+2k\pi}{4}\underline{\mathbf{u}}; k = \overline{0,1}; \\ &\pm\underline{\hat{\mathbf{q}}} = \frac{1}{1+|\underline{\mathbf{v}}|^2}\left[1 - |\underline{\mathbf{v}}|^2 + 2\underline{\mathbf{v}}\right]; \underline{\mathbf{R}} = \underline{\mathbf{I}} + \frac{4}{\left(1+|\underline{\mathbf{v}}|^2\right)^2}\left[\left(1-|\underline{\mathbf{v}}|^2\right)\underline{\tilde{v}} + 2\underline{\tilde{v}}^2\right]; \end{aligned} \tag{26}$$

$$\begin{aligned} &p_3(X) = 1 - 3X^2; q_3 = 3 - X^2; \underline{\mathbf{v}} = \tan\frac{\alpha+2k\pi}{6}\underline{\mathbf{u}}; \\ &k = \overline{0,2}; \pm\underline{\hat{\mathbf{q}}} = \frac{1}{\sqrt{\left(1+|\underline{\mathbf{v}}|^2\right)^3}}\left[1 - 3|\underline{\mathbf{v}}|^2 + \left(3-|\underline{\mathbf{v}}|^2\right)\underline{\mathbf{v}}\right]; \\ &\underline{\mathbf{R}} = \underline{\mathbf{I}} + \frac{2\left(3-|\underline{\mathbf{v}}|^2\right)}{\left(1+|\underline{\mathbf{v}}|^2\right)^3}\left[\left(1-3|\underline{\mathbf{v}}|^2\right)\underline{\tilde{v}} + \left(3-|\underline{\mathbf{v}}|^2\right)\underline{\tilde{v}}^2\right]. \end{aligned} \tag{27}$$

2.3. Poisson-Darboux problems in dual Lie algebra and vector parameterization

Consider the functions $\boldsymbol{Q} = \boldsymbol{Q}(t) \in SO_3^{\mathbb{R}}$ and $\boldsymbol{\rho} = \boldsymbol{\rho}(t) \in V_3^{\mathbb{R}}$ to be the parametric equations of any rigid motion. Thus, any rigid motion can be parameterized by a curve in $\underline{SO}_3$ where $\underline{\boldsymbol{R}}(t) = (I + \varepsilon\tilde{\boldsymbol{\rho}}(t))\boldsymbol{Q}(t)$, where t is time variable. Let $\underline{\mathbf{h}}_0$ embed the Plücker coordinates of a line feature at $t = t_0$. At a time stamp t the line is transformed into:

$$\underline{\mathbf{h}}(t) = \underline{\boldsymbol{R}}(t)\underline{\mathbf{h}}_0 \tag{28}$$

Theorem 10. *In a general rigid motion, described by an orthogonal dual tensor function* $\underline{\mathbf{R}}$*, the velocity dual tensor function* $\underline{\boldsymbol{\Phi}}$ *defined as*

$$\dot{\underline{\mathbf{h}}} = \underline{\boldsymbol{\Phi}}\underline{\mathbf{h}}, \forall \underline{\mathbf{h}} \in \underline{V}_3 \tag{29}$$

is expressed by

$$\underline{\boldsymbol{\Phi}} = \dot{\underline{\boldsymbol{R}}}\underline{\boldsymbol{R}}^{\mathrm{T}}. \tag{30}$$

Let $\underline{\boldsymbol{\Phi}} = \dot{\underline{\boldsymbol{R}}}\underline{\boldsymbol{R}}^{\mathrm{T}}$, then $\dot{\underline{\boldsymbol{R}}}\underline{\boldsymbol{R}}^{\mathrm{T}} + \underline{\boldsymbol{R}}\dot{\underline{\boldsymbol{R}}}^{\mathrm{T}} = \underline{\mathbf{0}}$, equivalent with $\underline{\boldsymbol{\Phi}} = -\underline{\boldsymbol{\Phi}}^{\mathrm{T}}$, which shows that $\underline{\boldsymbol{\Phi}} \in \underline{so}_3^{\mathbb{R}}$.

The dual vector $\underline{\boldsymbol{\omega}} = vect\dot{\underline{\boldsymbol{R}}}\underline{\boldsymbol{R}}^{\mathrm{T}}$ is called **dual angular velocity of the rigid body** and has the form:

$$\underline{\boldsymbol{\omega}} = \boldsymbol{\omega} + \varepsilon \mathbf{v} \tag{31}$$

where ω is the instantaneous angular velocity of the rigid body and $\mathbf{v} = \dot{\boldsymbol{\rho}} - \boldsymbol{\omega} \times \boldsymbol{\rho}$ represents the linear velocity of the point of the body that coincides instantaneously with the origin of the reference frame. The pair $(\boldsymbol{\omega}, \mathbf{v})$ is usually refereed as the **twist of the rigid body**.

2.3.1. Poisson-Darboux equation in dual Lie algebra

The next Theorem permits the reconstruction of the rigid body motion knowing in any moment the twist of the rigid body that is equivalent with knowing the dual angular velocity [5, 18].

Theorem 11. *For any continuous function* $\underline{\boldsymbol{\omega}} \in \underline{\mathbf{V}}_3^{\mathbb{R}}$ *a unique dual tensor* $\underline{\mathbf{R}} \in \underline{SO}_3^{\mathbb{R}}$ *exists so that*

$$\begin{gathered} \dot{\underline{\boldsymbol{R}}} = \tilde{\underline{\boldsymbol{\omega}}}\underline{\boldsymbol{R}} \\ \underline{\boldsymbol{R}}(t_0) = \underline{\mathbf{R}}_0, \underline{\mathbf{R}}_0 \in \underline{SO}_3 \end{gathered} \tag{32}$$

Due to the fact that orthogonal dual tensor $\underline{\boldsymbol{R}}$ completely models the six degree of freedom motion, we can conclude that the **Theorem 11** is the dual form of the **Poisson-Darboux problem** [28] for the case when the rotation tensor is computed from the instantaneous angular velocity. So, in order to recover $\underline{\boldsymbol{R}}$, it is necessary to find out how the dual angular velocity vector $\underline{\boldsymbol{\omega}}$ behaves in time and also the value of $\underline{\boldsymbol{R}}$ at time $t = t_0$.

The dual tensor $\underline{\boldsymbol{R}}$ can be derived from $\underline{\boldsymbol{\omega}}$, when is positioned in space, or from $\underline{\boldsymbol{\omega}}^B$, which denotes the dual angular velocity vector to be positioned in the rigid body.

Remark 6. The dual angular velocity vector positioned in the rigid body can be recovered from $\underline{\boldsymbol{\omega}}^B = \underline{\mathbf{R}}^{\mathrm{T}}\underline{\boldsymbol{\omega}}$, thus transforming Eq. (32) into:

$$\begin{cases} \dot{\underline{\boldsymbol{R}}} = \underline{\boldsymbol{R}}\tilde{\underline{\boldsymbol{\omega}}}^B \\ \underline{\boldsymbol{R}}(t_0) = \underline{\mathbf{R}}_0, \underline{\mathbf{R}}_0 \in \underline{SO}_3 \end{cases} \tag{33}$$

Eqs. (32) and (33) represent the dual replica of the classical orientation Poisson-Darboux problem [17, 28, 29].

The tensorial Eqs. (32) and (33) are equivalent with 18 scalar differential equations. The previous parameterizations of the orthogonal dual tensors allow us to determine some solutions of smaller dimension in order to solve the dual Poisson- Darboux problem.

2.3.2. *Kinematic equation for Euler dual vector parameterization*

Consider $\underline{\boldsymbol{\Psi}} \in V_3^{\mathbb{R}}$ such that $\underline{\boldsymbol{R}} = \exp \widetilde{\underline{\boldsymbol{\Psi}}}$. According to the Eq. (10), the Poisson-Darboux problem (32) is equivalent to

$$\begin{cases} \dot{\underline{\boldsymbol{\Psi}}} = \underline{\boldsymbol{T}}\underline{\boldsymbol{\omega}} \\ \underline{\boldsymbol{\Psi}}(t_0) = \underline{\boldsymbol{\Psi}}_0 \end{cases} \tag{34}$$

where $\exp \widetilde{\underline{\boldsymbol{\Psi}}}_0 = \underline{\boldsymbol{R}}_0$, and $\underline{\boldsymbol{T}}$ is the following dual tensor:

$$\underline{\boldsymbol{T}} = \frac{|\underline{\boldsymbol{\Psi}}|}{2}\cot\frac{|\underline{\boldsymbol{\Psi}}|}{2}\underline{\boldsymbol{I}} - \frac{1}{2}\widetilde{\underline{\boldsymbol{\Psi}}} - \frac{1}{2|\underline{\boldsymbol{\Psi}}|}\cot\frac{|\underline{\boldsymbol{\Psi}}|}{2}\widetilde{\underline{\boldsymbol{\Psi}}}^2 \tag{35}$$

The representation of the Poisson-Darboux problem from Eq. (33) is equivalent to

$$\begin{cases} \dot{\underline{\boldsymbol{\Psi}}} = \underline{\boldsymbol{T}}^{\mathrm{T}}\underline{\boldsymbol{\omega}}^{\mathrm{B}} \\ \underline{\boldsymbol{\Psi}}(t_0) = \underline{\boldsymbol{\Psi}}_0 \end{cases} \tag{36}$$

2.3.3. *Kinematic equation for high order Rodrigues dual vector parameterization*

Let $\underline{\mathbf{v}} \in \underline{\mathbf{V}}_3^{\mathbb{R}}$ such that $\underline{\boldsymbol{R}} = \mathrm{cay}_n \underline{\mathbf{v}}$. The problems (32) and (33) are equivalent to:

$$\begin{cases} \dot{\underline{\mathbf{v}}} = \underline{\boldsymbol{S}}\underline{\boldsymbol{\omega}} \\ \underline{\mathbf{v}}(t_0) = \underline{\mathbf{v}}_0, \end{cases} \tag{37}$$

$$\begin{cases} \dot{\underline{\mathbf{v}}} = \underline{\boldsymbol{S}}^{\mathrm{T}}\underline{\boldsymbol{\omega}}^{B} \\ \underline{\mathbf{v}}(t_0) = \underline{\mathbf{v}}_0 \end{cases} \tag{38}$$

where $\mathrm{cay}_n \underline{\mathbf{v}}_0 = \underline{\boldsymbol{R}}_0$, and $\underline{\boldsymbol{S}}$ is the following dual tensor [29]:

$$\underline{\boldsymbol{S}} = \frac{p_n|\underline{\mathbf{v}}|}{2q_n|\underline{\mathbf{v}}|}\underline{\boldsymbol{I}} - \frac{1}{2}\widetilde{\underline{v}} + \frac{\left(1+|\mathbf{v}|^2\right)q_n\left(|\underline{\mathbf{v}}| - np_n|\underline{\mathbf{v}}|\right)}{2n|\underline{\mathbf{v}}|^2 q_n|\underline{\mathbf{v}}|}\widetilde{\underline{v}}^2 \tag{39}$$

and the polynomials p_n, q_n are given by the Eqs. (22)–(24).

Eqs. (34), (36)–(38) are equivalent with six scalar differential equations. This is a minimal parameterization of the Poisson-Darboux problem in dual algebra.

2.3.4. Kinematic equation for dual quaternion parameterization

Let $\underline{\widehat{\mathbf{q}}} \in \underline{\mathbb{U}}^{\mathbb{R}}$ such that $\Delta\left(\underline{\widehat{\mathbf{q}}}\right) = \underline{\mathbf{R}}$. According to Eq. (16), the Poisson–Darboux problems (32) and (33) are equivalent to:

$$\begin{cases} \dot{\underline{\widehat{\mathbf{q}}}} = \dfrac{1}{2}\underline{\boldsymbol{\omega}}\underline{\widehat{\mathbf{q}}} \\ \underline{\widehat{\mathbf{q}}}(t_0) = \underline{\widehat{\mathbf{q}}}_0 \end{cases} \tag{40}$$

and

$$\begin{cases} \dot{\underline{\widehat{\mathbf{q}}}} = \dfrac{1}{2}\underline{\widehat{\mathbf{q}}}\underline{\boldsymbol{\omega}}^B \\ \underline{\widehat{\mathbf{q}}}(t_0) = \underline{\widehat{\mathbf{q}}}_0 \end{cases} \tag{41}$$

where $\Delta\left(\underline{\widehat{\mathbf{q}}}_0\right) = \underline{\mathbf{R}}_0$

Eqs. (40) and (41) are equivalent to eight scalar differential equations.

3. Rigid body motion in arbitrary non-inertial frame revised

To the author's knowledge, in the field of astrodynamics there aren't many reports on how the motion of rigid body can be studied in arbitrary non-inertial frames. Next, we proposed a dual tensors based model for the motion of the rigid body in arbitrary non-inertial frame. The proposed method eludes the calculus of inertia forces that contributes to the rigid body relative state. So, the free of coordinate state equation of the rigid body motion in arbitrary non-inertial frame will be obtained.

Let $\underline{\boldsymbol{R}}_D$ and $\underline{\boldsymbol{R}}_C$ be the dual orthogonal tensors which describe the motion of two rigid bodies relative to the inertial frame.

If $\underline{\boldsymbol{R}}$ is the orthogonal dual tensor which embeds the six degree of freedom relative motion of rigid body C relative to rigid body D, then:

$$\underline{\boldsymbol{R}} = \underline{\boldsymbol{R}}_C^T \underline{\boldsymbol{R}}_D \tag{42}$$

Let $\underline{\boldsymbol{\omega}}_C$ denote the dual angular velocity of the rigid body C and $\underline{\boldsymbol{\omega}}_D$ the dual angular velocity of the rigid body D, both being related to inertial reference frame. In the followings, the inertial motion of the rigid body C is considered to be known. If $\underline{\boldsymbol{\omega}}$ is the dual angular velocity of the rigid body D relative to the rigid body C, then, conforming with Eq. (42):

$$\underline{\boldsymbol{\omega}} = \underline{\boldsymbol{\omega}}_D - \underline{\boldsymbol{\omega}}_C \tag{43}$$

Considering $\underline{\boldsymbol{\omega}}_D^B$ being the dual angular velocity vector of the rigid body D in the body frame, the dual form of the Euler equation given in [30] results that:

$$\underline{M}\underline{\dot{\omega}}_D^B + \underline{\omega}_D^B \times \underline{M}\underline{\omega}_D^B = \underline{\tau}^B \tag{44}$$

In Eq. (44) $\underline{\boldsymbol{\tau}}^B = \mathbf{F}^B + \varepsilon\boldsymbol{\tau}^B$, where $\mathbf{F}^B$ the force applied in the mass center and $\boldsymbol{\tau}^B$ is the torque. Also in Eq. (44), $\underline{M}$ represents the inertia dual operator, which is given by $\underline{M} = m_D \frac{d}{d\varepsilon}\boldsymbol{I} + \varepsilon\boldsymbol{J}$, where $\boldsymbol{J}$ is the inertia tensor of the rigid body D related to its mass center and m_D is the mass of the rigid body D. Combining $\underline{M}^{-1} = \boldsymbol{J}^{-1}\frac{d}{d\varepsilon} + \varepsilon\frac{1}{m_D}\boldsymbol{I}$ with Eq. (44) results:

$$\underline{\dot{\omega}}_D^B + \underline{M}^{-1}\left(\underline{\omega}_D^B \times \underline{M}\underline{\omega}_D^B\right) = \underline{M}^{-1}\underline{\tau}^B \tag{45}$$

Taking into account that $\underline{\omega}_D = \underline{R}\underline{\omega}_D^B$, the dual angular velocity vector can be computed from

$$\underline{\omega} = \underline{R}\underline{\omega}_D^B - \underline{\omega}_C \tag{46}$$

this through differentiation gives:

$$\underline{\dot{\omega}} + \underline{\dot{\omega}}_C = \underline{\dot{R}}\underline{\omega}_D^B + \underline{R}\underline{\dot{\omega}}_D^B \tag{47}$$

If the previous equation is multiplied by $\underline{R}^T$, then

$$\underline{R}^T(\underline{\dot{\omega}} + \underline{\dot{\omega}}_C) = \underline{R}^T\underline{\dot{R}}\underline{\omega}_D^B + \underline{\dot{\omega}}_D^B \tag{48}$$

which combined with $\underline{\dot{R}} = \underline{\tilde{\omega}}\underline{R}$ generates:

$$\underline{R}^T(\underline{\dot{\omega}} + \underline{\dot{\omega}}_C) = \underline{R}^T\underline{\tilde{\omega}}\underline{R}\underline{\omega}_D^B + \underline{\dot{\omega}}_D^B \tag{49}$$

After a few steps, Eq. (49) is transformed into

$$\underline{\dot{\omega}} + \underline{\dot{\omega}}_C = \underline{R}\underline{\dot{\omega}}_D^B + \underline{\omega} \times \underline{\omega}_C \tag{50}$$

which combined with Eq. (45) gives:

$$\underline{\dot{\omega}} + \underline{\dot{\omega}}_C = \underline{R}\underline{M}^{-1}\underline{\tau}^B - \underline{R}\underline{M}^{-1}\left(\underline{\omega}_D^B \times \underline{M}\underline{\omega}_D^B\right) + \underline{\omega} \times \underline{\omega}_C \tag{51}$$

Because $\underline{\omega}_D^B = \underline{R}^T(\underline{\omega} \times \underline{\omega}_C)$, the final equation is:

$$\underline{\dot{\omega}} + \underline{\dot{\omega}}_C = \underline{R}\underline{M}^{-1}\left[\underline{\tau}^B - \underline{R}^T(\underline{\omega} + \underline{\omega}_C) \times \underline{M}\underline{R}^T(\underline{\omega} + \underline{\omega}_C)\right] + \underline{\omega} \times \underline{\omega}_C \tag{52}$$

The system:

$$\begin{cases} \underline{\dot{R}} = \underline{\tilde{\omega}}\underline{R} \\ \underline{\dot{\omega}} + \underline{\dot{\omega}}_C = \underline{R}\underline{M}^{-1}[\underline{R}^T\underline{\tau} - \underline{R}^T(\underline{\omega} + \underline{\omega}_C) \times \\ \times \underline{M}\underline{R}^T(\underline{\omega} + \underline{\omega}_C)] + \underline{\omega} \times \underline{\omega}_C \\ \underline{\omega}(t_0) = \underline{\omega}_0, \underline{\omega}_0 \in \underline{V}_3 \\ \underline{R}(t_0) = \underline{R}_0, \underline{R}_0 \in \underline{\mathbb{SO}}_3 \end{cases} \tag{53}$$

is a compact form which can be used to model the six D.O.F relative motion problem. In the previous equation the state of the rigid body D in relation with the rigid body C is modeled by the dual tensor $\underline{\boldsymbol{R}}$ and the dual angular velocities field $\underline{\omega}$. This initial value problem can be used to study the behavior of the rigid body D in relation with the frame attached to the rigid body C. In Eq. (53), all the vectors are represented in the body frame of C, which shows that the proposed solution is onboard and has the property of being coupled in $\underline{\boldsymbol{R}}$ and $\underline{\omega}$.

Next, we present a procedure that allows the decoupling of the proposed solution.

In order to describe the solution to Eq. (53), we consider the following change of variable:

$$\underline{\omega}_* = \underline{\boldsymbol{R}}^T(\underline{\omega} + \underline{\omega}_C) \tag{54}$$

This change of variable leads to $\underline{\dot{\omega}}_* = \underline{\dot{\boldsymbol{R}}}^T(\underline{\omega} + \underline{\omega}_C) + \underline{\boldsymbol{R}}^T(\underline{\dot{\omega}} + \underline{\dot{\omega}}_C) = -\underline{\boldsymbol{R}}^T\underline{\tilde{\omega}}(\underline{\omega} + \underline{\omega}_C) + \underline{\boldsymbol{R}}^T(\underline{\dot{\omega}}$ $+\underline{\dot{\omega}}_C)$. The result is equivalent with $\underline{\dot{\omega}}_* = \underline{\boldsymbol{R}}^T(\underline{\omega}_C \times \underline{\omega} + \underline{\dot{\omega}} + \underline{\dot{\omega}}_C)$ or

$$\underline{\omega}_C \times \underline{\omega} + \underline{\dot{\omega}} + \underline{\dot{\omega}}_C = \underline{\boldsymbol{R}}\,\underline{\dot{\omega}}_* \tag{55}$$

After some steps of algebraic calculus, from Eqs. (54), (55) and (52), results that:

$$\begin{cases} \underline{\boldsymbol{M}}\,\underline{\dot{\omega}}_* + \underline{\omega}_* \times \underline{\boldsymbol{M}}\,\underline{\omega}_* = \underline{\tau}_* \\ \underline{\omega}_*(t_0) = \underline{\omega}_*^0 \end{cases} \tag{56}$$

Where $\underline{\tau}_* = \underline{\boldsymbol{R}}^T\underline{\tau}$ is the dual torque related to the mass center in the body frame of the rigid body D and $\underline{\omega}_*^0 = \underline{\boldsymbol{R}}_0^T(\underline{\omega}_0 + \underline{\omega}_C(t_0))$. Eq. (56) is a dual Euler fixed point classic problem.

For any $\underline{\boldsymbol{R}} \in \underline{SO}_3^{\mathbb{R}}$, the solution of Eq. (53) emerges from

$$\begin{cases} \underline{\dot{\boldsymbol{R}}} = \underline{\tilde{\omega}}\,\underline{\boldsymbol{R}} \\ \underline{\boldsymbol{R}}(t_0) = \underline{\boldsymbol{R}}_0 \end{cases} \tag{57}$$

Making use of Eq. (54), results that $\underline{\boldsymbol{R}}\underline{\omega}_* = \underline{\omega} + \underline{\omega}_C$. If $\sim$ operator used, the previous calculus is transformed into $\widetilde{\underline{\boldsymbol{R}}\,\underline{\omega}_*} = \underline{\tilde{\omega}} + \underline{\tilde{\omega}}_C \Leftrightarrow \underline{\boldsymbol{R}}\,\underline{\tilde{\omega}}_*\underline{\boldsymbol{R}}^T = \underline{\dot{\boldsymbol{R}}}\,\underline{\boldsymbol{R}}^T + \underline{\tilde{\omega}}_C$. After multiplying the last expression by $\underline{\boldsymbol{R}}$, we obtain the initial value problem:

$$\begin{cases} \underline{\dot{\boldsymbol{R}}} = \underline{\boldsymbol{R}}\,\underline{\tilde{\omega}}_* - \underline{\tilde{\omega}}_C\,\underline{\boldsymbol{R}} \\ \underline{\boldsymbol{R}}(t_0) = \underline{\boldsymbol{R}}_0 \end{cases} \tag{58}$$

Using the variable change Eq. (54), the initial value problem (53) has been decoupled into two distinct initial value problems (56) and (58).

Let $\underline{\boldsymbol{R}}_{-\underline{\omega}_C} \in \underline{SO}_3^{\mathbb{R}}$ be the unique solution of the following Poisson-Darboux problem:

$$\begin{cases} \dot{\underline{\boldsymbol{R}}} + \tilde{\underline{\omega}}_C \, \underline{\boldsymbol{R}} = 0 \\ \underline{\boldsymbol{R}}(t_0) = \boldsymbol{I} - \varepsilon \tilde{\mathbf{r}}_C(t_0) \end{cases} \tag{59}$$

Considering $\underline{\boldsymbol{R}} = \underline{\boldsymbol{R}}_{-\underline{\omega}_C} \underline{\boldsymbol{R}}_*$, a representation theorem of the solution of Eq. (53) can be formulated.

Theorem 12. (Representation Theorem). *The solution of Eq. (53) results from the application of the tensor* $\underline{\boldsymbol{R}}_{-\underline{\omega}_C}$ *from Eq. (59) to the solution of the classical dual Euler fixed point problem:*

$$\begin{cases} \dot{\underline{\boldsymbol{R}}}_* = \underline{\boldsymbol{R}}_* \tilde{\underline{\boldsymbol{\omega}}}_* \\ \underline{\boldsymbol{M}} \, \dot{\underline{\boldsymbol{\omega}}}_* + \underline{\boldsymbol{\omega}}_* \times \underline{\boldsymbol{M}} \, \underline{\boldsymbol{\omega}}_* = \underline{\boldsymbol{\tau}}_* \\ \underline{\boldsymbol{\omega}}_*(t_0) = \underline{\boldsymbol{\omega}}_{*0} \\ \underline{\boldsymbol{R}}_*(t_0) = \underline{\boldsymbol{R}}_{*0} \end{cases} \tag{60}$$

where $\underline{\boldsymbol{\omega}}_{*0} = \underline{\boldsymbol{R}}_0^T(\underline{\boldsymbol{\omega}}_0 + \underline{\boldsymbol{\omega}}_C(t_0)), \underline{\boldsymbol{R}}_{*0} = (\boldsymbol{I} + \varepsilon \tilde{r}_C(t_0))\underline{\boldsymbol{R}}_0, \underline{\boldsymbol{\tau}}_* = \underline{\boldsymbol{R}}^T \underline{\boldsymbol{\tau}}$.

Different representations can be considered for the problem (60).

Using dual quaternion representation $\underline{\boldsymbol{R}}_* = \Delta\left(\hat{\underline{\mathbf{q}}}_*\right)$, Eq. (60) is equivalent with the following one:

$$\begin{cases} \dot{\hat{\underline{\mathbf{q}}}}^* = \frac{1}{2} \hat{\underline{\mathbf{q}}}_* \underline{\boldsymbol{\omega}}_* \\ \underline{\boldsymbol{M}} \, \dot{\underline{\boldsymbol{\omega}}}_* + \underline{\boldsymbol{\omega}}_* \times \underline{\boldsymbol{M}} \, \underline{\boldsymbol{\omega}}_* = \underline{\boldsymbol{\tau}}_* \\ \underline{\boldsymbol{\omega}}_*(t_0) = \underline{\boldsymbol{\omega}}_{*0} \\ \hat{\underline{\mathbf{q}}}_*(t_0) = \hat{\underline{\mathbf{q}}}_{*\mathbf{0}} \end{cases} \tag{61}$$

For the n-th order of Cayley transform based representation $\underline{\boldsymbol{R}}_* = cay_n\left(\underline{\xi}\right), \underline{\xi} = \tan \frac{\alpha}{2n} \underline{\mathbf{u}}$, the Eq. (60) becomes:

$$\begin{cases} \dot{\underline{\xi}} = \underline{S}\left(\underline{\xi}\right) \underline{\boldsymbol{\omega}}_* \\ \underline{\boldsymbol{M}} \, \dot{\underline{\boldsymbol{\omega}}}_* + \underline{\boldsymbol{\omega}}_* \times \underline{\boldsymbol{M}} \, \underline{\boldsymbol{\omega}}_* = \underline{\boldsymbol{\tau}}_* \\ \underline{\boldsymbol{\omega}}_*(t_0) = \underline{\boldsymbol{\omega}}_{*0} \\ \underline{\xi}(t_0) = \underline{\xi}_0 \end{cases} \tag{62}$$

where the tensor $\underline{S}$ is:

$$\underline{S} == \frac{p_n\left(|\underline{\xi}|\right)}{2q_n\left(|\underline{\xi}|\right)} \underline{\boldsymbol{I}} + \frac{1}{2} \tilde{\underline{\xi}} + \frac{\left(1 + |\underline{\xi}|^2\right) q_n\left(\underline{\xi}\right) - n p_n\left(|\underline{\xi}|\right)}{2n |\underline{\xi}|^2 q_n\left(|\underline{\xi}|\right)} \underline{\xi} \otimes \underline{\xi} \tag{63}$$

when $p_n(X)$ and $q_n(X)$ are defined by Eqs. (22) and (23).

Different particular cases can be analyzed for Eq. (62):

1. Let $\underline{\xi} = \tan\frac{\alpha}{2}\underline{\mathbf{u}}$ be the **Rodrigues dual vector** for n = 1:

$$\underline{S} = \frac{1}{2}\underline{I} + \frac{1}{2}\underline{\tilde{\xi}} + \frac{1}{2}\underline{\xi}\otimes\underline{\xi}$$

2. Let $\underline{\xi} = \tan\frac{\alpha}{4}\underline{\mathrm{u}}$ be the modified Rodrigues dual vector (Wiener-Milenkovic dual vector) for n = 2:

$$\underline{S} = \frac{1-|\underline{\xi}|^2}{4}\underline{I} + \frac{1}{2}\underline{\tilde{\xi}} + \frac{1}{2}\underline{\xi}\otimes\underline{\xi}.$$

The initial value problem (62) is a minimum parameterization of the six degrees of freedom motion problem. The singularity cases can be avoided using the **shadow parameters** of the **n-th order Modified Rodrigues Parameter dual vector.**

4. A dual tensor formulation of the six degree of freedom relative orbital motion problem

The results from the previous paragraphs will be used to study the six degrees of freedom relative orbital motion problem.

The relative orbital motion problem may now be considered classical one considering the many scientific papers written on this subject in the last decades. Also, the problem is quite important knowing its numerous applications: rendezvous operations, spacecraft formation flying, distributed spacecraft missions [3, 4, 6–10].

The model of the relative motion consists in two spacecraft flying in Keplerian orbits due to the influence of the same gravitational attraction center. The main problem is to determine the pose of the Deputy satellite relative to a reference frame originated in the Leader satellite center of mass. This non-inertial reference frame, known as "LVLH (Local-Vertical-Local- Horizontal)" is chosen as following: the C_x axis has the same orientation as the position vector of the Leader with respect to an inertial reference frame with the origin in the attraction center; the orientation of the C_z is the same as the Leader orbit angular momentum; the C_y axis completes a right-handed frame. The angular velocity of the LVLH is given by vector $\boldsymbol{\omega}_C$, which has the expression:

$$\omega_C = \dot{f}_C\frac{\mathbf{h}_C}{h_C} = \frac{1}{r_C^2}\mathbf{h}_C = \left[\frac{1+e_C\cos f_C(t)}{p_C}\right]^2\mathbf{h}_C \tag{64}$$

where vector $\mathbf{r}_C$ is

$$\mathbf{r}_C = \frac{p_C}{1+e_C\cos f_C(t)}\frac{\mathbf{r}_C^0}{r_C^0} \tag{65}$$

where p_C is the conic parameter, $\mathbf{h}_C$ is the angular momentum of the Leader, $f_C(t)$ being the true anomaly and e_C is the eccentricity of the Leader.

We propose dual tensors based model for the motion and the pose for the mass center of the Deputy in relation with LVLH. Both, the Leader satellite and the Deputy satellite can be considered rigid bodies.

Furthermore, the time variation of $\mathbf{r}_C$ is:

$$\dot{\mathbf{r}}_C = \frac{e_C|\mathbf{h}_C|\sin f_C(t)}{p_C}\frac{\mathbf{r}_C^0}{r_C^0} \tag{66}$$

In order to a more easy to read list of notations, for $t = t_0$ there will be used the followings:

$$\omega_C^0 = \left[\frac{1 + e_C\cos f_C(t_0)}{p_C}\right]^2 \mathbf{h}_C \tag{67}$$

$$\dot{\mathbf{r}}_C^0 = \frac{e_C|\mathbf{h}_C|\sin f_C(t_0)}{p_C}\frac{\mathbf{r}_C^0}{r_C^0} \tag{68}$$

where $\frac{\mathbf{r}_C^0}{r_C^0}$ is the unity vector of the X-axis from LVLH.

The full-body relative orbital motion is described by Eq. (53) where the dual angular velocity of the Chief satellite is:

$$\underline{\omega}_C = \omega_C + \varepsilon(\dot{\mathbf{r}}_C + \omega_C \times \mathbf{r}_C) \tag{69}$$

and the dual torque related to the mass center of Deputy satellite is:

$$\underline{\tau} = -\frac{\mu}{|\mathbf{r}_c + \mathbf{r}|^3}(\mathbf{r}_c + \mathbf{r}) + \varepsilon\boldsymbol{\tau}. \tag{70}$$

The representation theorem (**Theorem 12**) is applied in this case using the conditions (66)–(69), the solution of the Poisson-Darboux problem (59) is:

$$\underline{\boldsymbol{R}}_{-\underline{\omega}_C} = (\boldsymbol{I} - \varepsilon\widetilde{\boldsymbol{r}}_C(t))\left(\boldsymbol{I} - \sin f_c^0\frac{\widetilde{\boldsymbol{h}}_C}{h_c} + (1 - \cos f_c^0)\frac{\widetilde{\boldsymbol{h}}_C^2}{h_c^2}\right). \tag{71}$$

In (71), we've noted $h_c = \|\mathbf{h}_c\|$ and $f_c^0 = f_c(t) - f_c(t_0)$.

Theorem 13. (Representation Theorem of the full body relative orbital motion). *The solution of Eq. (53) results from the application of the tensor $\underline{\mathbf{R}}_{-\underline{\omega}_C}$ from Eq. (71) to the solution of the classical dual Euler fixed point problem* (60).

4.1. The rotational and translational parts of the relative orbital motion

The complete solution of Eq. (53) can be recovered in two steps.

Consider first the real part of Eq. (53). This leads to an initial value problem:

$$\begin{cases} \dot{Q} = \widetilde{\omega}Q \\ \dot{\omega} + \dot{\omega}_c = QJ^{-1}[Q^T\tau - Q^T(\omega+\omega_c)\times \\ \times JQ^T(\omega+\omega_c)] + \omega\times\omega_c \\ \omega(t_0) = \omega_0,\, \omega_0 \in V_3 \\ Q(t_0) = Q_0,\, Q_0 \in S\mathbb{O}_3 \end{cases} \tag{72}$$

which has the solution $Q = Q(t)$, the real tensor Q being the attitude of Deputy in relation with LVLH. In Eq. (72), ω is the angular velocity of the Deputy in relation with LVLH, ω_c is the angular velocity of LVLH, τ is the resulting torque of the forces applied on the Deputy in relation with is mass center, J is the inertia tensor of the Deputy in relation with its mass center. The angular velocity of Deputy in respect to LVLH at time t_0 is denoted with ω_0 and Q_0 is the orientation of Deputy in respect to LVLH at time t_0.

Consider now the dual part of Eq. (53). Taking into account the internal structure of $\underline{R}$, which is given by Eq. (2), after some basic algebraic calculus we obtain a second initial value problem that models the translation of the Deputy satellite mass center with respect to the LVLH reference frame:

$$\begin{cases} \ddot{\mathbf{r}} + 2\omega_c \times \dot{\mathbf{r}} + \omega_c \times (\omega_c \times \mathbf{r}) + \dot{\omega}_c \times \mathbf{r} + \\ + \dfrac{\mu}{|\mathbf{r}_c+\mathbf{r}|^3}(\mathbf{r}_c+\mathbf{r}) - \dfrac{\mu}{r_c^3}\mathbf{r}_c = 0 \\ \mathbf{r}(t_0) = \mathbf{r}_0,\, \dot{\mathbf{r}}(t_0) = \mathbf{v}_0 \end{cases} \tag{73}$$

where $\mu > 0$ is the gravitational parameter of the attraction center and $\mathbf{r}_0, \mathbf{v}_0$ represent the relative position and relative velocity vectors of the mass center of the Deputy spacecraft with respect to LVLH at the initial moment of time $t_0 \geq 0$.

Based on the **representation theorem 12**, the following theorem results.

Theorem 14. *The solutions of problems Eqs. (72) and (73) are given by*

$$\begin{aligned} Q &= R_{-\omega_C} Q_* \\ \mathbf{r} &= R_{-\omega_C}\mathbf{r}_* - \mathbf{r}_c \end{aligned} \tag{74}$$

where Q_* and $\mathbf{r}_*$ *are the solutions of the the classical Euler fixed point problem and, respectively, Kepler's problem*:

$$\begin{cases} \dot{Q}_* = Q_*\widetilde{\omega}_* \\ J\dot{\omega}_* + \omega_* \times J\omega_* = \tau_* \\ \omega_*(t_0) = Q_0^T(\omega_0 + \omega_c(t_0)) \\ Q_*(t_0) = Q_0 \end{cases} \tag{75}$$

and

$$\begin{cases} \ddot{\mathbf{r}}_* + \dfrac{\mu}{r_*^3}\mathbf{r}_* = \mathbf{0}; \\ \mathbf{r}_*(t_0) = \mathbf{r}_c^0 + \mathbf{r}_0; \\ \dot{\mathbf{r}}_*(t_0) = \dot{\mathbf{r}}_C^0 + \mathbf{v}_0 + \boldsymbol{\omega}_C^0 \times (\mathbf{r}_C^0 + \mathbf{r}_0) \end{cases} \tag{76}$$

where

$$\boldsymbol{R}_{-\omega_C} = \boldsymbol{I} - \sin f_c^0 \frac{\widetilde{h}_C}{|\mathbf{h}_c|} + (1 - \cos f_c^0)\frac{\widetilde{h}_C^{\,2}}{|\mathbf{h}_c|^2}) \tag{77}$$

and $\mathbf{r}_c$ *is given by Eq. (65).*

Remark 7: The problems (72) and (73) are coupled because, in general case, the torque $\boldsymbol{\tau}$ depends of the position vector $\mathbf{r}$.

The relative velocity of the translation motion may be computed as:

$$\mathbf{v} = \boldsymbol{R}_{-\omega_C}\dot{\mathbf{r}}_* - \widetilde{\omega}_c \boldsymbol{R}_{-\omega_C}\mathbf{r}_* - \frac{e_c|\mathbf{h}_c|\sin f_c(t)}{p_c}\frac{\mathbf{r}_c^0}{r_c^0} \tag{78}$$

This result shows a very interesting property of the translational part of the relative orbital motion problem (73). We have proven that this problem is super-integrable by reducing it to the classic Kepler problem [11, 12, 31, 32]. The solution of the translational part of the relative orbital motion problem is expressed thus:

$$\mathbf{r} = \mathbf{r}(t, t_0, \mathbf{r}_0, \mathbf{v}_0); \mathbf{v} = \mathbf{v}(t, t_0, \mathbf{r}_0, \mathbf{v}_0) \tag{79}$$

The exact closed form, free of coordinate, solution of the translational motion can be found in [11, 12, 31, 32, 34].

5. Conclusions

The chapter proposes a new method for the determination of the onboard complete solution to the full-body relative orbital motion problem.

Therefore, the isomorphism between the Lie group of the rigid displacements $S\mathbb{E}_3$ and the Lie group of the orthogonal dual tensors $\underline{S\mathbb{O}}_3$ is used. It is obtained a Poisson-Darboux like problem written in the Lie algebra of the group $\underline{S\mathbb{O}}_3$, an algebra that is isomorphic with the Lie algebra of the dual vectors. Different vectorial and non-vectorial parameterizations (obtained with n-th order Cayley-like transforms) permit the reduction of the Poisson-Darboux problem in dual Lie algebra to the simpler problems in the space of the dual vectors or dual quaternions.

Using the above results, the free of coordinate state equation of the rigid body motion in arbitrary non-inertial frame is obtained.

The results are applied in order to offer a coupled (rotational and translational motion) state equation and a representation theorem for the onboard complete solution of full body relative orbital motion problem. The obtained results interest the domains of the spacecraft formation flying, rendezvous operation, autonomous mission and control theory.

Nomenclature

a	real number
$\underline{a}$	dual number
$\mathbf{a}$	real vector
$\underline{\mathbf{a}}$	dual vector
$\boldsymbol{A}$	real tensor
$\underline{\boldsymbol{A}}$	dual tensor
$\boldsymbol{V}_3$	real vectors set
$\underline{\mathbb{V}}_3$	dual vectors set
$\boldsymbol{V}_3^{\mathbb{R}}$	time depending real vectorial functions
$\underline{\mathbb{V}}_3^{\mathbb{R}}$	time depending dual vectorial functions
$\underline{\widetilde{a}}$	skew-symmetric dual tensor corresponding to the dual vector $\underline{\mathbf{a}}$
f_c	true anomaly
p_c	conic parameter
h_c	specific angular momentum of the leader satellite
$L(\underline{V}_3, \underline{V}_3)$	dual tensor set
$\widehat{\mathbf{q}}$	real quaternion
$\underline{\widehat{\mathbf{q}}}$	dual quaternion
$\mathbb{R}$	real numbers set
$\underline{\mathbb{R}}$	dual numbers set
$S\mathbb{O}_3$	orthogonal real tensors set
$\underline{S\mathbb{O}_3}$	orthogonal dual tensor set

$\mathbb{SO}_3^{\mathbb{R}}$ time depending real tensorial functions

$\underline{\mathbb{SO}}_3^{\mathbb{R}}$ time depending dual tensorial functions

$\mathbb{U}$ unit quaternions set

$\underline{\mathbb{U}}$ unit dual quaternions set

Author details

Daniel Condurache

Address all correspondence to: daniel.condurache@tuiasi.ro

Technical University of Iasi, Iaşi, Romania

References

[1] Condurache D, Burlacu A. Onboard exact solution to the full-body relative orbital motion problem. AIAA Journal of Guidance, Control, and Dynamics. 2016;**39**(12):2638-2648. DOI: 10.2514/1.G000316

[2] Condurache D, Burlacu A. On six D.O.F relative orbital motion parameterization using rigid bases of dual vectors. Advances in the Astronautical Sciences. 2013;**150**:2293-2312

[3] Filipe N, Tsiotras P. Adaptive model-independent tracking of rigid body position and attitude motion with mass and inertia matrix identification using dual quaternions. AIAA Guidance, Navigation, and Control (GNC) Conference, Boston, Massachusetts; 2013. DOI: 10.2514/6.2013-5173

[4] Segal S, Gurfil P. Effect of kinematic rotation-translation coupling on relative spacecraft translational dynamics. Journal of Guidance, Control, and Dynamics. 2009;**32**(3):1045-1050. DOI: 10.2514/1.39320

[5] Condurache D. Poisson-Darboux problems's extended in dual Lie algebra. AAS/AIAA Astrodynamics Specialist Conference, Stevenson, WA, USA; 2017

[6] Alfriend K, Vadali S, Gurfil P, How J, Breger L. Spacecraft Formation Flying. New York: Elsevier; 1999. pp. 227-232. DOI: 10.1016/B978-0-7506-8533-7.00214-1

[7] Carter TE. New form for the optimal rendezvous equations near a Keplerian orbit. Journal of Guidance, Control, and Dynamics. 1990;**13**(1):183-186. DOI: 10.2514/3.20533

[8] Gim D-W, Alfriend KT. State transition matrix of relative motion for the perturbed noncircular reference orbit. Journal of Guidance, Control, and Dynamics. 2003;**26**(6):956-971. DOI: 10.2514/2.6924

[9] Sinclair AJ, Hurtado JE, Junkins JL. Application of the Cayley form to general spacecraft motion. Journal of Guidance, Control, and Dynamics. 2006;**29**(2):368-373. DOI: 10.2514/1.9910

[10] Yamanaka K, Ankersen F. New state transition matrix for relative motion on an arbitrary elliptical orbit. Journal of Guidance, Control, and Dynamics. 2002;**25**(1):60-66. DOI: 10.2514/2.4875

[11] Condurache D, Martinusi V. Kepler's problem in rotating reference frames. Part 1: Prime integrals, vectorial regularization. Journal of Guidance, Control, and Dynamics. 2007;**30**(1):192-200. DOI: 10.2514/1.20466

[12] Condurache D, Martinusi V. Kepler's problem in rotating reference frames. Part 2: Relative orbital motion. Journal of Guidance, Control, and Dynamics. 2007;**30**(1):201-213. DOI: 10.2514/1.20470

[13] Condurache D, Martinusi V. Relative spacecraft motion in a central force field. Journal of Guidance, Control, and Dynamics. 2007;**30**(3):873-876. DOI: 10.2514/1.26361

[14] Condurache D, Martinusi V. A Novel Hypercomplex Solution to Kepler's Problem. Vol. 19. Publications of the Astronomy Department of the Eötvös University (PADEU); 2007. pp. 201-213. ISBN: 963 463 557

[15] Condurache D, Martinusi V. Hypercomplex eccentric anomaly in the unified solution of the relative orbital motion. Advances in the Astronautical Sciences. 2010;**135**:281-300. ISSN: 0065-3438

[16] Angeles J. The application of dual algebra to kinematic analysis. Computational Methods in Mechanical Systems. 1998;**161**:3-32. DOI: 10.1007/978-3-662-03729-4_1

[17] Condurache D, Burlacu A. Dual Lie algebra representations of the rigid body motion. AIAA/AAS Astrodynamics Specialist Conference, AIAA Paper, San Diego; 2014. pp. 2014–4347. DOI: 10.2514/6.2014-4347

[18] Condurache D, Burlacu A. Dual tensors based solutions for rigid body motion parameterization. Mechanism and Machine Theory. 2014;**74**:390-412. DOI: 10.1016/j.mechmachtheory.2013.12.016

[19] Condurache D, Burlacu A. Recovering dual Euler parameters from feature-based representation of motion. Advances in Robot Kinematics. 2014:295-305. DOI: 10.1007/978-3-319-06698-1_31

[20] Pennestri E, Valentini PP. Dual quaternions as a tool for rigid body motion analysis: A tutorial with an application to biomechanics. The Archive of Mechanical Engineering. 2010;**LVII**:184-205. DOI: 10.2478/v10180-010-0010-2

[21] Pennestri E, Valentini PP. Linear Dual Algebra Algorithms and their Application to Kinematics, Multibody Dynamics: Computational Methods and Applications. Vol. 122009. pp. 207-229. DOI: 10.1007/978-1-4020-8829-2_11

[22] Fischer I. Dual-Number Methods in Kinematics. Statics and Dynamics: CRC Press; 1998. pp. 1-9. ISBN: 9780849391156

[23] Condurache D, Burlacu A. Orthogonal dual tensor method for solving the AX= XB sensor calibration problem. Mechanism and Machine Theory. 2016;**104**:382-404. DOI: 10.1016/j.mechmachtheory.2016.06.002

[24] Angeles J. Fundamentals of Robotic Mechanical Systems. Springer; 2014

[25] Tsiotras P, Junkins JL, Schaub H. Higher-order Cayley transforms with applications to attitude representations. Journal of Guidance, Control, and Dynamics. 1997;**20**(3):528-534

[26] Vasilescu FH. Quaternionic Cayley transform. Journal of Functional Analysis. 1999;**164**: 134-162. MR 2000d:47015

[27] Selig JM. Cayley Maps for SE(3). 12th IFToMM World Congress, Besancon, 2007

[28] Darboux G. Lecons sur la Theorie Generale des Surfaces et les Applications Geometriques du Calcul Infinitesimal. Paris: Gauthier-Villars; 1887. pp. 175-179. ark:/13960/t2h70912j

[29] Condurache D, Burlacu A. Fractional order Cayley transforms for dual quaternions based pose representation. Advances in the Astronautical Sciences. 2016;**156**:1317-1339

[30] Brodsky V, Shoham M. Dual numbers representation of rigid body dynamics. Mechanism and Machine Theory. 1999;**34**(5):693-718. DOI: 10.1016/S0094-114X(98)00049-4

[31] Condurache D, Martinusi V. Foucault pendulum-like problems: A tensorial approach. International Journal of Non-Linear Mechanics. 2008;**43**(8):743-760. DOI: 10.1016/j.ijnonlinmec.2008.03.009

[32] Condurache D, Martinusi V. Exact solution to the relative orbital motion in eccentric orbits. Solar System Research. 2009;**43**(1):41-52. DOI: 10.1134/S0038094609010043

[33] Tanygin S. Attitude parameterizations as higher-dimensional map projections. Journal of Guidance, Control, and Dynamics. 2012;**35**(1):13-24. DOI: 10.2514/1.54085

[34] Condurache D, Martinusi V. Quaternionic exact solution to the relative orbital motion problem. Journal of Guidance, Control, and Dynamics. 2010;**33**(4):1035-1047. DOI: 10.2514/1.47782

[35] Condurache D, Martinusi V. Exact solution to the relative orbital motion in a central force field, 2nd international symposium on systems and control in aeronautics and astronautics, Shenzhen, PRC; 2008. DOI: 10.1109/ISSCAA.2008.4776296

[36] Gurfil P, Kasdin JN. Nonlinear modeling of spacecraft relative motion in the configuration space. Journal of Guidance, Control, and Dynamics. 2004;**27**(1):154-157. DOI: 10.2514/1.9343

2

Long-Life Technology for Space Flight Hall Thrusters

Yongjie Ding, Liqiu Wei, Hong Li and Daren Yu

Abstract

The vastly improved durability of spacecrafts, coupled with the simultaneous continuous development of thrusters for high power output, has created a strong demand for Hall thrusters (HT) with long service lives. However, erosion of the discharge channel walls by high-energy ions is the most impactful and visible process that limits the lifetime of the thruster. This process is very sensitive to the operation mode of the thruster and the corresponding power density. We hereby present the results of our investigation on the factors that limit the lifetime of Hall thrusters, and three proven techniques for improving longevity of use including magnetic shielding (MS), wall-less technology, and aft-magnetic fields with large gradient.

Keywords: Hall thruster, long life, magnetic shield, wall-less, aft-magnetic

1. Introduction

The development of space propulsion technology is the cornerstone of development in the aerospace industry. With the rapid development of a wide range of satellite and spacecraft technologies, the demand for space transportation systems is on the rise. Electric propulsion technology is widely used in spacecrafts due to its high specific impulse, compact structure, low propellant consumption, and other advantages. The Hall thruster is currently one of the most widely used electric propulsion technologies at a global level [1].

Hall thrusters (HT), also called stationary plasma thrusters (SPT), were invented in the 1960s, and an early model was first used to transport a Russian satellite (METEOR-18) on December 29, 1971 [2]. The number of SPTs used for scientific and commercial space missions in the United States, Russia, Europe, and Japan is on the rise. The United States involved some of the original work on Hall Thruster in the early and mid-1960s [3–6]. However, interest in that particular accelerator was considerably less than that in ion thrusters. Russia has played a dominant role in the development of SPTs until relatively recently, when the USA, Europe, and Japan began to develop a strong interest in SPTs in the early 1990s. This resurgence of interest has generated a strong recovery in related research and development.

Figure 1 shows a schematic of a common Hall thruster. The basic process of operation begins with the release of electrons from a cathode, which enter a chamber and are subjected to a circumferential Hall drift movement by an orthogonal, axial electric field, and a magnetic field that acts primarily in the radial direction. Neutral atoms that are injected through an anode/gas distributor collide with the electrons in the closed drift and are ionized. Although the magnetic field is strong enough to lock the electrons in a circumferential drift within the discharge channel, its intensity is not sufficiently strong to affect the ions, which are accelerated by the axial electric field. An axial electron flux equal to that of the ion reaches the anode due to the cross-field mobility that often exceeds classical values. The cathode can provide the same electron flux to neutralize the exhausted ions. Therefore, quasi-neutrality is maintained throughout the discharge channel and the plume, and there is consequently no space-charge limitation on the acceleration. Therefore, the thrust density of SPTs is relatively high, compared to that of conventional electrostatic propulsion devices [7].

At present, commercial spacecrafts require thrusters that are capable of trouble-free operation for over 8000 h; however, conventional HTs have a relatively short operational lifetime. Thus, the development of long-life technology for Hall thrusters is significant.

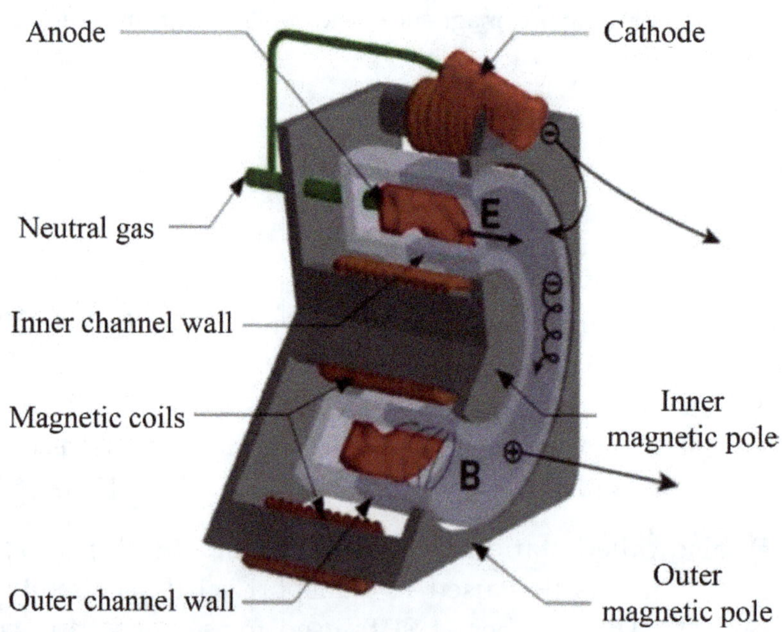

Figure 1. Schematic of a Hall thruster.

2. Long-life limitations of space flight Hall thrusters

Improvements in the operational lifetime of spacecrafts, and the continuous development of high-power thrusters, have resulted in an increasing demand for Hall thrusters with a long service life. There are several physical processes that limit the lifetime and reliability of Hall thrusters. These include [8] erosion of the cathode and magnetic system elements by the accelerated primary and secondary ions and the erosion of cathode's thermoemitter by ions which are accelerated in the near-cathode potential drop in the hollow cathode discharge plasma. Additional processes include oxidization of the getter, contaminated Xe gas flowing through the cathode, evaporation of the thermoemitter and heater materials. Finally, suboptimal temperatures under operation conditions, degradation of insulating and structural element materials, operation in space under increased temperature and radiation factor's impact, mechanical deformation and cracking of the heater, cathode and accelerator materials, due to the thermal shocks which occur when the thruster is started, can all have undesired effects.

The erosion of the discharge channel walls by high-energy ions is the most impactful and notable factor which limits the thruster's lifetime. This process is most sensitive to the thruster's operation mode and the corresponding power density [9, 10]. **Figure 2** shows photographs of the channel geometry of a PPS 1350-GQM thruster after 4200 h of operation. The interaction between the plasma and the wall causes power deposition on the channel wall and other structure components. The magnetic field topology leads directly to the large particle flux

Figure 2. Channel geometry of a PPS 1350-GQM thruster after 4200 h of operation.

with high energy, which is also directed toward the channel walls. In the discharge channel, the atoms undergo diffusion movement before ionization, which results in a radial motion component. The ions that are generated via the ionization process acquire the initial velocity of the atom and a radial velocity component. This results in an acceleration of the ion beam along the radial direction. In addition, the sheath and the presheath structures which are formed by the interaction of the plasma and channel walls also generate a radial electric field, which leads to radial ion divergence. Due to the influence of various physical factors mentioned above, the ion beam will diverge in the channel. In the acceleration zone, a portion of the high-energy ions will not be able to directly exit the channel. Instead, the wall material is sputtered and bombarded. When the bombardment energy is greater than the binding energy of the atoms in the wall, the wall material is sputtered and the geometrical morphology of the channel wall is altered [11, 12].

Long-term ion bombardment of the channel wall causes erosion, and the resulting change in the channel's geometry alters the optimum working condition of the thruster, which results in a decline in performance; more importantly, the breakdown of the channel's ceramic causes the magnetic pole to be exposed to the plasma, which would affect this field. Eventually, the performance of the Hall thruster is significantly affected, resulting in eventual failure. The end of the lifetime of a Hall thruster is generally accepted as the point of time when the channel is completely eroded by ion bombardment, and the magnetic pole is exposed to the plasma.

3. Magnetic shielding technology

During the years 2007 and 2009, Aerojet and Lockheed Martin Space Systems Company demonstrated the extension of the working hours of the qualification model (BPT-4000 4.5 kW HT) over 10,400 h. Most significantly, no measurable erosion of the insulator ring was observed from 5600 h to 10,400 h, which indicated that the thruster had achieved a "zero" erosion configuration [13, 14]. These improvements are the result of the topological structure of the magnetic field near the erosion surface. Jet Propulsion Laboratory (JPL) describes this process as "magnetic shielding (MS)." **Figure 3** shows the design principles involved in a magnetically shielded (MS) configuration, compared to an unshielded (US) configuration. In the US configuration, the magnetic lines near the channel's exit are almost perpendicular to the channel walls; however, the magnetic field lines of the MS configuration extend to the acceleration region deep within the channel and are arranged close to the ceramic walls without intersecting it. This is called the "grazing line," which effectively inhibits cavity wall erosion by high-energy ions.

The electron number density (*ne*) in HTs is so low that collisions between electrons and gases have little influence on the $E \times B$ drift (where E and B denote the electric and magnetic fields, respectively) or Hall drift, and an important current, the Hall current, is produced in a circumferential direction. The electron parameter, $\Omega_e \equiv \omega_{ce}/\upsilon_e \gg 1$, where ω_{ce} is the electron gyro-frequency and υ_e is the total collision frequency. Thus, electron temperature (*Te*) stays nearly constant along the magnetic field lines.

$$\nabla_{//} T_e \approx 0 \tag{1}$$

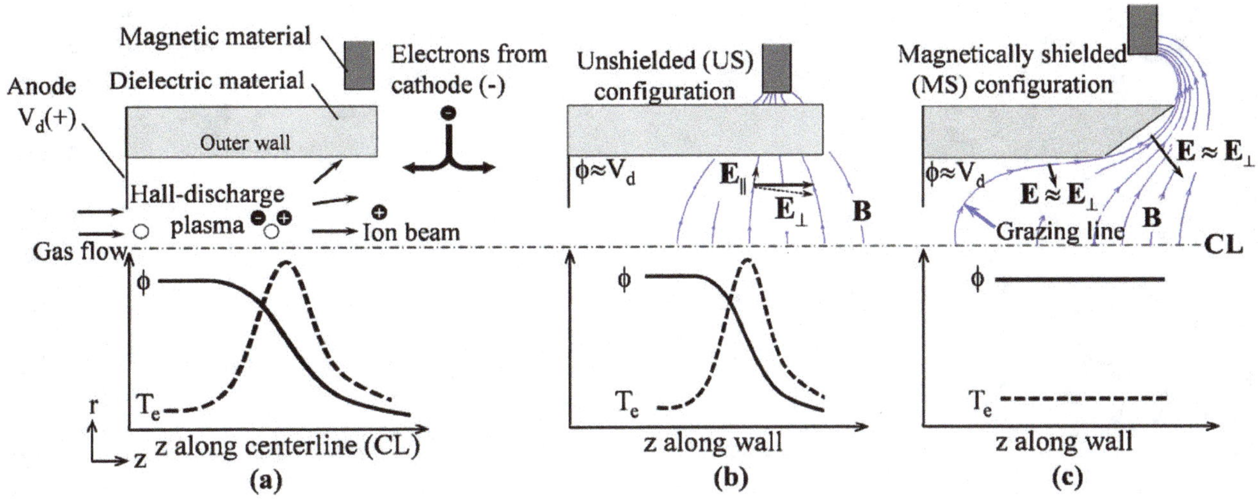

Figure 3. Schematics of the different structure of HTs (top) the potential (ϕ) and the electron temperature (*Te*) distribution along the center line. From left to right are traditional configuration, US configuration, and MS configuration, respectively.

Furthermore, the momentum equation of electrons can be simplified as

$$E_{//} \approx -T_e \nabla_{//} \ln n_e \tag{2}$$

and the resistive contribution to the electric field is negligibly small. Eqs. (1) and (2) contribute to two important properties of the force lines in HTs [15], that is, $T_e \approx T_{e0}$ and $\phi \approx \phi_0 + T_{e0} \ln(n_e/n_{e0})$ along a magnetic field lines, where T_{e0}, ϕ_0, and n_{e0} denote integration constants.

As shown in **Figure 3**, the MS configuration can be obtained by optimizing the magnetic field to realize a higher potential ϕ and a lower Te near the cavity surface. The parameter Te has its lowest value when the electrons are closet to the discharge voltage Vd, such that the kinetic energy of the injected ions and the sheath energy are reduced to values near or below the sputtering yield threshold. In addition, if the magnetic field is designed to appropriately match with the geometry of the discharge channel, the generated self-consistent electric field will be larger, and the field direction will be approximately perpendicular to the channel.

Therefore, the main principle when designing MS HTs is to recognize that the pressure of the electrons (yielding $T_e \times \text{In}\ (n_e)$ in Eq. (2)) is such that the electric field E is no longer orthogonal to the magnetic field, which can be clearly observed in **Figure 3**. Hence, if the magnetic field lines with convex curvature toward the anode [16] near the channel walls are not equipotential, then they are not able to effectively control the near-wall electric field.

These aforementioned ideas are consistent and provide some interesting insight into the theoretical development of magnetic shielding technology. The design of the H6MS Hall thruster in particular is based on this technological innovation [17, 18]. **Figure 4** (left) shows a photograph of the H6MS Hall thruster and its physical condition after operating continuously for 15 h. JPL demonstrated, using both numerical simulations and experiments, that the ion beam produced in a US HT can be controlled effectively, and the erosion rate on the walls is decreased by 2–3 orders [19].

Figure 4. H6MS Hall thruster before (left) and after (right) 15 h of testing. The ceramic walls were covered with a carbon film which was back-sputtered from the vacuum device's inner wall.

In addition, JPL also applied a magnetic shielding technique to a miniature Hall thruster. This investigation, which was performed with the cooperation of the University of California, led to the development of a magnetically shielded miniature HT (MaSMi HT), which was operated with a 275 V discharge voltage and a 325 W discharge power [20]. In Europe, CNRS (France) also realized a magnetic shielding technique for Hall thrusters which operated at a discharge power of 1.5 kW (200 W–PPS-flex and ISCT200-MS Hall thruster) [9, 21].

4. Wall-less technology

The relatively short lifetime of HTs due to plasma-surface interactions inside the discharge chamber is another drawback of conventional Hall thrusters. The underlying cause of this problem is channel wall erosion caused by the bombardment of high-energy electrons and ions. It is known that the choice of material of the channel wall influences the properties of the plasma discharge dynamics, which consequently influences the performance and the lifetime of the thruster. The plasma properties in a Hall thruster are also influenced by the secondary electron emission of the wall material.

Wall-less Hall thruster (WL-HT) was proposed to reduce the interaction between plasma and Hall thruster's channel walls. The objective is to limit the plasma-wall interaction by moving the ionization and acceleration regions to the exterior of the discharge channel. Such an unusual configuration was first proposed by Kapulkin et al. of Russia, during the 1990s. The concept was then proposed based on the idea of a Hall thruster, with an external electric field. Nevertheless, the assumption that limitations of the ion current are linked to the plasma instabilities led researchers to transition from a standard one-stage structure to a two-stage structure. However, the concept of a two-stage structure is less attractive because of its complicated design and operation. The concept of moving the electric field to the exterior of the channel was also investigated in Russia at TsNIIMASH, for thrusters with anode layer (TAL) in the late 1990s and early 2000s. The researchers demonstrated the possibility of stable operation at a high voltage with a high efficiency [22].

Figure 5 depicts the standard configuration of a conventional Hall thruster and a wall-less Hall thruster. The anode/gas distributor is usually positioned at the bottom of the discharge channel. The cathode is located on the outside of the channel and is the source of electrons for discharge balancing and neutralization of the ion beam in the plume area. A radial directed magnetic field with a bell-shaped intensity distribution along the center line is generally by coils or permanent magnets. As shown in **Figure 5**, the peak value of the magnetic field intensity is typically located near the discharge channel outlet. The ceramic channel constrains the propellant and thus maintains a higher atom density for subsequent ionization processes. The easiest way to move the ionization and the acceleration regions out of the discharge channel is to place the anode directly at the channel outlet plane, which is shown in **Figure 5**. This requires that the shape and size of the channel, as well as the magnetic field topology and discharge channel geometry, are unchanged. The proposed idea is the simplest way to transform Hall thrusters into WL-HTs.

Figure 6 depicts images of a low-power ion source working with Xenon propellant in the standard 200 W-class Hall thruster and WL configuration with a ring anode. The discharge voltage is 200 V, and the propellant mass flow rate (MFR) is 1 mg/s. The photograph with bright light near the channel exit (right) indicates that the discharge region was pushed outside the ceramic channel, as expected in the WL configuration. A distinct difference between the two methods is that the boundary of the ion beam with WL configuration is less distinct, which means that the divergence angle of the plume region in a wall-less configuration is much larger. The discharge current is also higher for WL compared to the standard configuration. Therefore, the thruster's performance will diminish and the erosion of external parts, such as the pole pieces, will be increased with time. Moreover, a large beam divergence means that the plasma from thrusters will have a negative effect on the spacecraft elements [23].

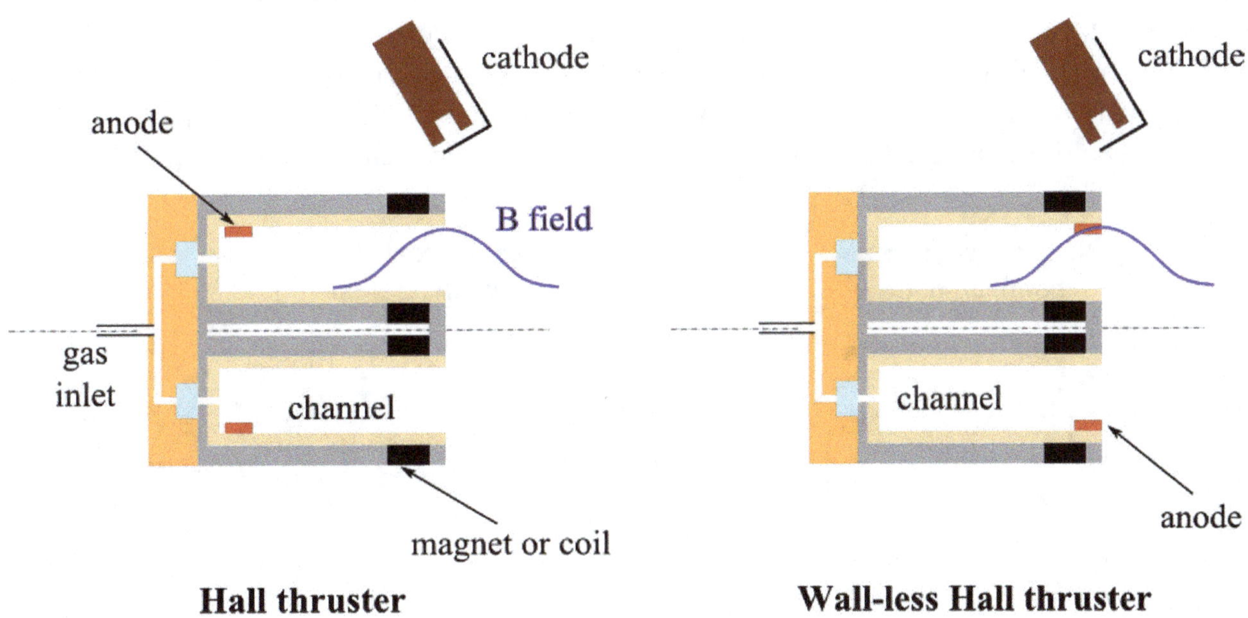

Figure 5. Configurations of a standard Hall thruster and wall-less Hall thruster.

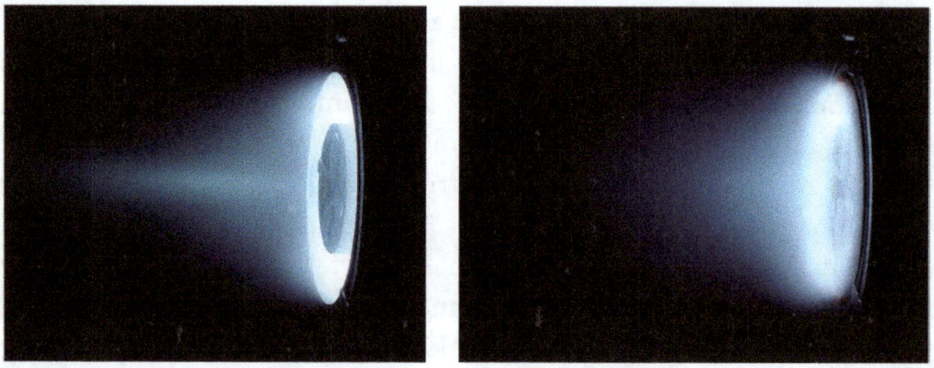

Figure 6. Photographs of the low-power PPI Hall thruster operating at the voltage of 200 V and a MFR of 1 mg/s in standard (left) and WLHT (right) configurations.

Figure 7 displays the interaction between the annular anode and the B-field lines. The magnetic circuit of the original WL-HT prototype, as shown in **Figure 7** (left), is based on the classical Hall thruster design. By shifting the anode from the bottom, to the channel outlet without any other changes, the magnetic field lines are roughly perpendicular to the ceramic wall and intersect with the anode located near the cavity outlet. This results in a decline in the efficiency of the electron confinement. Moreover, a large number of high-energy electrons emitted from the cathode will be trapped along the magnetic field lines and eventually arrive at the anode. Therefore, the electron current is relatively large, and the propellant utilization is low [24].

To solve the problem of excessive energy losses of the electrons at the anode, some optimized prototypes were proposed. The first optimization approach involves rotating the anode by 90 degrees to restore the magnetic barrier, while maintaining the topology of the magnetic field. However, this design does not perform satisfactorily due to two limitations. The first is that a large component of the magnetic field lines near the channel exit does not contribute to the trapping of electrons and the production of thrust. The other is that the ionization and acceleration region are too short for effective electron-atom collision. **Figure 7** (right) portrays a generally satisfactory design. The magnetic field lines are injected axially, and the peak of the magnetic field intensity is pushed downstream at the channel exit. The

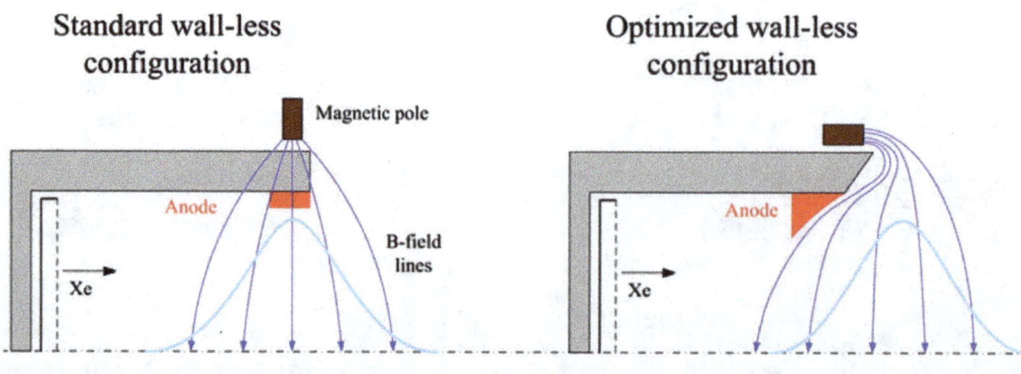

Figure 7. Schematics of original (left) and optimized (right) WL-HT prototype.

curved anode located at the exit is shaped so that it does not intersect with the field lines, which ensures that the magnetic field can trap electrons and effectively produce thrust. This type of optimization may appear similar to the MS Hall thruster, but the most striking difference is that it is not necessary for the field lines in the WL magnetic configuration to extend deep into the cavity to capture electrons. Therefore, it is quite easy to generate the required magnetic circuit.

Based on the 1.5-kW PPS-Flex HT, some experiments have also been performed. As expected, the discharge current is significantly reduced by the adjustment of the magnetic topology, and the positioning of the anode in parallel. In order to improve the utilization of propellant and achieve a satisfactory specific impulse, thrust level, and anode efficiency, the thruster was operated with a voltage of 500 V. However, the current magnetic does not allow the generation of WL topology with a peak magnetic field value above 90 G. This significantly impacts the operation at high voltage, and further optimization is necessary to reduce discharge current oscillations and increase the thruster efficiency. An improved Hall thruster based on PPS-Flex, which is capable of forming a stronger magnetic field intensity, is currently under development. The influence of the ceramic channel length of the thruster is an important factor that requires further study. It is possible that the channel length may be reduced as ionization takes place near the channel outlet plane. However, it should also be kept sufficiently long to ensure the homogenization of neutral gas.

Another proposed WL prototype was also based on the structure of the PPI thruster. To facilitate more effective and uniform distribution of the xenon gas, a 3-mm-thick gridded anode which covers the channel exit was designed, as shown in **Figure 4**. In order to limit the plasma diffusion in the discharge channel, the width of the anode was decreased. The gridded anode has a transparency of 68% with a 3-mm-diameter hole. Apart from the anode design, the second prototype is almost identical to the first one (**Figure 8**).

Figure 8. Photograph of the second wall-less thruster prototype with a gridded anode at the exit plane.

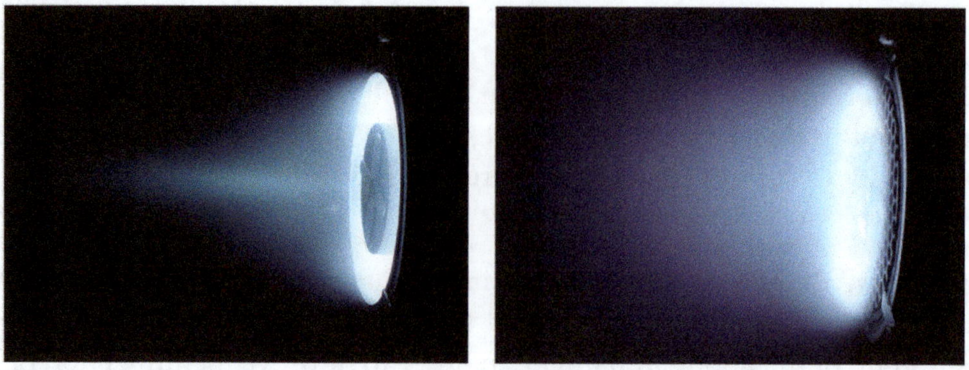

Figure 9. (Left) Photograph of the low-power PPI Hall thruster in standard configuration firing with Xe at 200 V and 1 mg/s. (Right) Photograph of the second prototype of WL-HT with a gridded anode firing with Xe under same conditions.

Figure 9 shows two plume region photographs of the PPI thruster operating with a Xe propellant; the left photograph shows the thruster in a standard configuration, and the right one is the wall-less Hall thruster with a gridded anode. The discharge voltage is 200 V, and the MFR is 1 mg/s. Compared to the anode ring, the discharge area of the Hall thruster with the gridded anode is repositioned outside the discharge channel, which is indicated by the bright light in front of the outlet. The ion beam boundaries of the WL Hall thruster are also less defined in this prototype, which implies that there is a degradation in performance.

Hall thrusters in WL configuration generally experience significant benefit in integration, lifetime, operating envelope, and propellant options. Since the acceleration zone is outside the discharge channel, the channel wall can be substantially shortened, thus reducing the mass and improving the economy of volume. The interaction between the plasma and the walls is also significantly reduced. Therefore, the impact of the channel material on the thrusters' performance is reduced. More importantly, it is presumably possible for the thruster to operate at a higher voltage and with an extended lifetime. In addition, the reduction of the plasma-wall interaction can lead to higher electron temperatures and positive points, which should result in efficient ionization of the propellants such as krypton and argon.

5. Aft-magnetic field with large gradient technology

To address the problems associated with power losses, and the low lifetime associated with the high surface-to-volume ratio of low power Hall thrusters, Harbin Institute of Technology proposed an aft-magnetic field with large gradient technique. In this approach, the maximum magnetic field strength is located on the outside of the channel with a large gradient. Harbin Institute of Technology developed a Hall thruster using a focused magnetic field of low power, which was excited using only two permanent magnet rings, such that the maximum magnetic field strength is outside the channel (Brexit/Brmax = 0.75 can be achieved). The magnetic field gradient in this configuration is much larger than that of a conventional Hall thruster, which can achieve a value of 20 G/mm [25, 26]. **Figure 10** shows the magnetic structure and configuration of the aft-magnetic field setup.

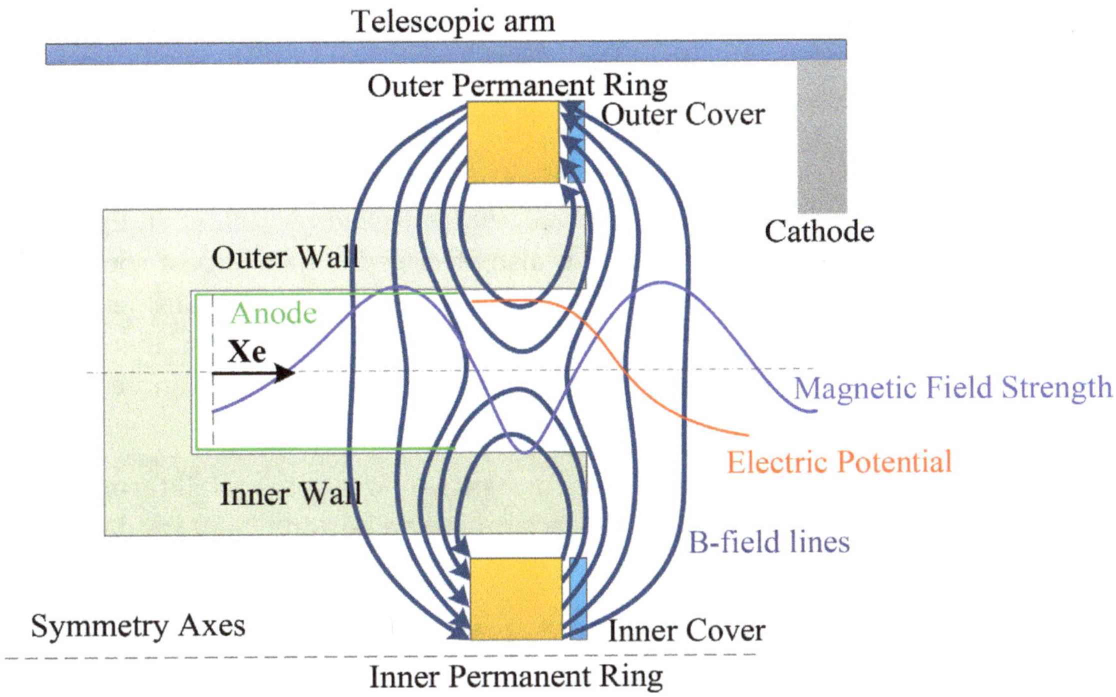

Figure 10. Magnetic structure and configuration.

An approach to push down the magnetic field and the channel can be adjusted accordingly and can achieve plasma discharge without wall loss. The result of calculations based on simulation has confirmed that the abovementioned approach causes acceleration processes to occur outside the channel, but ionization occurs in the channel. The temperature of the walls remains relatively low, since the resulting power deposition on this structure is minimal. This is because the wall is only bombarded with low-energy ions and electrons. Therefore, the channel erosion is effectively reduced, and the operational life of the thruster is extended. In addition, the overall efficiency of the system is improved because additional coil power is not consumed.

Based on this research, Harbin Institute of Technology designed a 200-W Hall thruster with two permanent magnet rings, to facilitate an in-depth investigation of the effects of the aft-magnetic field with a large gradient, on discharge properties and device performance. This thruster has five noteworthy features. First, the magnetic field is only excited by an inner and an outer permanent magnetic ring. Second, the gas distributor and the anode are made of nonmagnetic stainless steel, while the other metal structures of the thruster are made of titanium. Therefore, the other parts of the entire thruster are nonmagnetic, and a magnetic screen is not necessary. Third, the anode's front end-face is at the internal magnetic separatrix position, and it has a hollow structure. Compared to the traditional Hall thrusters, the distance from the channel outlet to the zero-magnetic region is shorter, which implies that the magnetic field gradient is larger than that of traditional Hall thrusters. Fourth, by using various sets of ceramic rings, the channel length can be easily changed while keeping the width of the channel fixed. Finally, 50% of the thruster's shell components are hollow. To effectively reduce the discharge channel temperature, they are directly exposed. References [27, 28] highlight the visual preliminarily evidence, which confirms the feasibility of the proposed thrusters. The thruster is able to discharge with lower wall energy loss and eliminate wall erosion both in a straight channel and in an oblique arrangement (Brexit/Brmax = 0.75). The maximum anode efficiency is 29.1% (straight channel) and 34.2% (oblique channel) with a discharge power of 200 W. When the channel is enlarged to Brexit/Brmax = 0.9, the anode efficiency can be improved to 42% [27]. **Figure 11** depicts photographs of the ceramic channel after a discharge with Brexit/Brmax = 0.75. It is observed that there is a 1-mm-long area, which is slightly yellow, in the outlet area of the inner ceramic wall. A black deposition is also observed, which almost completely covers the entire outer ceramic wall. Neither of these observations indicate that the whiteness of the ceramic bottom is caused by a bombardment of high-energy ions. It can therefore be concluded that there are very few high-energy ions which bombard the wall and cause erosion. The resulting ions are mainly low-energy ions [26].

In order to extend the life and improve the performance of low-power Hall thrusters, Harbin Institute of Technology has done further research on anode design [29–31] and channel wall material analysis [32].

Unlike conventional Hall thrusters, the peak of the magnetic field strength is outside the discharge channel, for Hall thrusters which adopt an aft-magnetic field with a large gradient and double peak. Therefore, the distance from the channel outlet to the zero magnetic field region is relatively short. However, if the thruster adopts a traditional anode configuration and anode location, it will experience a drop in its performance, as this configuration may cause an inadequate homogenization of neutral gas. Hence, a comparative study was performed for a U-shaped hollow anode with the front end-face and the flat plate anodes in the zero magnetic field region, with the first magnetic peak (corresponding to the rear and front end-faces of the U-shaped anode, respectively). The research shows that under the same operating conditions, the highest overall performance is achieved for thrusters with a hollow anode. For an anode positioned at the magnetic peak, its ionization rate is at a maximum. However, most of the ionized ions produced bombarded the walls, resulting in energy loss and reduced performance. For an anode in the zero magnetic field region, the voltage and propellant utilization are lower than those of the hollow anode. Thus, although the maximum ionization rate is higher than that of the hollow anode, the wall power loss is slightly smaller. In addition, due to its shorter ionization region and relatively shorter channel, it also has a poor overall performance compared to that of the hollow anode [28].

Figure 11. Ceramics rings after discharge.

Due to the large gradient of the magnetic field, matching the magnetic field to the anode's position is very important, which when carried out to a very large extent determines the performance of the thruster. Simulation and experimental results demonstrate that when the anode is placed between the outer and inner magnetic separatrices, both the efficiency and the thrust are at a maximum. The significant energy losses on the walls result in a low efficiency and thrust, despite the high degree of ionization, when the anode is placed at the inner magnetic separatrix. Thus, the performance of the thruster is at its lowest when the anode is at the outer magnetic separatrix, because of the lower ionization level and larger divergence angle of the plume, as the ionization zone is shifted toward the plume region [24].

A hollow indented anode is proposed to increase the neutral gas density in the discharge channel, so that the performance of the thruster can be improved. The experimental results to date indicate that this structure can effectively improve the performance (in terms of anode efficiency, ionization rate, propellant utilization, and thrust) compared to the hollow straight anode, under similar operating conditions. Simulation results indicate that the neutral gas density can be effectively increased by the utilization of an indented anode in a discharge channel and on the centerline of the channel. Furthermore, the ionization rate in the channel and the preionization in the anode can also be increased. Therefore, the hollow indented anode can be considered as an important design concept for improving the thruster's performance [30].

As acceleration occurs in the plume area and ionization occurs in the channel, the simulation and experimental results indicate that the maximum electron temperature can be found in the plume zone, while the electron temperature in the channel is relatively low. The secondary electron emission yield of the channel material will have a small but measurable effect on the thruster's performance. This assertion was experimentally verified. It was confirmed that materials with low sputtering yield could be used to further increase the life of the low-power thrusters while discharging through a channel with walls of titanium and graphite. **Figure 12** shows a picture of the 200-W prototype Hall thruster and plume discharge with titanium wall material [31].

Two additional low-power Hall thrusters were designed by Harbin Institute of Technology, with power ratings of 10–20 W and 50–100 W. The maximum anode efficiency was about 30%,

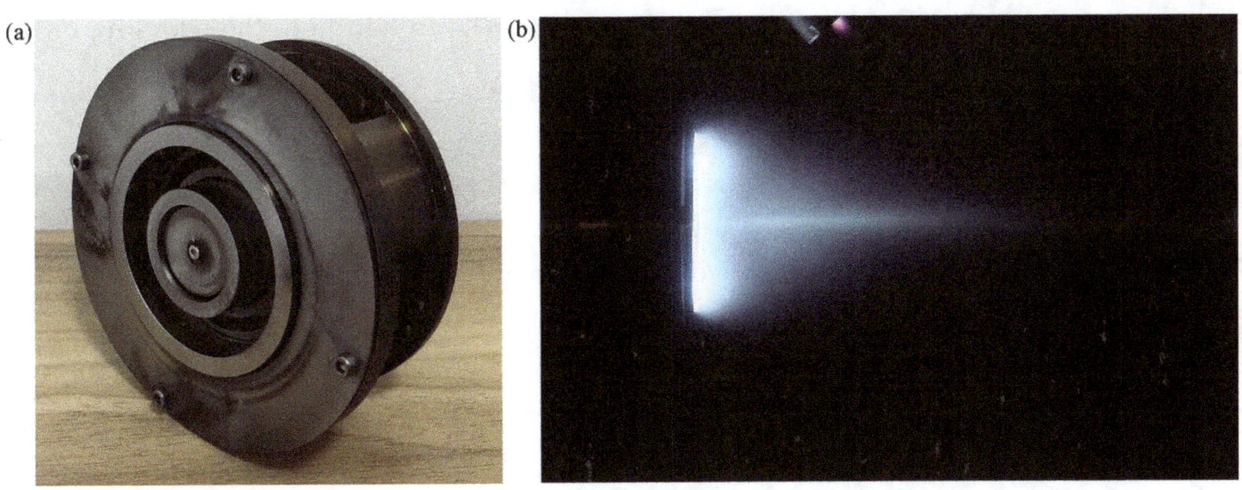

Figure 12. 200 W Hall thruster discharge with titanium wall material (a) prototype of the Hall thruster and (b) picture of discharge plume.

when the design was based on an aft-magnetic field with large gradient technique. High-power thrusters which operate at 1.35 and 5 kW have been designed and tested at the Harbin Institute of Technology using the aft-magnetic field with large gradient technique. The maximum efficiency attained was 65%. Therefore, the aft-magnetic field with large gradient technique can be widely used in Hall thrusters to achieve different power outputs.

6. Outstanding problems

The purpose of the MS technology is to facilitate the equipotentialization of the near-wall magnetic field lines. The topology of these lines reaches deeply into the near anode region and eliminates the influence on the potential originating from electron pressure. On the basis of the isothermal principle of magnetic field lines, $E_{//}$ is negligibly small. Meanwhile, the induced $E\perp$ prevents ion bombardment of the ceramic walls, which significantly reduces channel erosion. Nevertheless, there are still two primary problems that need to be addressed: (1) the large excitation power consumption and the relatively low thruster efficiency. As a result, an additional component for heat dissipation is required, especially for low-power HTs; (2) Ioannis et al. [33–35] first discovered that the pole erosion of the magnetic shield of Hall thrusters is a by-product of magnetic shielding. Although the erosion rate is small, it will affect the lifetime of thrusters over long periods of time.

The wall-less technology involves moving the anode to the channel exit, which entirely shifts the ionization and acceleration region to the outside of the channel defined by the wall-less Hall thrusters. The ionization of neutrals occurs in the plume region, where the neutrals spread radially without the control of the channel wall, thus resulting in a larger plume divergence (55°–62°). Thus, the performance of this device is relatively lower.

The aft-magnetic field with large gradient technique causes the maximum magnetic field strength to be generated on the outside of the channel with a large gradient. Primary ionization can be maintained inside the channel, and the primary acceleration can be directed toward the plume region, which can maintain a high level of propellant utilization while decreasing the energy, flux of electrons, and the ions that bombard the ceramic channel wall. In the future, the channel and the magnetic field should be the two main considerations while attempting to optimize the discharge performance of HTs. In addition, the coupling of the cathode with the thrusters should be studied.

Acknowledgements

The authors want to gratefully acknowledge the financial support from the National Natural Science Foundation of China (Grant Nos. 51777045 and 51477035).

Conflict of interest

This chapter has no conflicts of interest.

Author details

Yongjie Ding[1*], Liqiu Wei[2], Hong Li[1] and Daren Yu[1]

*Address all correspondence to: dingyongjie@hit.edu.cn

1 School of Energy Science and Engineering, Harbin Institute of Technology, Harbin, People's Republic of China

2 Academy of Fundamental an Interdisciplinary Science, Harbin Institute of Technology, Harbin, People's Republic of China

References

[1] Boeuf JP. Tutorial: Physics and modeling of Hall thrusters. Journal of Applied Physics. 2017;**121**(1):011101. DOI: 10.1063/1.4972269

[2] Morozov AI. The conceptual development of stationary plasma thrusters. Plasma Physics Reports. 2003;**29**(3):235-250. DOI: 10.1134/1.1561119

[3] Lary E, Meyerand R, Salz F. Ion acceleration in a gyro-dominated neutral plasma: Theory and experiment. Bulletin of the American Physical Society. 1962;**7**:441

[4] Seikel GR, Reshotko E. Hall current ion accelerator. Bulletin of the American Physical Society, Series II. June 1962;**7**:414

[5] Banas CM, Brown CO, Pinsley EA. Hall-current accelerator utilizing surface contact ionization. Journal of Spacecraft & Rockets. 1971;**1**(5):525-531. DOI: 10.2514/3.27692

[6] Janes GS, Lowder RS. Anomalous electron diffusion and ion acceleration in a low-density plasma. The Physics of Fluids. 1966;**9**(6):1115-1123. DOI: 10.1063/1.1761810

[7] Choueiri EY. Plasma oscillations in Hall thrusters. Physics of Plasmas. 2001;**8**(4):1411-1426. DOI: 10.1063/1.1354644

[8] Clauss C, Day M, Kim V, et al. Preliminary study of possibility to ensure large enough lifetime of SPT operating under increased powers. In: Proceedings of the 33rd AIAA/ASME/SAE/ASEE Joint Propulsion Conference and Exhibit; 11-14 July 1997; Seattle, WA, U.S.A. AIAA; 1997. p. 2789

[9] Mazouffre S, Vaudolon J, Largeau G, et al. Visual evidence of magnetic shielding with the PPS-flex Hall thruster. IEEE Transactions on Plasma Science. 2014;**42**(10):2668-2669. DOI: 10.1109/tps.2014.2331180

[10] Mikellides IG, Katz I, Hofer RR, et al. Magnetic shielding of a laboratory Hall thruster. I. Theory and validation. Journal of Applied Physics. 2014;**115**(4):203. DOI: 10.1063/1.4862313

[11] Cai N. Influence of magnetic lens on wall erosion of Hall effect thruster [thesis]. Harbin: Harbin Institute of Technology; 2010

[12] Dumazert P, Marchandise F, Jolivet L, Estublier D, Cornu N. PPS-1350-G qualification status. In: Proceedings of the 40th AIAA/ASME/SAE/ASEE Joint Propulsion Conference; 11-14 July 2004; Fort Lauderdale. AIAA; 2004. p. 3604

[13] De Grys K, Mathers A, Welander B, Khayms V. Demonstration of 10,400 hours of operation on 4.5 kW qualification model Hall Thruster. In: Proceedings of the 46th AIAA/ASME/SAE/ASEE Joint Propulsion Conference; 25-28 July 2010; Nashville. AIAA; 2010. p. 6698

[14] Mikellides I, Katz I, Hofer R, Goebel D, de Grys K, Mathers A. Magnetic shielding of the acceleration channel walls in a long-life Hall thruster. In: Proceedings of the 46th AIAA/ASME/SAE/ASEE Joint Propulsion Conference; 25-28 July 2010; Nashville. AIAA; 2010. p. 6942

[15] Morozov AI, Esipchuk YV, Tilinin GN, Trofimov AV, Sharov YA, Shchepkin GY. Plasma accelerator with closed electron drift and extended acceleration zone soviet physics: Technical physics. 1972;**17**:38

[16] Morozov AI, Savelyev VV. Fundamentals of stationary plasma thruster theory. Reviews of Plasma Physics. 2000;**21**:203-391. DOI: 10.1007/978-1-4615-4309-1_2

[17] Mikellides I, Katz I, Hofer R. Design of a laboratory Hall thruster with magnetically shielded channel walls, Phase I: Numerical simulations. In: Proceedings of the 47th AIAA/ASME/SAE/ASEE Joint Propulsion Conference; 31 July–03 August 2011; San Diego. AIAA; 2011. p. 5809

[18] Hofer R, Goebel D, Mikellides I, Katz I. Design of a laboratory Hall thruster with magnetically shielded channel walls, Phase II: Experiments. In: Proceedings of the 48th AIAA/ASME/SAE/ASEE Joint Propulsion Conference; 30 July–01 August 2012; Atlanta. AIAA; 2012. p. 3788

[19] Mikellides IG, Katz I, Hofer RR, et al. Magnetic shielding of walls from the unmagnetized ion beam in a Hall thruster. Applied Physics Letters. 2013;**102**(2):4906-4911. DOI: 10.1063/1.4776192

[20] Conversano R, Goebel D, Hofer R, Matlock T, Wirz R. Magnetically shielded miniature Hall thruster: Development and initial testing. In: Proceedings of 33rd International Electric Propulsion Conference (IEPC); October 2013; Washington, IEPC-2013-201

[21] Grimaud L, Vaudolon J, Mazouffre S, Boniface C. Design and characterization of a 200W Hall thruster in "magnetic shielding" configuration. In: Proceedings of the 52nd AIAA/ASME/SAE/ASEE Joint Propulsion Conference; 25-27 July 2016; Salt Lake City. AIAA; 2016. p. 4832

[22] Mazouffre S, Tsikata S, Vaudolon J. Development and experimental characterization of a wall-less Hall thruster. Journal of Applied Physics. 2014;**116**:243302. DOI: 10.1063/1.4904965

[23] Mazouffre S, Tsikata S, Vaudolon J. Development and characterization of a wall-less Hall thruster. In: Proceedings of the 50th AIAA/ASME/SAE/ASEE Joint Propulsion Conference; 28-30 July 2014; Cleveland. AIAA; 2014. p. 3513

[24] Vaudolon J, Mazouffre S, Hénaux C, Harribey D, Rossi A. Optimization of a wall-less Hall thruster. Applied Physics Letters. 2015;**107**:174103. DOI: 10.1063/1.4932196

[25] Ding Y, Peng W, Wei L, et al. Computer simulations of Hall thrusters without wall losses designed using two permanent magnetic rings. Journal of Physics D Applied Physics. 2016;**49**(46):465001. DOI: 10.1088/0022-3727/49/46/465001

[26] Ding Y, Peng W, Sun H, et al. Performance characteristics of no-wall-losses Hall thruster. European Physical Journal Special Topics. 2017;**226**(13):2945-2953. DOI: 10.1140/epjst/e2016-60247-y

[27] Ding Y, Peng W, Sun H, et al. Visual evidence of suppressing the ion and electron energy loss on the wall in Hall thrusters. Japanese Journal of Applied Physics. 2017;**56**(3):038001. DOI: 10.7567/jjap.56.038001

[28] Ding Y, Peng W, Sun H, et al. Effect of oblique channel on discharge characteristics of 200-W Hall thruster. Physics of Plasmas. 2017;**24**:023507. DOI: 10.1063/1.4976104

[29] Ding Y, Sun H, Li P, et al. Application of hollow anodes in a Hall thruster with double-peak magnetic fields. Journal of Physics D Applied Physics. 2017;**50**(33):335201. DOI: 10.1088/1361-6463/aa7bbf

[30] Ding Y, Sun H, Li P, et al. Influence of hollow anode position on the performance of a Hall-effect thruster with double-peak magnetic field. Vacuum. 2017;**143**:251-261. DOI: 10.1016/j.vacuum.2017.06.030

[31] Ding Y, Sun H, Wei L, et al. A 200W Hall thruster with hollow indented anode. Acta Astronautica. 2017;**139**:521-527. DOI: 10.1016/j.actaastro.2017.08.001

[32] Ding Y, Sun H, Peng W, et al. Experimental test of 200 W Hall thruster with titanium wall. Japanese Journal of Applied Physics. 2017;**56**(5):050312. DOI: 10.7567/jjap.56.050312

[33] Mikellides IG, Lopez Ortega A, Jorns B. Assessment of pole erosion in a magnetically shielded Hall thruster. In: Proceedings of the 50th AIAA/ASME/SAE/ASEE Joint Propulsion Conference; 28-30 July 2014. Cleveland AIAA; 2014. p. 3897

[34] Goebel DM, Jorns B, Hofer RR, Mikellides I G, Katz I. Pole-piece interactions with the plasma in a magnetically shielded Hall thruster. In: Proceedings of the 50th AIAA/ASME/SAE/ASEE Joint Propulsion Conference; 28-30 July 2014; Cleveland. AIAA; 2014. p. 3899

[35] Jorns B, Dodson CA, Anderson JR, Goebel DM, Hofer RR, Sekerak MJ, et al. Mechanisms for pole piece erosion in a 6-kW magnetically-shielded Hall thruster. In: Proceedings of the 52nd AIAA/ASME/SAE/ASEE Joint Propulsion Conference; 25-27 July 2016; Salt Lake City. AIAA; 2016. p. 4839

3

Mars Networks-Based Navigation: Observability and Optimization

Zhengshi Yu, Pingyuan Cui, Rui Xu and
Shengying Zhu

Abstract

In order to achieve more scientific returns for Mars, future Mars landers will be required to land at certain landing point with special scientific interest. Therefore, autonomous navigation is indispensable during the Mars approach, entry, and landing phase. However, the number of beacons or the Mars orbiters which can provide the navigation service is so limited and the line-of-sight visibility cannot be guaranteed during the landing period. So the navigation scheme especially the beacon configuration has to be optimized in order to efficiently use the limited navigation information. This chapter aims to analyze the feasibility and optimize the performance of the Mars Networks-based navigation scheme for the Mars pinpoint landing. The observability of navigation system is used as an index describing the navigation capability. Focusing on the relationship between the configuration of radio beacons and observability, the Fisher information matrix is introduced to analytically derive the degree of observability, which gives valuable conclusions for navigation system design. In order to improve the navigation performance, the navigation scheme is optimized by beacon configuration optimization, which gives the best locations of beacons (or the best orbit of navigation orbiters). This is the main approach to improve the navigation capability.

Keywords: Mars networks, navigation, observability, optimization

1. Introduction

As the most similar planet to the Earth in the Solar system, Mars is considered as an ideal target for planetary exploration [1, 2]. Since the 1960s, humans have investigated the Mars exploration missions in the near distance. With the development of aerospace science and technology, the manner of Mars exploration has shifted from flyby/orbiting to landing and roving explorations. Considering scientific returns and exploration capabilities, Mars landing exploration is also essential and is one of the most popular tasks of human deep space exploration in the near future. The representative Mars landing missions including NASA's Viking 1 and 2, Mars Pathfinder (MPF), Mars Exploration Rovers (MER, including the Spirit and Opportunity rovers), Phoenix, Mars Science Laboratory (MSL, including the Curiosity rover), and ESA's Mars Express/Beagle 2 mission. All of these greatly inspire the development of advanced guidance, navigation, and control (GNC) technologies.

During the past 50 years of Mars exploration, 46 Mars exploration spacecraft have been launched. The overall success rate is only 41.3% though. Furthermore, among the 20 Mars landing attempts, only 7 robotic rovers were successful. The success rate for Mars landing missions is only 35%. Among the failed landing missions, most failures occur during the landing phase. The pinpoint landing has to be based on the precise autonomous navigation technology.

In the entry phase of a Mars landing, the lander is covered by a heat shield which blocks the optical sensor measurement, causing that all landers relied on the Inertial Measurement Unit (IMU) recursion. The initial errors of the lander cannot be corrected by IMU data. Even worse, the recursion errors using IMU are accumulated due to the sensor bias and noise. To overcome the incapability of IMU, the Mars Network-based Mars entry navigation is developed based on high frequency radio communication between the lander with ground or orbiting radio beacons [3–5]. Involving the radio measurement date into a navigation filter, the position and velocity of the lander can be optimally estimated.

The Mars Network-based Mars entry navigation is faced with two challenging. One is that the geometric configuration of the radio beacons affects the navigation performance. The other is that the available beacons at present are very limited. Considering these two factors, effort should be devoted to optimizing the configuration of radio beacons to maximize the function of the limited beacons. In [7], the navigation accuracy from the Extend Kalman Filter (EKF) by processing the radio measurements is analyzed, and the optimal configuration of ground beacons is selected among potential beacon position. Yu focused on the navigation observability and take it as a performance index to optimize the configuration of radio beacons [8]. The research on ground beacons, to some extent, inspired the future Mars landing navigation. However, the practice application of ground beacon-based navigation is hardly applied in practice. The first concern is that no ground beacon is available. Even if several beacons are distributed on Mars surface, it's still a tough job to place them exactly at the optimal locations. Moreover, the accurate positions of the beacons are hardly obtained accurately. Considering the immovability of ground beacons, the potential location areas are constrained by the line-of-sight visibility, resulting in an unsatisfactory beacon configuration during the entry phase.

As a substitution of ground radio beacons, the Mars orbiters which can also serve as beacons for Mars Network-Based Navigation are of more practice value. Currently, the operational orbiter around Mars includes 2001 Mars Odyssey and 2005 Mars Reconnaissance Orbiter. With another forthcoming spacecraft Mars Atmosphere and Volatile Evolution (MAVEN) [9], the capability of Mars network can be further increased. Focusing on how to fulfill the function of a Mars network, Ely firstly established the basic principle to design a constellation for navigation [10]. Then, taking the Mean of the Position Accuracy Response Time (MPART) as the performance index, the constellation configuration was optimized [11]. In [12], the number of orbiters and the coverage was considered to design the Martian navigation constellations envisaged in the ESA's Martian Constellation for Precise Object Location program. The optimization method of the above researches is inherited from the Global Positioning System (GPS). The global navigation performance was emphasized. For the limited amount of Mars orbiters, global coverage is difficult to realized, and local navigation performance should be investigated thoroughly for specific missions. Moreover, the effect of geometric configuration of the Mars network on the navigation performance should be revealed clearly. Inspired by these requirements, Yu et al. optimized the orbits of Mars orbiters in the observability point of view, and tried to explain the relationship between the configuration of beacons and orbiters and the navigation capability [13, 14].

To optimize the configuration of the radio beacons, a performance index should be firstly setup. The observability of the navigation system is selected as the performance index since it reflects the navigation capability directly. A lot of work has investigated the observability of linear and nonlinear dynamic systems [6, 15–17]. However, the analytic relationship between geometric configuration and observability has never been revealed. According to Cramér-Rao inequality [18], the inverse of the Fisher Information Matrix (FIM) estimates the lower bound of the estimation error. Therefore, FIM can be used to quantify the observability of the navigation system [19–21]. In this circumstance, some valuable analytic conclusions about the navigation design can thus be obtained.

Based on the requirement of the navigation optimization for Mars pinpoint landing, this chapter discusses the design and optimization of the Mars Networks-based navigation during Mars entry phase. Firstly, the Mars Networks-based navigation scheme is introduced, and the dynamic model and the observation model are given. Based on the navigation system, the observability of the Mars entry navigation analysis, and the analysis methods based on the quadratic approximation and Fisher information matrix are proposed. The relationship between the observability and the beacon configuration is derived, and the theoretically optimal configuration is given. Considering the constraints of Mars entry scenario, the ground beacons and the orbit of Mars orbiters are optimized based on observability based on an entry trajectory. The simulations also indicate the improved navigation performance.

2. Mars networks-based navigation scheme

2.1. Dynamic model of Mars entry phase

In the dynamical model with respect to a stationary atmosphere of a rotating planet, the 6 dimensional states x of the entry vehicle include r (radius from the center of Mars to the vehicle's center of mass), θ (longitude), ϕ (latitude), V (relative velocity), γ (flight path angle), and Ψ (heading angle, with $\Psi = 0$ as due east). The motion of the entry vehicle is governed by the following state equations:

$$\begin{aligned}
\dot{r} &= V\sin\gamma \\
\dot{\theta} &= V\cos\gamma\cos\Psi/(r\cos\phi) \\
\dot{\phi} &= V\cos\gamma\sin\Psi/r \\
\dot{V} &= -d - g\sin\gamma \\
\dot{\gamma} &= \left[l\cos\sigma - \left(g - V^2/r\right)\cos\gamma\right]/V + 2\omega\left(\tan\gamma\sin\Psi\cos\phi - \sin\phi\right) \\
\dot{\Psi} &= -\left(l\sin\sigma + V^2\cos^2\gamma\cos\Psi\tan\phi/r\right)/(V\cos\gamma) + 2\omega\cos\Psi\cos\phi
\end{aligned} \tag{1}$$

In the equation, σ is the banking angle, which is fixed at 0 in the following analysis. ω refers to the rotation rate of Mars. For simplicity, the second order terms of ω are neglected, which is feasible because the value of ω is quite small. Then the gravity acceleration g, lift and drag accelerations l and d are given by

$$g = \mu/r^2 \tag{2}$$

$$l = 0.5\rho V^2 C_l S/m \tag{3}$$

$$d = 0.5\rho V^2 C_d S/m \tag{4}$$

where μ is the Martian gravitational constant. S and m denotes the reference area and mass of the entry vehicle, and C_l and C_d are the lift and drag coefficients respectively. Furthermore, the Mars atmospheric density ρ is approximated by the conventional exponential model

$$\rho = \rho_0\exp[(r_0 - r)/h_s] \tag{5}$$

where $\rho_0 = 2 \times 10^{-4}$ kg/m3 is the reference density, $r_0 = 3437.2$ km is the reference radial position, and $h_s = 7500$ m refers to the atmospheric scale height. The dynamical model of the entry vehicle is abbreviated as $\dot{\boldsymbol{x}} = \boldsymbol{f}(\boldsymbol{x})$.

2.2. Observation model

The radio ranging and velocity data between the lander and the radio beacon can be measured through radio communication, given by

$$\begin{aligned}
y_{R_i} &= R_i + \varepsilon_R^i \\
&= \sqrt{\left(x_B^i - x\right)^2 + \left(y_B^i - y\right)^2 + \left(z_B^i - z\right)^2} + \varepsilon_R^i \\
x = r\cos\phi\cos\theta,&\quad y = r\cos\phi\sin\theta,\quad z = r\sin\phi
\end{aligned} \tag{6}$$

where R_i is the real range between the lander and the ith beacon, x_B^i, y_B^i, and z_B^i represent respectively the triaxial position components of the beacon, and ε_R^i is the radio ranging measurement noise.

The relative velocity model is given by

$$y_{V_i} = V_i + \varepsilon_V^i = dR_i/dt + \varepsilon_V^i \tag{7}$$

where V_i is the real line-of-sight relative velocity between the lander and the ith radio beacon, and ε_V^i is the velocity measurement noise.

With different radio beacons come different navigation scenarios. Without losing the generality, the observation model can be summarized as $\boldsymbol{y} = \boldsymbol{h}(\boldsymbol{x})$. Obviously, both radio measurements in Eqs. (6) and (7) are nonlinear. Moreover, the navigation performance is closely related to the geometric configuration of radio beacons. Therefore, the beacon configuration needs to be optimized based on the observability analysis.

3. Observability of the navigation system

3.1. Observability analysis based on the quadratic approximation

Consider the following nonlinear system:

$$\Sigma : \begin{cases} \dot{\boldsymbol{x}} = \boldsymbol{f}(\boldsymbol{x}) \\ \boldsymbol{y} = \boldsymbol{h}(\boldsymbol{x}) \end{cases} \tag{8}$$

where $\boldsymbol{x} \in \mathbb{R}^n$ is the n-dimensional state vector and $\boldsymbol{y} \in \mathbb{R}^m$ is the m-dimensional observation vector. Define $\boldsymbol{h} : \mathbb{R}^n \to \mathbb{R}^m$ as the nonlinear measurement operator.

The Lie algebra is an efficient tool for observability analysis. For the kth order Lie derivative of the jth measurement function, which can be expressed as $L_f^k h_j$, the $k+1$th order Lie derivative $L_f^{k+1} h_j$ with respect to the state equation $\boldsymbol{f}$ can be computed as:

$$L_f^{k+1} h_j = \sum_{i=1}^{n} \frac{\partial L_f^k h_j}{\partial x_i} f_i = \nabla L_f^k h_j \boldsymbol{f}, \quad k = 0, 1, \cdots \quad j = 1, 2 \cdots, m \tag{9}$$

The differential of $L_f^k h_j$ is defined as

$$\nabla L_f^k h_j = \left[\frac{\partial L_f^k h_j}{\partial x_1}, \quad \cdots, \quad \frac{\partial L_f^k h_j}{\partial x_n} \right] \tag{10}$$

Regarding the zero-order Lie derivative of the jth measurement function h_j as h_j itself, the matrix$\nabla L_f^k \boldsymbol{h}$ is given as

$$\nabla L_f^k \boldsymbol{h} = \left[\left(\nabla L_f^k h_1 \right)^T, \quad \cdots, \quad \left(\nabla L_f^k h_m \right)^T \right]^T \tag{11}$$

It is proven that the dynamical system $\boldsymbol{\Sigma}$ at state x_0 is locally observable if the observability matrix $\boldsymbol{O}_{\Sigma}$ given below has the rank of n.

$$\boldsymbol{O}_{\Sigma} = \left[\left(\nabla L_f^0 \boldsymbol{h}\right)^T, \quad \left(\nabla L_f^1 \boldsymbol{h}\right)^T, \quad \dots, \quad \left(\nabla L_f^{n-1} \boldsymbol{h}\right)^T \right]^T \Bigg|_{x=x_0} \tag{12}$$

It's a heavy burden to calculate the observability matrix in Eq. (12) due to the existence of high order differential, especially for the 6-dimensional dynamics of Mars entry phase which requires the calculation of 5th order Lie derivatives. Next, a quadratic approximation method is developed to simplify the computation of the observability matrix.

First of all, the quadratic approximation of the kth order Lie derivative $L_f^k h_j$ is given as

$$L_f^k h_j \approx L_f^k h_{j0} + \boldsymbol{J}_{Lj}^k (\boldsymbol{x} - \boldsymbol{x}_0) + \frac{1}{2} (\boldsymbol{x} - \boldsymbol{x}_0)^T \boldsymbol{H}_{Lj}^k (\boldsymbol{x} - \boldsymbol{x}_0) \tag{13}$$

where $L_f^k h_{j0}$ is the value of $L_f^k h_j$ at x_0, and $\boldsymbol{J}_L^k$and $\boldsymbol{H}_L^k$ refer to, respectively, the Jacobian and Hessian matrix of $L_f^k h_j$ at x_0. The linearized state equation is given by

$$\boldsymbol{f} \approx \boldsymbol{f}_0 + \boldsymbol{J}_f (\boldsymbol{x} - \boldsymbol{x}_0) \tag{14}$$

in which $\boldsymbol{f}_0$ refers to the value of $\boldsymbol{f}$ at x_0, and $\boldsymbol{J}_f$ is the Jacobi matrix of $\boldsymbol{f}$ at x_0.

According to Eq. (9) and Eq. (13), the relationship between the kth and k + 1th order Lie derivative can be rewritten as

$$\begin{aligned}
L_f^{k+1} h_j &= \nabla L_f^k h_j \cdot \boldsymbol{f} \\
&= \left[\boldsymbol{J}_{Lj}^k + \frac{1}{2} (\boldsymbol{x} - \boldsymbol{x}_0)^T \left(\boldsymbol{H}_{Lj}^k + \left(\boldsymbol{H}_{Lj}^k\right)^T \right) \right] \left[\boldsymbol{f}_0 + \boldsymbol{J}_f (\boldsymbol{x} - \boldsymbol{x}_0) \right] \quad = \boldsymbol{J}_{Lj}^k \boldsymbol{f}_0 \\
&+ \left[\boldsymbol{J}_{Lj}^k \boldsymbol{J}_f + \frac{1}{2} \left(\left[\boldsymbol{H}_{Lj}^k + \left(\boldsymbol{H}_{Lj}^k\right)^T \right] \boldsymbol{f}_0 \right)^T \right] (\boldsymbol{x} - \boldsymbol{x}_0) + \frac{1}{2} (\boldsymbol{x} - \boldsymbol{x}_0)^T \left[\boldsymbol{H}_{Lj}^k + \left(\boldsymbol{H}_{Lj}^k\right)^T \right] \boldsymbol{J}_f (\boldsymbol{x} - \boldsymbol{x}_0) \\
&= L_f^{k+1} h_{j0} + \boldsymbol{J}_{Lj}^{k+1} (\boldsymbol{x} - \boldsymbol{x}_0) + (\boldsymbol{x} - \boldsymbol{x}_0)^T \boldsymbol{H}_{Lj}^{k+1} (\boldsymbol{x} - \boldsymbol{x}_0)
\end{aligned} \tag{15}$$

This leads to

$$\begin{aligned}
L_f^{k+1} h_{j0} &= \boldsymbol{J}_{Lj}^k \boldsymbol{f}_0 \\
\boldsymbol{J}_{Lj}^{k+1} &= \boldsymbol{J}_{Lj}^k \boldsymbol{J}_f + \frac{1}{2} \left(\left[\boldsymbol{H}_{Lj}^k + \left(\boldsymbol{H}_{Lj}^k\right)^T \right] \boldsymbol{f}_0 \right)^T \\
\boldsymbol{H}_{Lj}^{k+1} &= \frac{1}{2} \left[\boldsymbol{H}_{Lj}^k + \left(\boldsymbol{H}_{Lj}^k\right)^T \right] \boldsymbol{J}_f
\end{aligned} \tag{16}$$

The observability matrix can be computed as

$$\begin{aligned} \boldsymbol{O}_\Sigma &= \left[\left(\nabla L_f^0 \boldsymbol{h}\right)^T, \quad \left(\nabla L_f^1 \boldsymbol{h}\right)^T, \quad \dots, \quad \left(\nabla L_f^{n-1} \boldsymbol{h}\right)^T \right]^T \Bigg|_{x=x_0} \\ &= \left[(\boldsymbol{J}_L^0)^T, \quad (\boldsymbol{J}_L^1)^T, \quad \dots, \quad (\boldsymbol{J}_L^{n-1})^T \right]^T \end{aligned} \tag{17}$$

where $\boldsymbol{J}_L^k = \left[(\boldsymbol{J}_{L1}^k)^T, \quad \dots, \quad (\boldsymbol{J}_{Lm}^k)^T \right]^T$.

Obtaining $\boldsymbol{J}_{Lj}^0$ and $\boldsymbol{H}_{Lj}^0$, the observability matrix can be iteratively calculated. Only 2nd order differential of $\boldsymbol{h}$ is needed here to compute the Jacobian and Hessian matrices, reducing largely the computation cost.

Linearize the dynamical and observation model by first-order approximation

$$\begin{aligned} \boldsymbol{f} &\approx \boldsymbol{f}_0 + \boldsymbol{J}_f(\boldsymbol{x} - \boldsymbol{x}_0) \\ \boldsymbol{h} &\approx \boldsymbol{h}_0 + \boldsymbol{J}_h(\boldsymbol{x} - \boldsymbol{x}_0) \end{aligned} \tag{18}$$

Construct the observability matrix according to the linear system theory

$$\boldsymbol{O}_\Sigma^l = \left[(\boldsymbol{J}_h)^T, \quad \left(\boldsymbol{J}_h \boldsymbol{J}_f\right)^T, \quad \dots, \quad \left(\boldsymbol{J}_h^{n-1} \boldsymbol{J}_f\right)^T \right]^T \tag{19}$$

The Hessian matrix is involved in the quadratic approximation, improving the accuracy of observability analysis compared with the linearized observability analysis. However, the higher order terms of $\boldsymbol{x} - \boldsymbol{x}_0$ may appear when computing $L_f^{k+1} h_j$ in Eq. (9) if the state equation is approximated to a higher order. In this case, the predetermined presentation form in Eq. (13) is no longer valid. One way to defeat this case is to increase the approximation order of Lie derivatives. Note that tensor calculus can be involved and the computation complexity is increased. Thus, the trade between accuracy and computation cost is balanced by the quadratic approximation of Lie derivatives and the linearization of state equation.

In the optimization of observability, the condition number of observability matrix is selected as the performance index, given by

$$cond(\boldsymbol{M}) = \frac{\sigma_{\max}(\boldsymbol{M})}{\sigma_{\min}(\boldsymbol{M})} \tag{20}$$

where $\sigma_{\max}$ and $\sigma_{\min}$ are, respectively, the maximum and minimum singular value of the matrix. The condition number measures the singularity of the matrix. A larger condition number means a more singular matrix. Here we take the inverse of the condition number to quantify the system observability.

$$\delta = \frac{1}{cond(\boldsymbol{O}_\Sigma)} = \frac{\sigma_{\min}(\boldsymbol{O}_\Sigma)}{\sigma_{\max}(\boldsymbol{O}_\Sigma)} \tag{21}$$

Obviously, the observability degree δ is in the interval [0, 1]. When $\delta = 0$, the observability matrix is rank defect, and the navigation system is locally unobservable. When $\delta > 0$, the observability is full rank, indicating an observable navigation system.

3.2. Observability analysis based on the fisher information matrix

Without loss of generality, we will consider the nonlinear observation models

$$y_i = h_i(\boldsymbol{x}) + \varepsilon_i, \quad i = 1, \cdots, N \tag{22}$$

This equation may describe the measurement of relative range and range-rate according to Eq. (6) and (7). Meanwhile, in order to investigate the impact of different measurement methods on the observability of position and velocity of the entry vehicle separately, the 3-dimensional state x may be r or v of the entry vehicle. The likelihood function of x is defined as the joint probability density function of multiple measurements given by

$$L(y_1, \cdots, y_3|\boldsymbol{x}) = \prod_{i=1}^{N} \frac{1}{\sqrt{2\pi}\sigma_i} \exp\left(-\frac{1}{2}\sigma_i^{-2}\|y_i - h_i(\boldsymbol{x})\|^2\right) \tag{23}$$

Then, take the negative of the natural log of Eq. (23) and omitting the terms not related to x, and the loss function can be derived as

$$J(\boldsymbol{x}) = \frac{1}{2}\sum_{i=1}^{N} \sigma_i^{-2}\|y_i - h_i(\boldsymbol{x})\|^2 \tag{24}$$

Find a state vector to minimize $J(\boldsymbol{x})$ and the state vector is the optimal estimation of the lander's states. The FIM of the state is given by

$$\boldsymbol{F} = E\left\{\frac{\partial^2}{\partial \boldsymbol{x} \partial \boldsymbol{x}^T} J(\boldsymbol{x})\right\} = \sum_{i=1}^{N} \sigma_i^{-2} \frac{\partial h_i(\boldsymbol{x})}{\partial \boldsymbol{x}} \left(\frac{\partial h_i(\boldsymbol{x})}{\partial \boldsymbol{x}}\right)^T \tag{25}$$

The estimate error covariance and FIM satisfy the following equation

$$\boldsymbol{P} \geq \boldsymbol{F}^{-1} \tag{26}$$

where $\boldsymbol{P}$ is the estimate error covariance, and "≥" means that $(\boldsymbol{P} - \boldsymbol{F}^{-1})$ is positive semidefinite. According to Eq. (26), the FIM can be used to evaluate the lower bound of the estimation error covariance, and further the system observability. Give the trace of $\boldsymbol{F}^{-1}$ in Eq. (27).

$$\mathrm{tr}\left(\boldsymbol{F}^{-1}\right) = \sum_{i=1}^{3} \frac{1}{\lambda_i} \tag{27}$$

where λ_i $(i = 1,\ 2,\ 3)$ are the eigenvalues of F. It's illustrated from Eq. (27) that larger eigenvalues of the FIM leads to smaller trace of estimation error covariance and stronger system observability. Quantify the observability by the determinant of FIM $\det(\boldsymbol{F}) = \prod_{i=1}^{3} \lambda_i$. The following relationship can be obtained.

$$\mathrm{tr}(\boldsymbol{P}) \geq \mathrm{tr}\left(\boldsymbol{F}^{-1}\right) = \sum_{i=1}^{3} \frac{1}{\lambda_i} > \frac{3}{\sum_{i=1}^{3} \lambda_i} = \frac{3}{\mathrm{tr}(\boldsymbol{F})} \tag{28}$$

Eq. (28) means that the trace of FIM measures the lower bound of estimation errors.

4. Observability analysis of Mars networks-based navigation

4.1. Observability analysis using only range measurements

In this subsection, the system observability using only range measurements between the lander and ground beacons is analyzed. Since no velocity information is included in Eq. (6), only the observability of the position vector is studied. The cases with different amount of beacons are studied.

4.2. One-beacon case

In this case, the FIM is given by

$$\boldsymbol{F}_1 = \sigma_{R1}^{-2} \frac{\partial R_1(\boldsymbol{r})}{\partial \boldsymbol{r}} \left(\frac{\partial R_1(\boldsymbol{r})}{\partial \boldsymbol{r}}\right)^T = \sigma_{R1}^{-2} \boldsymbol{n}_1 \boldsymbol{n}_1^T \equiv \sigma_{R1}^{-2} \boldsymbol{N}_1 \tag{29}$$

The rank of the matrix $\boldsymbol{N}_1$is only one. Solving the following equation

$$\det(\lambda \boldsymbol{I}_{3\times3} - \boldsymbol{N}_1) = 0 \tag{30}$$

Clearly, the eigenvalues of $\boldsymbol{N}_1$ are given by twice repeated 0 and $n_{1x}^2 + n_{1y}^2 + n_{1z}^2 = 1$. Therefore, the eigenvalues of $\boldsymbol{F}_1$ are given by $\lambda_1 = \lambda_2 = 0, \quad \lambda_3 = \sigma_{R1}^{-2}$.

Next, we have the eigenvector corresponding to λ_3

$$\boldsymbol{w}_3 = \frac{1}{n_{1z}} \left[n_{1x}, \quad n_{1y}, \quad n_{1z} \right]^T = \frac{1}{n_{1z}} \boldsymbol{n}_1 \tag{31}$$

The vector $\boldsymbol{w}_3$ corresponds to the observable state combination, and means that only the state component along the vector $\boldsymbol{n}_1$ can be observable.

According to Eq. (28), the lower bound of estimation errors can be obtained as

$$\frac{3}{\mathrm{tr}(\boldsymbol{F}_1)} = 3\sigma_{R1}^2 \tag{32}$$

Eq. (32) means the lower bound of estimation errors is higher than the estimation accuracy. In another word, the estimation accuracy cannot be higher than the measurement accuracy. Note that, even if multiple beacons are involved in the navigation system, the observability is still deteriorated if the beacons are located in similar direction.

4.3. Two-beacon case

Assume two non-collinear beacons, the FIM in Eq. (25) is derived by

$$F_2 = \sum_{i=1}^{2} \sigma_{Ri}^{-2} \frac{\partial R_i(\boldsymbol{r})}{\partial \boldsymbol{r}} \left(\frac{\partial R_i(\boldsymbol{r})}{\partial \boldsymbol{r}} \right)^T = \sum_{i=1}^{2} \sigma_{Ri}^{-2} \boldsymbol{n}_i \boldsymbol{n}_i^T \tag{33}$$

Involving one more measurement, the rank of F_2 is increased to two. The observable state combinations can be obtained by solving the eigenvalue and eigenvector. In this case, the eigenvalues of F_2 are given by $\lambda_1 = \lambda_2 \neq 0$, $\lambda_3 = 0$. The eigenvector corresponding to the zero eigenvalue is obtained as

$$w_3 = \left[\frac{n_{1y}n_{2z} - n_{1z}n_{2y}}{n_{1x}n_{2y} - n_{1y}n_{2x}}, \quad \frac{n_{1z}n_{2x} - n_{1x}n_{2z}}{n_{1x}n_{2y} - n_{1y}n_{2x}}, \quad 1 \right]^T = \frac{1}{n_{1x}n_{2y} - n_{1y}n_{2x}} \boldsymbol{n}_1 \times \boldsymbol{n}_2 \tag{34}$$

The vector w_3 gives the unobservable state component which is in the direction perpendicular to the plane constructed by $\boldsymbol{n}_1$ and $\boldsymbol{n}_2$. From an opposite view, all state components in plane are observable.

Since, in this case, the observability matrix is still zero, the navigation system is unobservable. According to Eq. (28), the lower bound of the estimation errors can be obtained as

$$\frac{3}{\mathrm{tr}(F_2)} = \frac{3}{\sum_{i=1}^{2} \sigma_{Ri}^{-2} \left(n_{ix}^2 + n_{iy}^2 + n_{iz}^2 \right)} = \frac{3}{\sum_{i=1}^{2} \sigma_{Ri}^{-2}} \geq \frac{3\sigma_{R\min}^2}{2} \tag{35}$$

where $\sigma_{R\min}$ is the smaller standard deviation among σ_{R1} and σ_{R2}. It's known by comparing Eqs. (32) and (35) that the estimation accuracy can be improved by using one more radio beacon.

4.4. More-than-two-beacon case

In this case, the FIM is given by

$$F_N = \sum_{i=1}^{N} \sigma_{Ri}^{-2} \frac{\partial R_i(\boldsymbol{r})}{\partial \boldsymbol{r}} \left(\frac{\partial R_i(\boldsymbol{r})}{\partial \boldsymbol{r}} \right)^T = \sum_{i=1}^{N} \sigma_{Ri}^{-2} \boldsymbol{n}_i \boldsymbol{n}_i^T, \quad N \geq 3 \tag{36}$$

The matrix F_N has a full rank, indicating an observable system. The determinant of F_N is given in Eq. (37).

$$\det(\boldsymbol{F}_N) = \sum_{1 \le k_3 < k_2 < k_1 \le N} \sigma_{Rk_1}^{-2} \sigma_{Rk_2}^{-2} \sigma_{Rk_3}^{-2} [\boldsymbol{n}_{k_1} \cdot (\boldsymbol{n}_{k_2} \times \boldsymbol{n}_{k_3})]^2 \tag{37}$$

The detailed derivation can be found in Ref. [14]. From Eq. (7), we can know that more radio beacons, no matter where they are, increase the determinant of the FIM, thus increase the system observability. To analyze the maximum value of $\det(\boldsymbol{F}_N)$, Eq. (37) is reorganized as

$$\det(\boldsymbol{F}_N) \le \sigma_{R\min}^{-6} \sum_{1 \le k_3 < k_2 < k_1 \le N} [\boldsymbol{n}_{k_1} \cdot (\boldsymbol{n}_{k_2} \times \boldsymbol{n}_{k_3})]^2 \tag{38}$$

where $\sigma_{R\min}$ is the minimum value among σ_{Ri}. The selection of the direction of radio beacons to maximize the observability can be described by the following optimization problem

$$\begin{aligned} &\max && \sum_{1 \le k_3 < k_2 < k_1 \le N} [\boldsymbol{n}_{k_1} \cdot (\boldsymbol{n}_{k_2} \times \boldsymbol{n}_{k_3})]^2 \\ &\text{subject to} && \|\boldsymbol{n}_i\| = 1, \quad i = 1, \cdots, N \end{aligned} \tag{39}$$

Note that the locations of radio beacons are not constrained. In cases with three beacons, the determinant of $\boldsymbol{F}_3$ is maximized if and only if $\boldsymbol{n}_1$, $\boldsymbol{n}_2$, and $\boldsymbol{n}_3$ are orthogonal to each other. However, no analytic results can be obtained when there are more than three beacons. Thus, a Genetic Algorithm is exploited to solve the optimization problem. The maximum determinants are listed in **Table 1**.

According to the results in **Table 1**, the relationship between the maximum determinant and the number of beacons can be induced by an exponential formulation, given by

$$\det(\boldsymbol{F}_N)_{\max} = \left(\frac{N}{3\sigma_{R\min}^{-2}}\right)^3 \tag{40}$$

The lower bound of estimation errors is derived as

$$\frac{3}{\operatorname{tr}(\boldsymbol{F}_N)} = \frac{3}{\sum_{i=1}^{N} \sigma_{Ri}^{-2} \left(n_{ix}^2 + n_{iy}^2 + n_{iz}^2\right)} \ge \frac{3\sigma_{R\min}^2}{N} \tag{41}$$

The change of lower bound of estimation errors with number of beacons is shown in **Figure 1**. It's shown that with more beacons comes more accurate estimation. However, the increasing rate of accuracy is slowed down, indicating that the navigation accuracy cannot be improved endlessly by only increasing the number of beacons.

4.5. Observability analysis of the navigation using range-rate measurements

4.5.1. Observability analysis of vehicle's velocity

The FIM of vehicle's velocity using range-rate data is given by

$$F_N = \sum_{i=1}^{N} \sigma_{Vi}^{-2} \frac{\partial V_i(\boldsymbol{r}, \boldsymbol{v})}{\partial \boldsymbol{v}} \left(\frac{\partial V_i(\boldsymbol{r}, \boldsymbol{v})}{\partial \boldsymbol{v}} \right)^T = \sum_{i=1}^{N} \sigma_{Vi}^{-2} \boldsymbol{n}_i \boldsymbol{n}_i^T, \quad N \geq 1 \tag{42}$$

Eq. (42) has a similar form with Eq. (36) which describes the FIM of position. The only difference lies in the measurement deviation. Hence the same conclusion of the observability of velocity can be obtained as that in Section 4.1. The detailed analysis is omitted here.

Number of beacons	Maximum determinant of FIM
3	$1.000\ \sigma_{Rmin}^{-6}$
4	$2.3704\ \sigma_{Rmin}^{-6}$
5	$4.6296\ \sigma_{Rmin}^{-6}$
6	$8.0000\ \sigma_{Rmin}^{-6}$
7	$12.7037\ \sigma_{Rmin}^{-6}$
8	$18.9630\ \sigma_{Rmin}^{-6}$

Table 1. Maximum determinants of FIM related to different number of beacons.

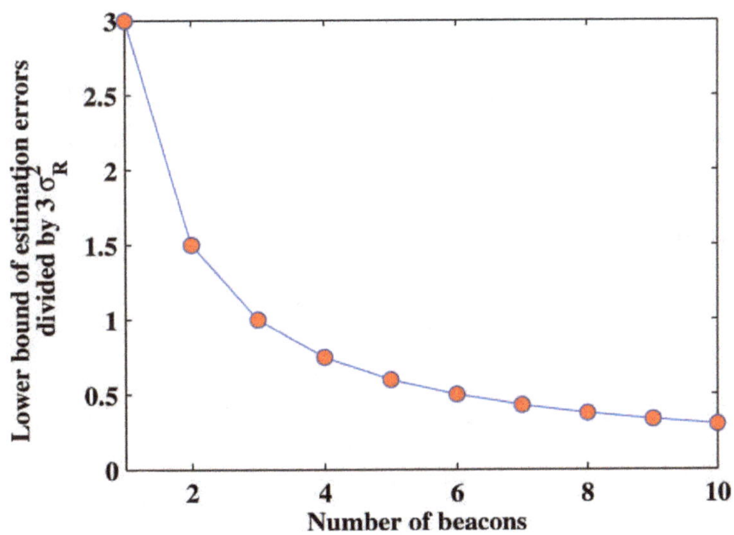

Figure 1. Lower bound of estimation errors with beacon number.

4.6. Observability analysis of vehicle's position

Using the range-rate measurements, the FIM of the lander's position is derived as

$$F_N = \sum_{i=1}^{N} \sigma_{Vi}^{-2} \frac{\partial V_i(\boldsymbol{r}, \boldsymbol{v})}{\partial \boldsymbol{r}} \left(\frac{\partial V_i(\boldsymbol{r}, \boldsymbol{v})}{\partial \boldsymbol{r}} \right)^T = \sum_{i=1}^{N} \sigma_{Vi}^{-2} \boldsymbol{L}_i \boldsymbol{v} \boldsymbol{v}^T \boldsymbol{L}_i^T, \quad N \geq 1 \tag{43}$$

where $\boldsymbol{L}_i$ is given by

$$L_i = \frac{1}{R_i}\begin{bmatrix} n_{iy}^2 + n_{iz}^2 & -n_{ix}n_{iy} & -n_{ix}n_{iz} \\ -n_{iy}n_{ix} & n_{iz}^2 + n_{ix}^2 & -n_{iy}n_{iz} \\ -n_{iz}n_{ix} & -n_{iz}n_{iy} & n_{ix}^2 + n_{iy}^2 \end{bmatrix} \tag{44}$$

The FIM here is much more complicated than that in Section 4.1 due to the involvement of both range and velocity information in FIM. Define $V_i = L_i v$, the following equation can be obtained.

$$F_N = \sum_{i=1}^{N} \sigma_{Vi}^{-2} V_i V_i^T, \quad N \geq 1 \tag{45}$$

When there are one or two beacons, the FIM is rank defect, and the navigation system is also unobservable. With three or more beacons comes the full-rank FIM. In this section, only the observable cases are focused on.

It is also concluded that the determinant of FIM will be zero if only one or two beacons is used, which indicates that the position of entry vehicle will be observable if more than two beacons are used. Furthermore, we focus on three-beacon and more-than-three beacon cases. Compare Eq. (46) with Eq. (36), we can find that the determinant of FIM for range-rate measurement cases has a similar format as Eq. (37)

$$\det(F_N) = \sum_{1 \leq k_3 < k_2 < k_1 \leq N} \sigma_{Vk_1}^{-2}\sigma_{Vk_2}^{-2}\sigma_{Vk_3}^{-2}[V_{k_1} \cdot (V_{k_2} \times V_{k_3})]^2 \leq \sigma_{V\min}^{-6} \sum_{1 \leq k_3 < k_2 < k_1 \leq N} [V_{k_1} \cdot (V_{k_2} \times V_{k_3})]^2, N \geq 3 \tag{46}$$

It's shown that Eq. (46) has the similar format with Eq. (37). Thus, the change of the observability with the number of radio beacons is similar with the results in **Table 1**. However, due to involving relative range and velocity information, the optimal geometric configuration is different with the cases using only range measurements.

The lower bound of estimation errors in this case is evaluated by

$$\begin{aligned}
\frac{3}{\mathrm{tr}(F_N)} &= \frac{3}{\sum_{i=1}^{N} \sigma_{Vi}^{-2} V_i^T V_i} \\
&= \frac{3}{\sum_{i=1}^{N} \frac{\sigma_{Vi}^{-2}}{R_i^2} \left\{ \begin{array}{l} \left[v_x\left(n_{iy}^2 + n_{iz}^2\right) - v_y n_{ix} n_{iy} - v_z n_{ix} n_{iz}\right]^2 + \left[-v_x n_{ix} n_{iy} + v_y\left(n_{ix}^2 + n_{iz}^2\right) - v_z n_{iy} n_{iz}\right]^2 + \\ \left[-v_x n_{ix} n_{iz} - v_y n_{iy} n_{iz} + v_z\left(n_{ix}^2 + n_{iy}^2\right)\right]^2 \end{array} \right\}} \\
&> \frac{3}{\sum_{i=1}^{N} \frac{\sigma_{Vi}^{-2}}{R_i^2} \left\{ \left[v_x\left(1 + \frac{1}{2}\left(n_{iy}^2 + n_{iz}^2\right)\right) + v_y\left(1 + \frac{1}{2}\left(n_{ix}^2 + n_{iz}^2\right)\right) + v_z\left(1 + \frac{1}{2}\left(n_{ix}^2 + n_{iy}^2\right)\right)\right]^2 \right\}} \\
&> \frac{3}{\frac{3}{2}\sum_{i=1}^{N} \frac{\sigma_{Vi}^{-2}}{R_i^2} (v_x + v_y + v_z)^2} \\
&\geq \frac{2}{3\sum_{i=1}^{N} \sigma_{Vi}^{-2} \frac{v^2}{R_i^2}}
\end{aligned} \tag{47}$$

where $v = \sqrt{v_x^2 + v_y^2 + v_z^2}$ is the lander's velocity value. Obviously, more radio beacons lead to more accurate estimation. Since the value of relative range is much bigger than relative velocity, the lower bound of estimation errors using range-rate date is larger than that using range data. Besides, it's concluded that more accurate range-rate measurement, closer relative range, and slower velocity can realizer more accurate position estimation.

5. Orbit optimization based on observability analysis

5.1. Optimization of navigation using ground beacons

The configuration radio beacons is expressed by the following set

$$C = \{p_B^i | i = 1, \cdots, l\} \tag{48}$$

where $p_B^i = \begin{bmatrix} x_B^i, & y_B^i, & z_B^i \end{bmatrix}^T$ is the position of the ith beacon. Considering the time-varying observability, the minimum value of the observability in the entry phase is taken as the optimization performance index.

$$D(C) = \min_{x \in T_x} \delta \tag{49}$$

To realize the Mars network-based navigation, the visibility of the beacons to the lander should be guaranteed. Define two unit vectors as follows

$$n_{Bi} = \frac{\begin{bmatrix} x_B^i, & y_B^i, & z_B^i \end{bmatrix}^T}{\sqrt{(x_B^i)^2 + (y_B^i)^2 + (z_B^i)^2}}, \quad n_C = \frac{[\tilde{x}, \tilde{y}, \tilde{z}]^T}{\sqrt{\tilde{x}^2 + \tilde{y}^2 + \tilde{z}^2}} \tag{50}$$

where $[\tilde{x}, \tilde{y}, \tilde{z}]^T$ is the relative position vector from the lander to the radio beacon, obtained as

$$[\tilde{x}, \tilde{y}, \tilde{z}]^T = [x, \quad y, \quad z]^T - \begin{bmatrix} x_B^i, & y_B^i, & z_B^i \end{bmatrix}^T \tag{51}$$

To guarantee the visibility, the two vectors in Eq. (50) should satisfy

$$\arccos (n_{Bi} \cdot n_C) < \frac{\pi}{2}, \quad x \in T_x \tag{52}$$

The schematic of visibility is shown in **Figure 2**.

The optimization problem of beacon configuration is given as

$$\begin{aligned} \max \quad & D(C) \\ s.t. \quad & p_B^i \in \Omega, \quad i = 1, \cdots, l \end{aligned} \tag{53}$$

where Ω is the set of the areas of radio beacons that satisfy the visibility during the whole entry phase. In this optimization problem, the global optimization algorithm is selected to obtain the optimal beacon configuration.

The initial states of the lander are listed in **Table 2**.

It is assumed that the Mars entry phase lasts approximately 240 seconds. The entry trajectory and the corresponding visible area are shown in **Figure 3**.

Three radio ranging measurements at a certain time can geometrically determine the position of the lander. Thus the navigation scenario with three beacons is first analyzed with respect to the observability. The optimal locations of beacons are displayed in **Figure 4**.

The optimal three beacons are located close to the edge of both sides of the visible area. The beacon on the east side is almost along the entry trajectory, while the west two beacons are separated on the north and south side of the entry trajectory. The observability degree in this situation calculated by different methods is illustrated in **Figure 5**, and the computation time for each method is listed in **Table 3**.

Figure 5 shows a huge undulation in observability degree during the Marts entry phase. The maximum and minimum value are 1.413×10^{-8} and 2.945×10^{-7} respectively. Considering the machine precision, the navigation system is observable only if the observability exceeds

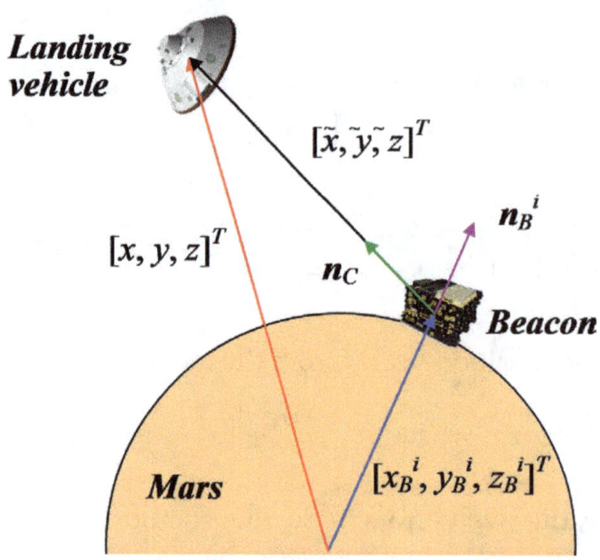

Figure 2. Principle of the line-of-sight visibility.

Initial state	$r(0)$km	$\theta(0)$deg	$\varphi(0)$deg	$V(0)$m/s	$\gamma(0)$deg	$\psi(0)$deg
Value	3518.2	−89.872	−28.02	5515	−11.8	5.156

Table 2. Initial states of the lander.

1×10^{-16}. The observability degree during the entire entry phase passes through the threshold, and thus, the navigation system is observable. The minimum degree of observability occurs at the beginning of the entry phase when the entry vehicle is at its greatest distance from radio beacons, while the maximum degree of observability occurs when the entry vehicle approaches two beacons on the west side. In order to explain the evolution of the degree of observability. An

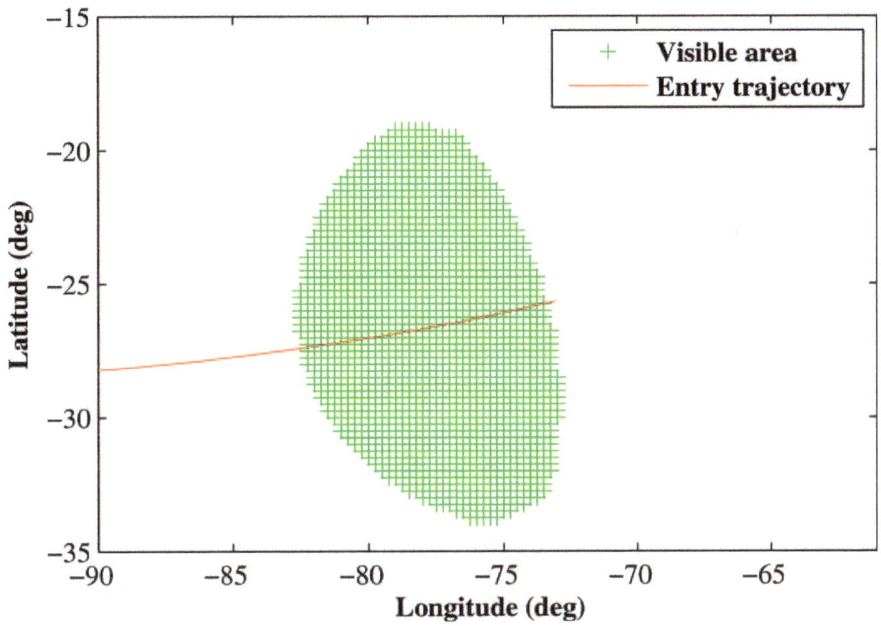

Figure 3. Entry trajectory and the visible area.

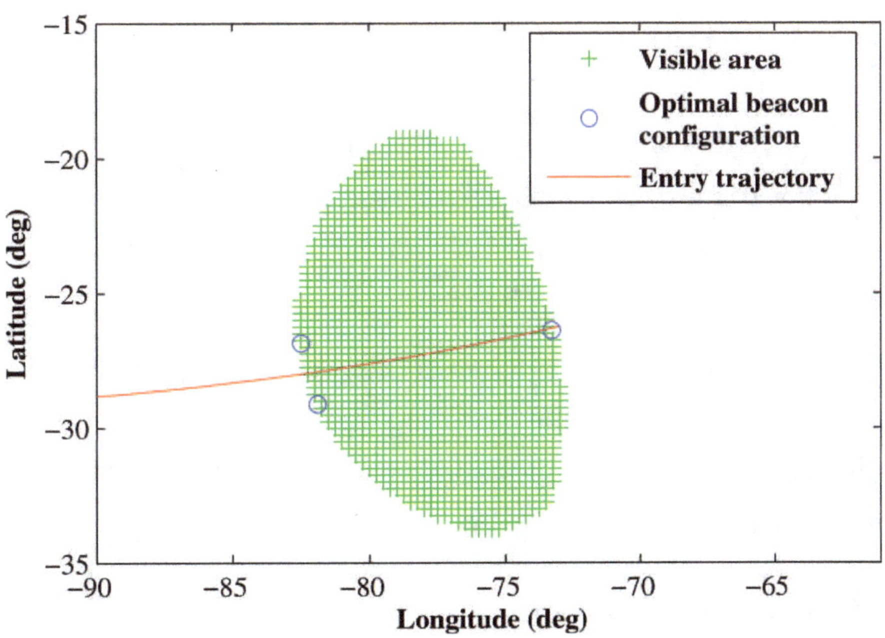

Figure 4. Optimal configuration for the scenario with three beacons.

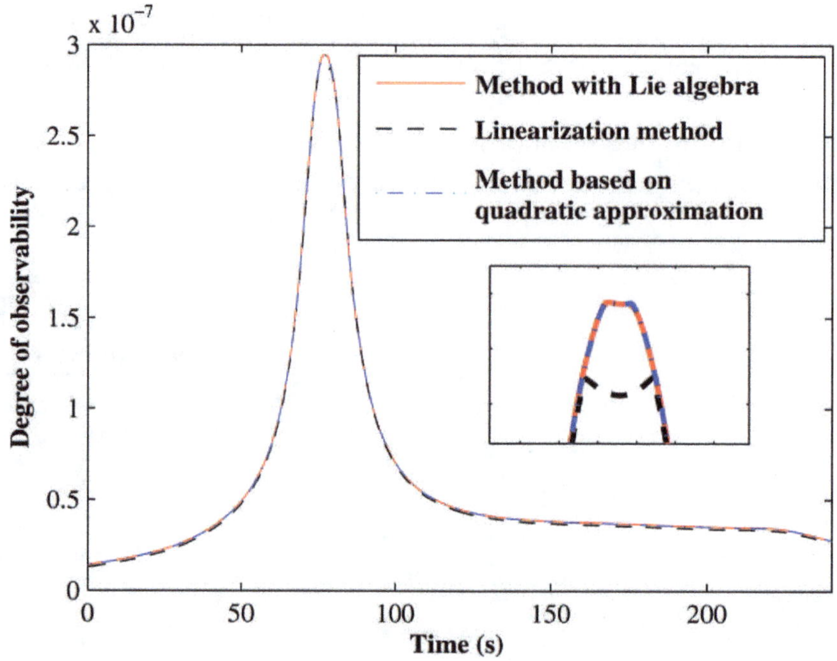

Figure 5. Degree of observability for the optimal scenario with three beacons.

Analysis approach	Computation time, s
Method with Lie algebra	>10,000
Linearization method	1.3987
Method based on quadratic approximation	2.1558

Table 3. Computation time for each approach.

index related to the geometric configuration of the lander and radio beacons is given in Eq. (54) to explain the evolution of the degree of observability.

$$I = \sum_{1 \leq i<j<k \leq N} \left[\boldsymbol{n}_i \cdot \left(\boldsymbol{n}_j \times \boldsymbol{n}_k \right) \right]^2 \tag{54}$$

where $\boldsymbol{n}_i$, $\boldsymbol{n}_j$, and $\boldsymbol{n}_k$ are the unit vectors from the beacon to the lander, N is the number of beacons. The evolution of index I is displayed in **Figure 6**, showing an identical variation trend with observability degree and backing up the observability analysis conclusion.

The observability degree obtained from the three methods is quite close to each other. However, the method based on Lie algebra consumes the most time. The linearization method provides the largest deviations, especially at the peak time, indicating a relatively low accuracy. The proposed quadratic approximation method achieves a performance balance in accuracy and complexity. To analyze the navigation accuracy, the Extended Kalman Filter (EKF) is

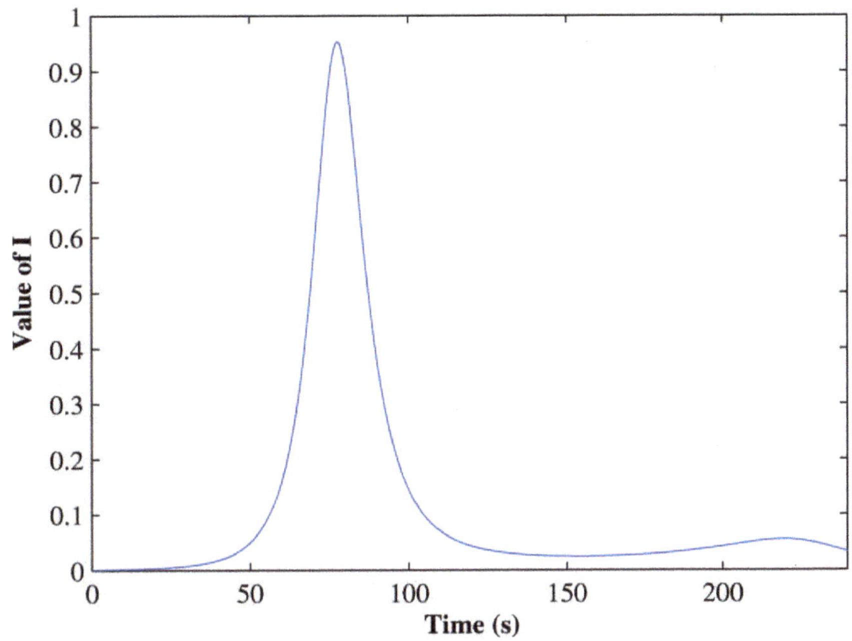

Figure 6. The value of I for the optimal scenario with three beacons.

Initial state	$r(0)$m	$\theta(0)$deg	$\varphi(0)$deg	$V(0)$m/s	$\gamma(0)$deg	$\psi(0)$deg
Error	1000	0.2	0.2	10	0.2	0.2

Table 4. Errors of initial states.

used to estimate the lander's states. The range measurement error is assumed to be Gaussian white noise with a standard deviation of 100 m. The initial errors are listed in **Table 4**. The estimation errors and the 1-sigma uncertainty bounds are depicted in **Figure 7**.

It's illustrated that θ and ϕ have the most accurate estimation and the fastest convergence. The convergence of the states V, γ, and Ψ is relatively slow at the beginning of the Mars entry phase due to the weak observability. With the increase of the observability degree comes the rapid convergence of the uncertainty bounds and the state estimation errors from about 90 to 115 seconds. The max deceleration of the lander also contributes to the rapid convergence.

5.2. Optimization of navigation using Mars orbiters

Compared with ground beacons, the Mars obiters are constrained by the orbital dynamics, which is considered to be two-body dynamics here. In this subsection, the initial states of the Mars orbiters are considered as the optimized variables. Furthermore, assuming that the Mars orbiters moves in a circular orbit, the variables to be optimized are simplified as inclination i, longitude of ascending node Ω, and the true anomaly f. The initial states of the orbiter can be expressed by the optimized variables, given by

$$\begin{cases} \boldsymbol{r}_{i0}^{B} = (R_M + a_i)\cos f_i \boldsymbol{P}_i + (R_M + a_i)\sin f_i \boldsymbol{Q}_i \\ \boldsymbol{v}_{i0}^{B} = -\sqrt{\mu/(R_M + a_i)}\sin f_i \boldsymbol{P}_i + \sqrt{\mu/(R_M + a_i)}\cos f_i \boldsymbol{Q}_i \end{cases} \tag{55}$$

where R_M is the radius of Mars, a_i is the orbit altitude, and $\boldsymbol{P}_i$ and $\boldsymbol{Q}_i$ are given by

$$\begin{cases} \boldsymbol{P}_i = [\cos\Omega_i, \quad \sin\Omega_i, \quad 0]^T \\ \boldsymbol{Q}_i = [-\sin\Omega_i\cos i_i, \quad \cos\Omega_i\cos i_i, \quad \sin i_i]^T \end{cases} \tag{56}$$

Given the initial states of the Mars obiter, the subsequent states can be obtained by propagating the two-body dynamics. Likewise, the trajectory of the lander can be also obtained by propagating the entry dynamics. To evaluate the overall performance of the observability of the entry phase, the integration of the observability is taken as the performance index, given by

$$I(\boldsymbol{e}) = \int_{t=0}^{t_f} O(t)dt \tag{57}$$

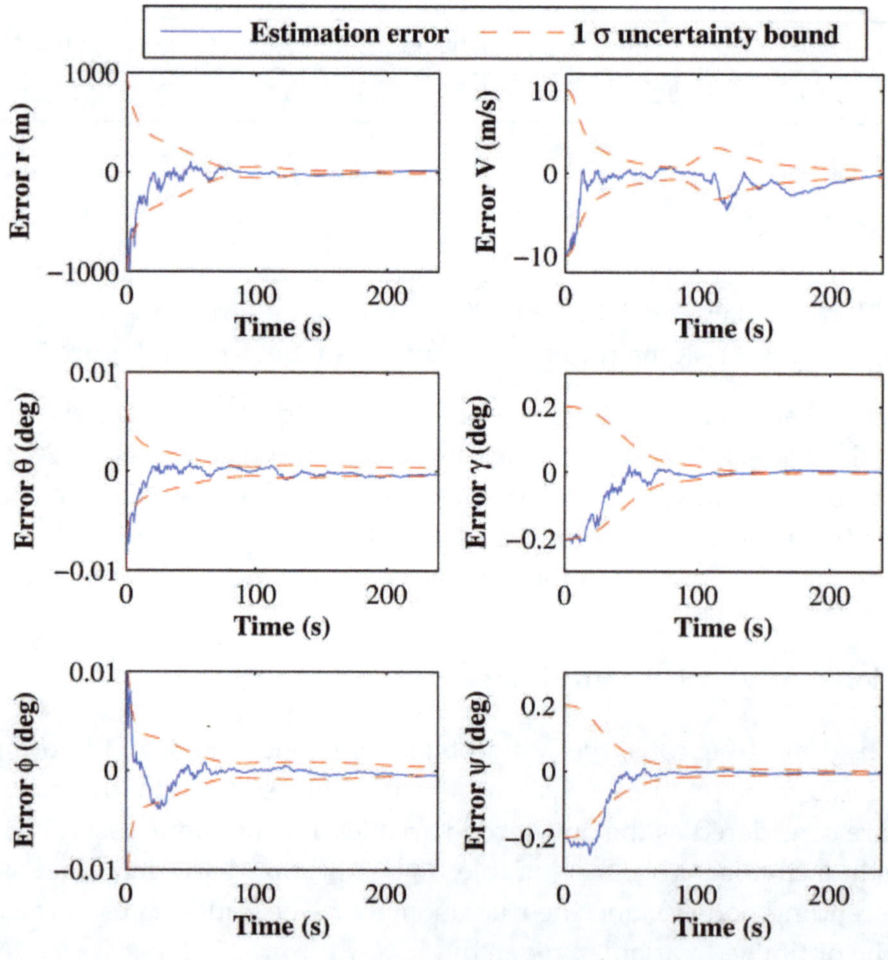

Figure 7. Navigation results for the optimal scenario with three beacons.

where t_f is the final time of entry phase, $\boldsymbol{e} = \{\boldsymbol{e}_1, \quad \cdots, \quad \boldsymbol{e}_n\}$, $\boldsymbol{e}_i = \begin{bmatrix}\Omega_i, & i_i, & f_i\end{bmatrix}^T$ denotes the optimization variables. Similar to ground beacon-based navigation, the visibility between the lander and the Mars obiters should be also guaranteed. Define two angles as follows:

$$\begin{cases} \theta_0 = \arccos\left(\dfrac{R_M}{\|\boldsymbol{r}\|}\right) \\ \theta_{1i} = \arccos\left(\dfrac{R_M}{\|\boldsymbol{r}_i^B\|}\right) \end{cases} \tag{58}$$

The angle between the position vectors $\boldsymbol{r}$ and $\boldsymbol{r}_i^B$ is given by

$$\theta_i = \arccos\left(\frac{\boldsymbol{r}_i^B \cdot \boldsymbol{r}}{\|\boldsymbol{r}_i^B\|\|\boldsymbol{r}\|}\right) \tag{59}$$

The visibility requires that

$$\theta_i < \theta_0 + \theta_{1i}, \quad t \in \begin{bmatrix}0, & t_f\end{bmatrix} \tag{60}$$

The schematic of the visibility is illustrated in **Figure 8**. The gray part represents the area in which the Mars orbiter is invisible to the lander.

Then the orbit optimization problem is given by

$$\begin{aligned} &\max \quad I(\boldsymbol{e}) = \int_{t=0}^{t_f} O(t)dt \\ &\text{subject to} \quad \theta_i < \theta_0 + \theta_{1i}, \quad t \in \begin{bmatrix}0, & t_f\end{bmatrix}, \quad i = 1, \cdots, n \end{aligned} \tag{61}$$

In the optimization problem, the performance index cannot be expressed explicitly by the optimization variables, and the gradient cannot be obtained. Thus, the heuristic global optimization algorithm is chosen to solve the optimization problem. The lander's initial states are listed in **Table 5** with the assumption of a ballistic entry having a banking angle of zero. The duration of entry phase is setup as 240 seconds.

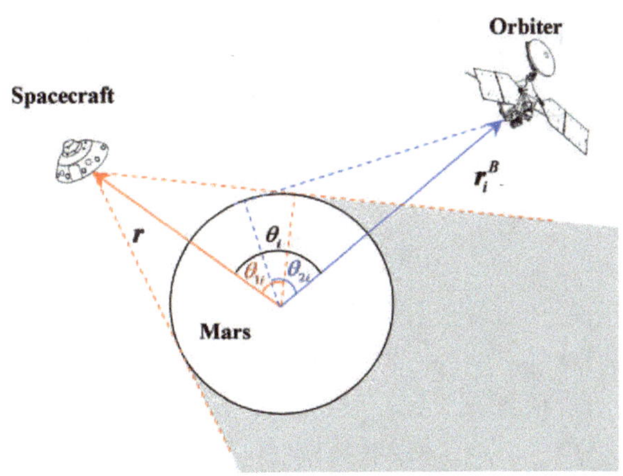

Figure 8. The schematic of the visibility.

The navigation scenario with three Mars orbiters is analyzed. The nominal orbit altitude of the three orbiters is 725 km. The observability is quantified by

$$O = \det(N_3) = [\boldsymbol{n}_1 \cdot (\boldsymbol{n}_2 \times \boldsymbol{n}_3)]^2 \tag{62}$$

At a certain epoch, the maximum value of O is 1 when and only when three unit vectors $\boldsymbol{n}_1$, $\boldsymbol{n}_2$, and $\boldsymbol{n}_3$ are orthogonal to each other. Considering the overall observability of the entry phase, the orbits of the Mars orbiters are optimized and shown in **Figure 9**, and the optimal initial elements are listed in **Table 6**.

It's shown that the three orbiters keep a relatively stable configuration, and stays orthogonal approximately to each other. The value of maximized performance index is 237.963. The observability almost reaches the maximum value all the time during the Mars entry phase. The comparison of Mars obiters-based navigation and ground beacon-based navigation is performed. The observability degree of these two scenarios is shown in **Figure 10**.

The fixed ground beacons have limited locations due to the visibility constrain and the geometric configuration cannot remain optimal during the entry phase. Thus, the observability is undulated to a large extent. The Mars orbiters overcome this defect with its moving property. To show straightforward the geometric configuration., the observability degree is close to maximum value at each epoch during the Mars entry phase. The angles between the vectors $\boldsymbol{n}_1$, $\boldsymbol{n}_2$, and $\boldsymbol{n}_3$ are depicted in **Figure 11**.

It's shown that, using the ground beacons, the angles between the three vectors change dramatically in the entry phase. The optimal configuration can be met only at the epoch of 75 s. However, for the orbiter-based navigation scheme, $\boldsymbol{n}_1$, $\boldsymbol{n}_2$, and $\boldsymbol{n}_3$ are almost orthogonal throughout the entry phase. The advantages of orbiter-based navigation scheme in the configuration and observability performance improve the navigation capability.

Next, 500-time Monte Carlo simulations of navigation systems based on EKF are carried out. The initial position and velocity have standard deviations of 1 km and 0.5 m/s respectively. The measurement error is set to be 50 m, and considered as Gaussian white noise. The simulation results are shown in **Figure 12**.

Since no information of entry vehicle's velocity is provided from range measurements, the convergence of velocity estimation is not as quick as position estimation. A much better navigation performance can be achieved by the Mars orbiter-based navigation. It can be

State	Value	Unit
x	−3.92	km
y	−3103.37	km
z	−1665.41	km
v_x	5775.31	m/s
v_y	1124.27	m/s
v_z	1175.48	m/s

Table 5. Initial states of the lander.

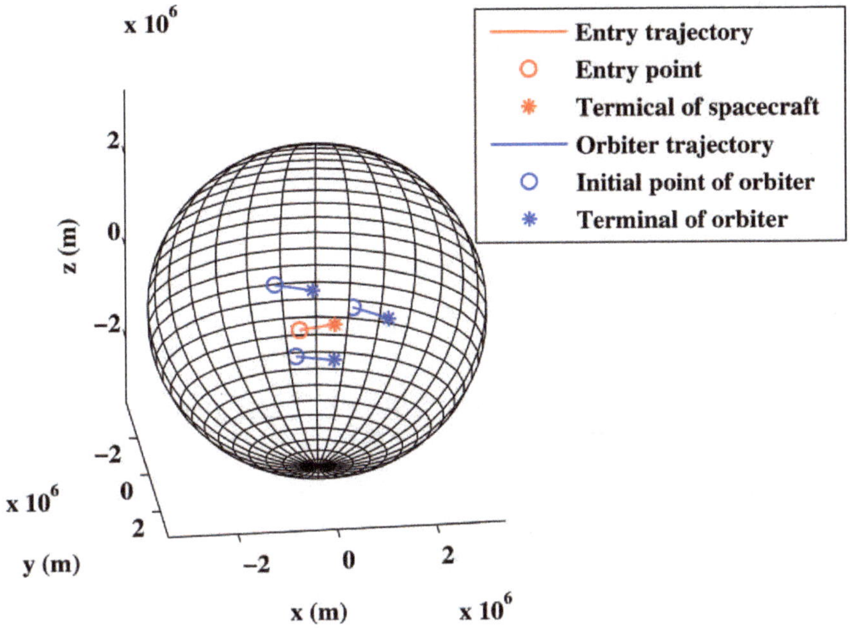

Figure 9. The optimal orbits of three orbiters.

Orbit element	Obiter 1	Orbiter 2	Orbiter 3
Ω (deg)	49.329	16.136	36.562
i (deg)	24.209	35.889	18.901
f (deg)	240.219	256.141	229.294

Table 6. Initial orbit elements of three orbiters.

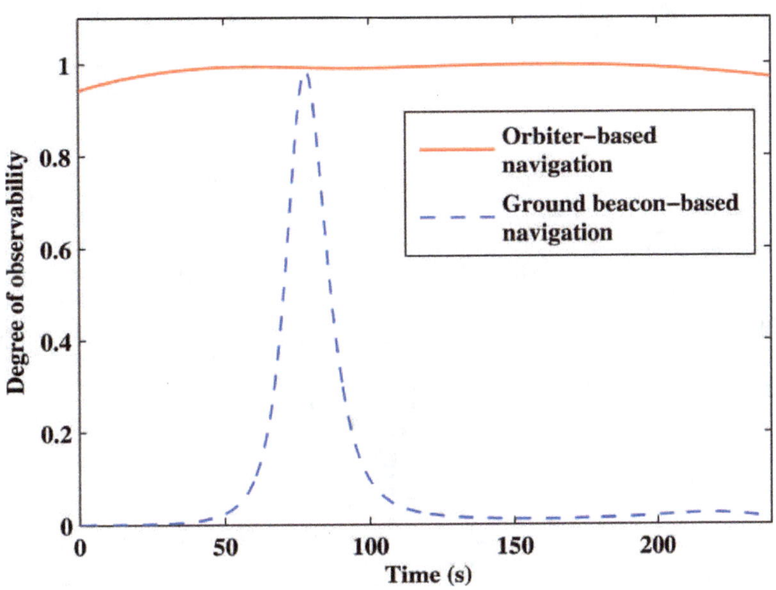

Figure 10. Degree of observability in two navigation schemes.

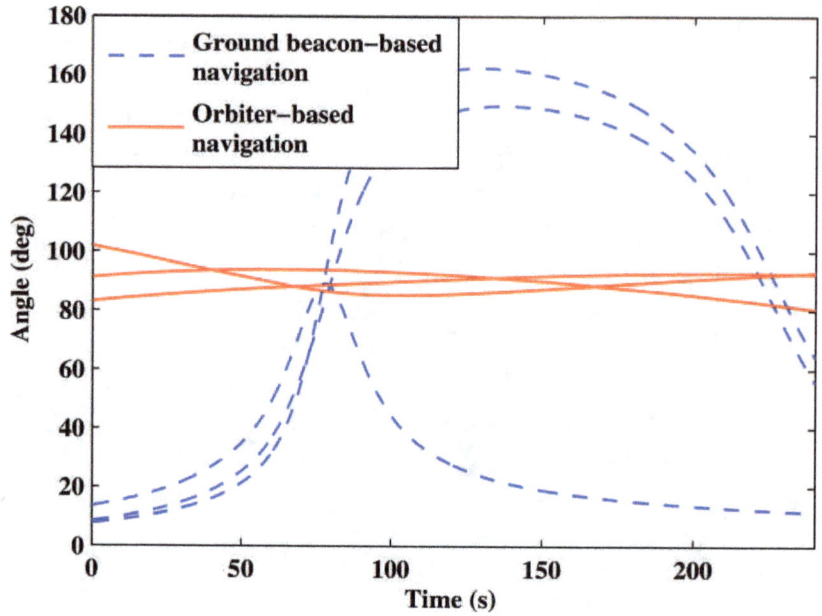

Figure 11. Angles between three unit vectors in two navigation schemes.

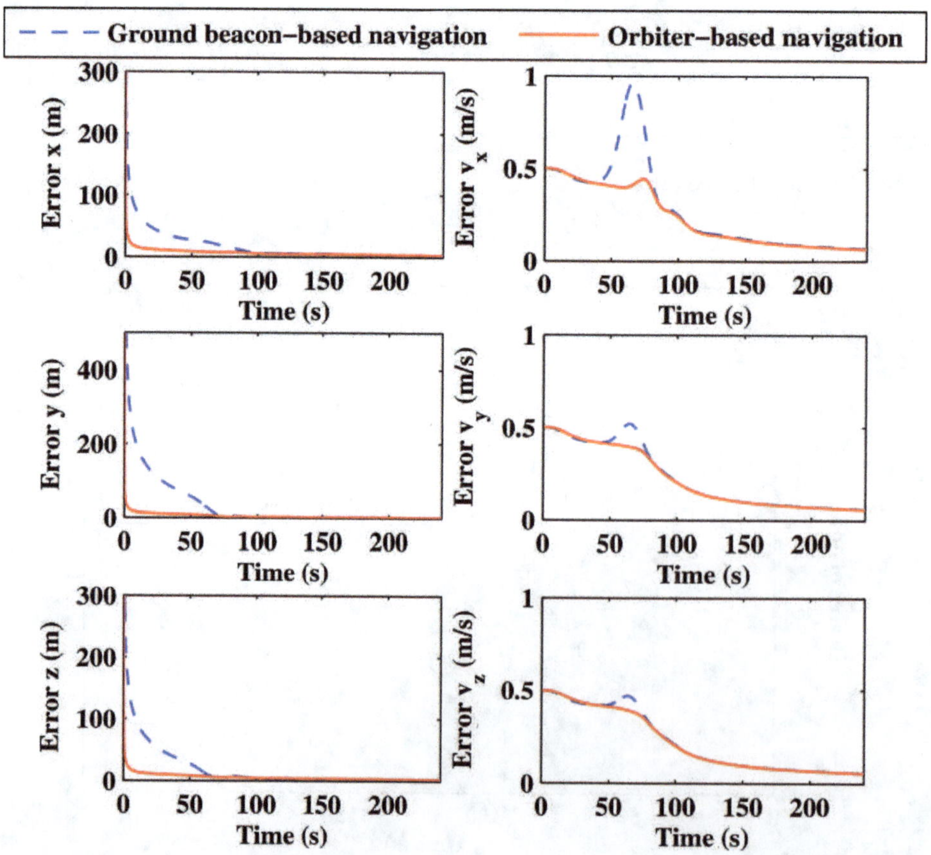

Figure 12. 1σ error bounds of states in two navigation schemes.

concluded that the configuration of orbiters is a main contributor to the navigation performance. The Mars orbiter-based navigation, which can achieve a better configuration, is more practical for Mars entry navigation.

6. Conclusions

This chapter introduced the Mars Networks-based navigation for the Mars entry phase. Based on the navigation scheme, the observability of the navigation system was analyzed using the proposed two novel observability analysis methods. Furthermore, the beacon configuration was optimized based on observability considering the line-of-sight constraints were concluded that the beacon configuration is a main contributor to the Mars Networks-based navigation. The observability analysis showed that an improved behavior of observability and more flexibility of beacon configuration determination can be achieved using more beacons. Navigation also demonstrated this conclusion. Meanwhile, compared with the ground beacons, Mars orbiters may be a better choice as Mars Network which gives a more accurate navigation result.

Author details

Zhengshi Yu[1,2]*, Pingyuan Cui[1,2], Rui Xu[1,2] and Shengying Zhu[1,2]

*Address all correspondence to: yuzhengshibit@qq.com

1 School of Aerospace Engineering, Beijing Institute of Technology, Beijing, China

2 Key Laboratory of Autonomous Navigation and Control for Deep Space Exploration, Ministry of Industry and Information Technology, Beijing, China

References

[1] Yu Z, Cui P, Crassidis J. Design and optimization of navigation and guidance techniques for Mars pinpoint landing: Review and prospect. Progress in Aerospace Sciences. 2017; **94C**:82-94

[2] Cui P, Yu Z, Zhu S. Research progress and prospect of autonomous navigation techniques for Mars entry phase. Journal of Astronautics. 2013;**34**(4):447-456

[3] Edwards CD, Adams JT, Bell DJ, et al. Strategies for telecommunications and navigation in support of Mars exploration. Acta Astronautica. 2001;**48**:661-668

[4] Hastrup RC, Bell DJ, Cesarone RJ, et al. Mars network for enabling low-cost missions. Acta Astronautica. 2003;**52**:227-235

[5] Lightsey EG, Mogensen A, Burkhart PD, Ely TA, Duncan C. Real-time navigation for Mars missions using the Mars network. Journal of Spacecraft and Rockets. 2008;**45**:519-533

[6] Lévesque JF, de Lafontaine J. Innovative navigation schemes for state and parameter estimation during Mars entry. Journal of Guidance, Control, and Dynamics. 2007;**30**(1): 169-184

[7] Pastor R, Bishop RH, Gay RS, Striepe SA. Mars entry navigation from EKF processing of beacon data. In: AIAA/AAS Astrodynamics Specialist Conference; 14–17 August; Denver, CO; 2000. p. AIAA 2000-4426

[8] Yu Z, Cui P, Zhu S. Observability-based beacon configuration optimization for Mars entry navigation. Journal of Guidance, Control, and Dynamics. 2015;**38**(4):643-650

[9] Chamberlain N, Gladden R, Bruvold K. MAVEN relay operations concept. In: 2012 IEEE Aerospace Conference; 3–10 March; Bigsky, MT; 2012

[10] Ely TA, Anderson R, Bar-Sever YE, et al. Mars network constellation design drivers and strategies. In: AAS/AIAA Astrodynamics Specialist Conference; 16–19 August; Girwood, AK; 1999. p. AAS 99-301

[11] Ely TA. Optimal orbits for sparse constellations of mars navigation satellites. In: AAS/AIAA Spaceflight Mechanics Meeting; 11–15 February; Santa Barbara, CA; 2001

[12] Pirondini F, Fernández AJ. A new approach to the design of navigation constellations around Mars: The MARCO POLO evolutionary system. In: the 57th International Astronautical Congress; 02–06 October; Valencia, Spain; 2006

[13] Yu Z, Zhu S, Cui P. Orbit optimization of mars orbiters for entry navigation: From an observability point of view. Acta Astronautica. 2015;**111**:136-145

[14] Yu Z, Cui P, Zhu S. On the observability of Mars entry navigation using radiometric measurements. Advances in Space Research. 2014;**54**(8):1513-1524

[15] Maessen DC, Gill E. Relative state estimation and observability for formation flying satellites in the presence of sensor noise. Acta Astronautica. 2013;**82**:129-136

[16] Hermann R, Krener AJ. Nonlinear controllability and observability. IEEE Transactions on Automatic Control. 1977;**22**(5):728-740

[17] Lall S, Marsden JE, Glavaški S. A subspace approach to balanced truncation for model reduction of nonlinear control systems. International Journal of Robust and Nonlinear Control. 2002;**12**(6):519-535

[18] Crassidis J, Junkins J. Optimal Estimation of Dynamic Systems. 2nd ed. Chapman & Hall/CRC: Boca Raton, FL; 2011

[19] Sun D, Crassidis J. Observability analysis of six-degree-of-freedom configuration determination using vector observations. Journal of Guidance, Control, and Dynamics. 2002; **25**(6):1149-1157

[20] Lee W, Bang H, Leeghim H. Cooperative localization between small UAVs using a combination of heterogeneous sensors. Aerospace Science and Technology. 2013;**27**(1): 105-111

[21] Cui P, Yu Z, Zhu S, Ai G. Real-time navigation for Mars final approach using X-ray pulsars. In: AIAA Guidance, Navigation, and Control Conference and Exhibit; 19–22 August; Boston, MA; 2013. p. AIAA 2013-5204

4

Suborbital Flight: An Affordable and Feasible Option for Mexican Aerospace Development

Barbara Bermudez-Reyes, Frederic Trillaud, Fernando Velazquez-Villegas, Jonathan Remba-Uribe, Ana M. Arizmendi-Morquecho, Alberto Caballero-Ruíz, Mario A. Mendoza-Barcenas, Rafael Prieto-Melendez, Leopoldo Ruiz-Huerta and Lauro Santiago-Cruz

Abstract

Suborbital flights are a low-cost option for universities. To perform suborbital missions, it is necessary to design, plan, test, verify, and validate each and every one of the subsystems that integrate the payload without leaving the Earth. In Mexico, some experiments have been carried out since the 1990s to test communication systems in case of disaster and emergency. The Mexican Service Gondola (CSM) from 2015 to date has made suborbital flights in conjunction with the National Polytechnic Institute and the group of Protective Coatings Resistant to Thermal Changes and Cosmic Radiation (CRTCR) to test communication systems and glass-ceramic coatings. Suborbital flights are a great opportunity to explore the national territory and test new communication systems, structures, and materials.

Keywords: suborbital flight, mexican service gondola, mission design, communication systems, glass-ceramic coatings

1. Introduction

Every day, there are new advances in the space area, specifically in satellites. Each time it becomes more complex to access space, due to requirements, restrictions (mass, volume, structural, etc.), cost and regulations, integration tests, and orbit assignment. For this, suborbital flights have become a viable option for probes of nanosatellite systems [1]. A suborbital flight can exceed 100 km in height and will not orbit the earth; that is, it will not leave the Earth's atmosphere [2]. Therefore, suborbital flights are viable for testing various subsystems and segments that are composing picosatellite and nanosatellite systems [3]. A nanosatellite system is characterized by its mass (1–10 kg), and its geometry can be cubic (CubeSat) or cylindrical (TubeSat). It should be noted that these nanosatellites can be composed by units of 1 kg (1 U) up to 10 units (10 U) [4]. These characteristics allow you to adjust and perform specific experiments or test subsystems in a timely manner and recover the nanosatellite, only if it does not go into space [3]. In addition, one of the advantages of performing suborbital flights is that you have a wide range of launch platforms as rockets, UAVs, and stratospheric balloons to climb into the high atmosphere [5]. The stratospheric balloons are an affordable platform for uploading nanosatellites and allow measurements during the ascent (infrared and ultraviolet radiation, X-rays, gamma rays, photographic recognition, and video capture) [6]. It also allows testing deployable systems (parachutes) for recovery of payloads without major damage [3]. For all the above, in Mexico, suborbital flights have become an alternative for sensors, communication, attitude, electrical subsystems, new materials resistant to thermal changes and cosmic radiation, etc., and the most important thing is that the universities allow the formation of human resources in the space area.

2. Suborbital flight: a window into space

In terms of altitude, a suborbital flight is limited by the Kármán line, which is a line 100 km above sea level. The Fédération Aéronautique Internationale defines this limit because it is roughly the point where a vehicle flying fast enough to support itself with aerodynamic lift from the Earth's atmosphere would be flying faster than orbital speed [7]. The Kármán line covers the troposphere (until 20 km), the stratosphere (until 50 km), the mesosphere (until 85 km), and a little section of the thermosphere, which extends until 690 km. **Figure 1** shows some physical properties of the atmosphere vs. altitude; this demonstrates that suborbital flight implies interesting conditions to implement scientific experiments [8].

Perhaps, the most attractive characteristic of suborbital flight is its cost, which is very low in comparison with space flight. According to NASA, today, it costs $10,000 to put a pound of payload in Earth orbit [9]. However, it costs about $1000 per pound to make a suborbital mission and that includes launching, tracking, and recovering the payload.

3. Suborbital flight: historical panorama

Mexico has been entering the area of suborbital flights in the late 1990s. Since then, a group of amateur radio operators from the "Radio Club Satélite" has sent several high-altitude balloons for experimental communication projects such as TSAT-1, TSAT-2, TSAT-3, and TSAT-4

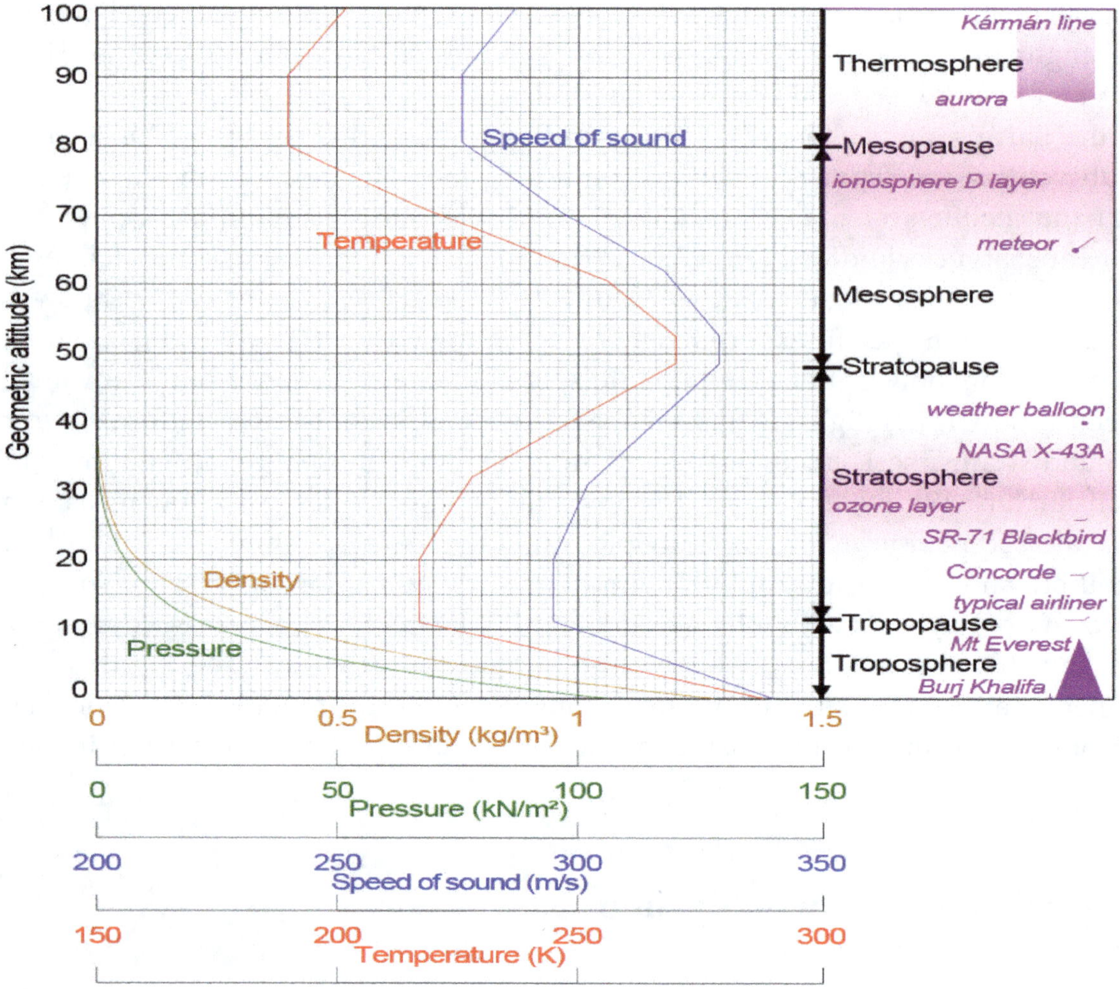

Figure 1. Comparison of the 1962 US standard atmosphere graph of geometric altitude against density, pressure, the speed of sound, and temperature with approximate altitudes of various objects [8].

and RCS-V, RCS-VI, RCS-VII, and RCS-VIII completing two test flights and eight full flights over the past two decades. In generally, flights only the RCS-V gondola was lost. The rate at recovery is over 90% owing to the use of onboard GPS that was not available at the time of the RCS-V project. The payloads were typically a 40 m band 4 W transmitter, 2 m band 0.7 W transmitter, altimeter, two temperature sensors, and a flight computer (**Figure 2**) [10].

In 2009, the CRAEG (Club de Radio Amateur del Estado de Guanajuato A.C.) launched their first project and one of the most complexes to date, SARSEM-ICARUS 1 (Mexican Aerostatic Subspace Repeater System). The objective of this project was to provide wide area communications in case of disaster or emergency for handheld and mobile VHF and UHF radios. The system carried a flight computer, a camera, temperature sensors, an onboard radio repeater, a dual GPS system, and a data communication system to download data and upload commands for controlling remotely specific subsystems. The gondola reached 28.8 km, had a radio coverage of 800 km, and was successfully recovered. In the following years (2010, 2011, 2013), SARSEM II, SARSEM III, and SARSEM IV were improved and redesigned, reaching their nominal altitudes and coverage as expected (**Table 1** and **Figure 3**) [11].

Figure 2. Projects (a) TSAT-1 (1992) and (b) RCS-V, RCS AC (1996).

Mission	Altitude (Km)
SARSEM II (2010)	28.7–800 km
SARSEM III (2011)	33.6–900 km
SARSEM IV (2013)	32.4–900 km

Table 1. Mexican suborbital missions [11].

Figure 3. SARSEM-ICARUS III, CRAEG AC (2011).

4. Mexican missions: CSM

Following the achieved successes over the past two decades, the "Carga de Servicio Mexicana" (CSM) or Mexican Service Gondola was born in 2014. It should be mentioned that this project was initiated as a spin-off of the Pixqui payload which flew in a NASA gondola on August 2013 with the participation of the Engineering Faculty of the Institute of Engineering under the leadership of the Institute of Nuclear Sciences of the National Autonomous University of Mexico (UNAM) [12]. The idea was to build a suborbital platform to promote aerospace technologies by means of stratospheric flights between the altitudes of 25 and 35 km to test electronic systems and detectors used in satellites. This project followed the need to increment the aerospace technologies following the creation of the Mexican Space Agency around 2010, merging the effort of the national academic and industrial sectors [13]. From the start, the participation of undergrad and graduate students was considered essential for the future development of the technology in Mexico. It should be noted that suborbital platforms provide a cheap and easy-to-handle test facility based on sounding stratospheric balloons, which have flight durations of the order of a few hours. They are useful tools to test in near-space conditions of various aerospace systems and subsystems. In 2015 and 2016, two suborbital flights were successfully carried out over the state of Guanajuato, Mexico. A first gondola of 2.5 kg was launched, the CSM-1 housed one of the subsystems of Ulises 2.0, a nanosatellite developed by the Unidad de Alta Tecnología (UAT) of the Faculty of Engineering of the UNAM. It reached an altitude of nearly 31 km monitoring the thermal behavior of the electronic payload in addition to the temperatures inside and outside the gondola. This first experience was achieved because of a successful collaboration between the Institute of Engineering, the Engineering Faculty of the UNAM, the company, Remtronic Telecomunicaciones, and the Amateur Radio Club of the state of Guanajuato (CRAEG). The second gondola, CSM-2, with a weight of 2.1 kg, was an improvement of the first version, CSM-1, which had a crude structural design. Indeed, during the first flight, the gondola reached a velocity of about 8 m/s at landing despite the use of a parachute. The force of the impact deformed the structure, but the payload did not suffer any damages thanks to a custom-designed floating structure holding the payload inside the gondola. However, it appeared necessary to improve the impact absorbers, and a new design was tested during the second flight. This second flight carried two payloads, a monitoring atmospheric electronic system (SADM-1) and an experiment to try out a ceramic coating for satellites [14, 15]. Indeed, the new design allowed lowering the overall mass of the gondola while ensuring a better absorption at impact. This flight reached an altitude of 34 km. The external temperature of the gondola reached −70°C over a flight duration of about 2 h. For this second flight, the original collaboration included a new member, the National Laboratory for Additive Manufacturing, 3D Digitization and Computed Tomography (MADiT) of the Center of Applied Sciences and Technological Development (CCADET) of the UNAM. MADiT has strong capabilities and experience in design and manufacturing by means of additive manufacturing technologies. This national laboratory collaborates in the development of the structure of the gondola.

Part of the development of the gondola, CSM, a series of studies covering mechanical and thermal aspects have been conducted to improve the reliability of the structure and to diminish the risks, mainly associated with cold temperatures (<−60°C) and impact at landing [16, 17]. **Figure 4** shows a comparison between experimental and numerical values of the altitude of the balloon versus time of flight for both flights [17]. **Figure 5** is a photograph of the CSM-2 team just before the launch from the Explora Science Center, León, Guanajuato, Mexico, on November 2016. A new flight of CSM-2 is scheduled on April 2017 with additional clients.

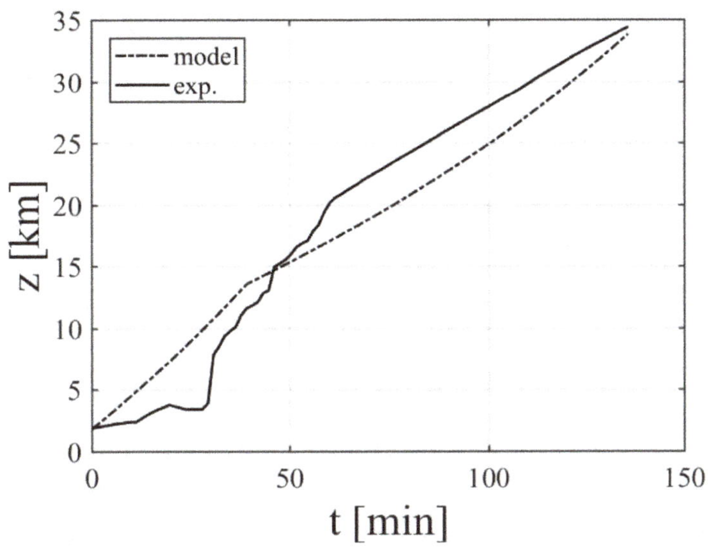

Figure 4. Comparison between the recorded altitude and the modeling one as a function of time [17].

Figure 5. CSM-2 team just before the launch of CSM-2 in November 2016 at the Explora Science Center, León, Guanajuato, Mexico.

The development of space technologies requires the use of stratospheric balloons since they offer a cheap, near-space environmental platform to test electronic systems of any kind. For the Mexican team developing CSM, it is the belief that it can lead to a sustainable development of the aerospace sector in the country involving the academic, governmental, military, and industrial sector. The first step has been carried out in that direction, and the future goal is to lift greater mass of a few tens of kilograms to improve the service already provided by CSM.

On the other hand, in Mexico, they have been designing glass-ceramic coverings to protect satellite systems of cosmic radiation and thermal changes. Aerospace materials must be lightweight and resistant to structural stresses, as well as to conditions in space [18, 19].

In the manufacture of satellites, aluminum alloys maintain the leadership among lightweight and relatively low-cost structural metal materials, followed by alloys based on titanium [20, 21] for their low density and high resistance to corrosion.

Glass-ceramic coatings have proven to be a key technology in thermal stability and protection coatings in extreme environments such as aerospace gas turbines, which together with the engine operate at high temperatures (1370–1425°C) [22, 23]. So, the selection of materials for thermal barrier coatings is restricted by some basic requirements: (1) high melting point of the ceramics, (2) they must not present phase transformation between the ambient temperature and the operating temperature, (3) low thermal conductivity, (4) chemically inert in space environment, (5) the thermal expansion must be related to that of the metallic substrate,

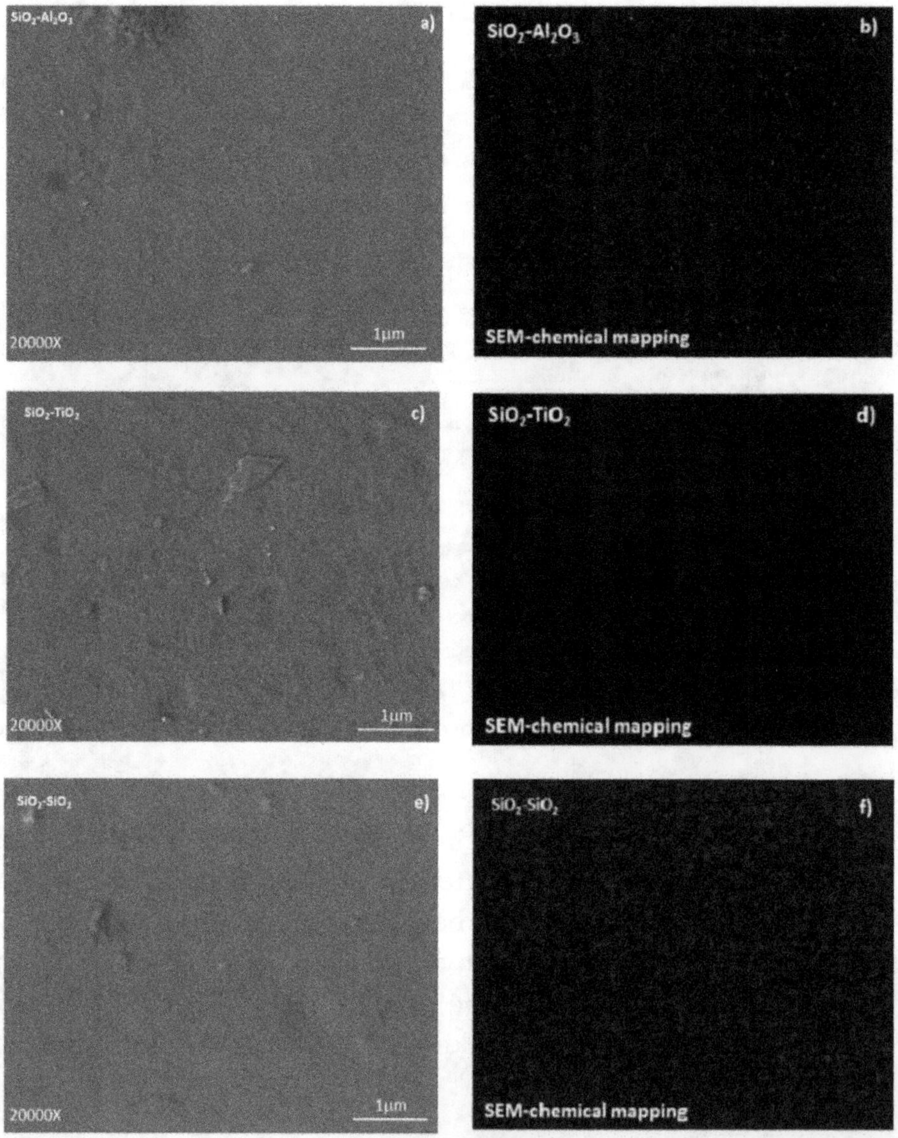

Figure 6. SEM images coatings: (a) SiO_2-Al_2O_3 surface, (b) SiO_2-Al_2O_3 chemical mapping, (c) SiO_2-TiO_2 surface, (d) SiO_2-TiO_2 chemical mapping, (e) SiO_2-SiO_2 surface, and (f) SiO_2-SiO_2 chemical mapping.

(6) good adhesion to the metallic substrate, and (7) low heating rate during the sintering process [24, 25]. The materials that can be used as glass-ceramic coatings is very limited; so far, only some materials have been found that meet these requirements [26]: aluminum oxide (Al_2O_3), mullite ($3Al_2O_3$-$2SiO_2$), cordierite ($2MgO$-$2Al_2O_35SiO_2$), zirconium oxide (ZrO_2), and zirconia stabilized with yttrium (ZrO_2-Y_2O_3). These coatings have been initially proposed for use as thermal control surfaces in aircraft because lightweight coatings with good adhesive properties can be obtained under thermal shock conditions [27]. This is why it could be considered that glass-ceramic coatings applied in satellite systems could resist ultraviolet, cosmic, and high-energy particles over a wide temperature range [28].

Therefore, the team of Coatings Resistant to Thermal Changes and Cosmic Radiation (CRTCR) from Space Science and Technology Network (REDCyTE) has designed glass-ceramic coatings reinforced with nanometric ceramic particles of Al_2O_3, SiO_2, and TiO_2, to obtain a multilayer system that is highly reflective and thin and homogeneous through the sol-gel. **Figure 6** shows images of scanning electron microscopy (SEM) in topography mode, in which the surfaces of the coatings are shown. In the same figure, SEM images are shown in chemical mapping mode in which the uniform dispersion of the ceramic nanoparticles immersed in the vitreous matrix is shown.

5. Challenges in medium and long term

In 2010, the Mexican Space Agency was born [29]. It followed a few decades of academic work and the need for an institution to formalize and federate projects toward the development of national space technologies. In 2015, a specific fund allocated to space technology was created by the government through the Federal Funding Agency (CONACYT) [30]. A few projects have been so far benefitting from this fund, and among them, one project dedicated to suborbital flights using stratospheric balloon was funded (ATON) [30]. Unfortunately, it was the only project funded to promote suborbital flights, and it is clear that more incentives are needed to allow the development of this sector in a near future and also to sustain the development of the space technologies in Mexico. In medium term, one can prospect for a need to increase the gondola size to provide services to larger payloads with greater mass. Typical light payloads range from less of a few kilograms to tens of kilograms. To lower the cost of flights and provide a service throughout the year, it is essential to be able to carry up to a ton if not a few hundreds of kilograms. At those masses, the risk associated with structural damages and health safety that such a gondola can create in semi-urban centers is too high. Therefore, flights over the oceans or the deserts, both available in Mexico, should be planned under the resources of the marine, air force, and the government for proper permits. It seems unlikely that flights over the jungle as one could plan in south of Mexico is realistic, due to the complexity to recover a gondola in a harsh, difficult-to-access environment with dense vegetation. States such as Sonora, Chihuahua, and Durango for desert lands or Baja California, Sinaloa, and Veracruz for oceans can be explored as possible sites for launches. Additionally, specific balloons of larger flight durations should be investigated and tried out in conjunctions with large masses of a few hundreds of kilograms to a ton.

Besides the practicability of the launch, tracking and recovery of the gondola, and the payload, it is important that the structure, which the gondola offers, houses onboard electronics and power services. The system should be able to monitor the environment and itself and to supply the payloads with the necessary power for their operation. Multidisciplinary teams of engineers in mechanics, aerospace technology, telecommunication, electrical and electronic systems, and aeronautics should be part of this effort, and likely a national center for aerospace technology can be the right place to gather those specialties. There is an important risk that allocating resources to small individual projects may not be fructuous on the long term and it may be sensible to lay out one or a couple of projects of national interest that can lead the effort with a sustainable funding.

6. Conclusions

Mexico has been involved into suborbital flights since the past two decades through amateur impulse. Lately, the academic sector has been starting to promote the use of stratospheric sounding balloons to provide the scientific and technological communities and inexpensive, easy-to-use facility to test their aerospace technologies. A few suborbital flights over the country have been carried out by different institutions with a large participation of students. Among the different projects, CSM is the only one involving mostly researchers and engineers from the academic and private sectors with the objective to professionalize this area of knowledge following the example of leading countries. This platform has provided two flights of 2-h duration for altitudes over 30 km to three clients. It is expected to provide yearly flights to future clients accommodating payloads of larger masses.

Overall, suborbital flights are the first step for Mexico to get involved in the aerospace development. Increasing funding, engineering schools, and national laboratories dedicated to aerospace studies and technological and scientific projects with the participation of the government, the military and the private sector is considered mandatory to achieve a significant contribution to the field in Latin America.

Acknowledgements

The CSM team would like to thank the Institute of Engineering and the Faculty of Engineering of the UNAM for sponsoring CSM.

They also thank the DGAPA for the support through the project IN113315. Finally, the authors really appreciate the participation and enthusiasm of Eduardo Amaro Calderon, Diego Dominguez Baez and Alfredo Sanchez Labra.

Author details

Barbara Bermudez-Reyes[1*], Frederic Trillaud[2], Fernando Velazquez-Villegas[3], Jonathan Remba-Uribe[4], Ana M. Arizmendi-Morquecho[5], Alberto Caballero-Ruíz[6], Mario A. Mendoza-Barcenas[7], Rafael Prieto-Melendez[7], Leopoldo Ruiz-Huerta[6] and Lauro Santiago-Cruz[2]

*Address all correspondence to: barbara.bermudezry@uanl.edu.mx

1 Faculty of Mechanical and Electrical Engineering, Autonomous University of Nuevo Leon, Nuevo León, México

2 Institute of Engineering, National Autonomous University of Mexico, México City, México

3 Center of Advanced Engineering, Faculty of Engineering, National Autonomous University of Mexico, México City, México

4 Remtronic Telecommunications, Guanajuato, México

5 Center of Research in Advanced Materials-Monterrey, Nuevo León, México

6 National Laboratory of Additive Manufacturing, 3D digitalization and Computed Tomography, Center of Applied Sciences and Technological Development, National Autonomous University of Mexico, México City, México

7 Aerospace Development Center, National Polytechnic Institute, México City, México

References

[1] Welti CR. Satellite Basics for Everyone. Blomington: iUniverse, Inc.; 2012. pp. 10-16

[2] Thom Stone, Marcus Murbach, Richard Alema, Ray Gilstrap. SOAREX- Suborbital Experiments 2015 – A New Paradigm for Small Spacecraft Communication. In: Aerospace Conference 2015. Montana, USA: IEEE

[3] Sako N, Tsuda Y, Ota S, Eishima T, Yamamoto T, Ikeda I, Li H, Yamamoto H, Tanaka H, Tanaka A, Nakasuka S. Cansat suborbital launch experiment-university educational space program using can sized Pico-satellite. Acta Astronautica. 2001;**48**(5-12):767-766

[4] Briess K. Space design process. In: Ley W, Wittmann K, Hallmann W, editors. Handbook of Space Technology. 1st ed. United Kingdom: Jonh Wiley and Sons, Ltd; 2009

[5] Selva D, Dingwall B, Altunc S. A concept for an Agile Mission Development Facility for Cubesat and suborbital Missions. In: Aerospace Conference. Montana, USA: IEEE; 2016

[6] Antunes S. DIY Instruments for Amateur Space. 1st ed. California, USA: O'Reilly Media Inc.; 2013

[7] Sanz Fernández de Córdoba S. Karman separation line, used as the boundary separating Aeronautics and Astronautics. Fédération Aéronautique Internationale; 2004. https://www.webcitation.org/618QHms8h?url=http://www.fai.org/astronautics/100km.asp [accessed: 12/10/2017]

[8] Advanced Space Transportation Program: Paving the Highway to Space, Marshall Space Flight Center, NASA. Available from: https://www.nasa.gov/centers/marshall/news/background/facts/astp.html [Accessed: Dec 11, 2017]

[9] U. S. Centennial Commission. Geometric altitude vs. temperature, pressure, density, and the speed of sound derived from the 1962 U.S. Standard Atmosphere. http://www.centennialofflight.net/essay/Theories_of_Flight/atmosphere/TH1G1.htm [Accessed: 12/07/2017]

[10] Solana J. Radio Club satellite A.C. BOLETIN XE1RCS. Historical Achieves (2008-2015). Mexico; 2017

[11] Club de Radio Amateur del Estado de Guanajuato A.C.-XE1CRG. Historical Archives (2009-2015). Guanajuato, Mexico; 2017

[12] CONACYT Press. Pixqui: plataforma mexicana de pruebas satelitales, Aug 18 2015. Available from: www.conacytprensa.mx [Accessed: Aug 11, 2017]

[13] Agencia Espacial Mexicana (AEM) Antecedentes de la AEM. Available from: www.gob.mx/aem/acciones-y-programas/antecedentes-de-la-aem [Accessed: Apr 12, 2017]

[14] Mendoza Bárcenas MA, Prieto Meléndez R, Santiago Cruz L, Trillaud F, Espinosa Calderón A, Herraiz Sarachaga M, Velázquez Villegas F. Módulo experimental de carga útil "SADM-1" para fines de exploración atmosférica. Sociedad Mexicana de Instrumentacion. In: XXXII Congreso de Instrumentación. Guerrero, Mexico; October 2017

[15] Herrera-Arroyave JE, Bermúdez-Reyes B, Ferrer-Pérez JA, Colín A. CubeSat system structural design. In: 67th International Astronautical Congress. Guadalajara, Mexico; September 2016. pp. 1-5

[16] Sebastian Rosas Contreras. Diseño de la Estructura Mecánica de una Carga de Servicio Ligera de Globos Estratosféricos Nacionales [undergraduate thesis]. UNAM; Oct 1, 2016

[17] Singer Genovese R, Trillaud F, Velazquez Villegas F, Santiago Cruz L, Remba J. Model and simulations of high altitude sounding balloons: Dynamics, stress-strain and thermal analysis. In: 67th International Astronautical Congress. Guadalajara, Mexico; Sep 2016. pp. 1-11

[18] DiCarlo JA, Yun H-M. Non-oxide (Silicon Carbide) Fibers. In: Bansal NP, editor. Handbook of Ceramics Composites, NASA Glenn Research Center. United States of America: Kluwer Academic Publisher; 2005. pp. 33-38

[19] ECSS Secretariat ESA-ESTEC. Space Engineering. Structural Materials Handbook. Part 4: Integrity control, verification guidelines and manufacturing. European Cooperation for Space Standardization. Noordwijk, The Netherlands: Requirements & Standards Division; 2011. pp. 219-230

[20] ASM International, editor. Metals Handbook. Vol. 2. Properties and Selection: Nonferrous Alloys and Special Purpose Materials. 10 ed. Metals Park, OH: ASM; 1999

[21] Hussey R, Wilson J, editors. Light Alloys Directory and Handbook. London: Chapman and Hall; 1998

[22] Berndt CC, Brindley W, Goland AN, Herman H, Houck DL, Jones K, Miller RA, Neiser R, Riggs W, Sampath S, Smith M, Spanne P. Journal of Thermal Science and Technology. 1992;**1**:1

[23] Cortese B, Caschera D, de Caro T, Ingo GM. Micro-chemical and -morphological features of heat treated plasma sprayed zirconia-based thermal barrier coatings. Thin Solid Films. 2013;**549**:321-329

[24] Cernuschi F, Bianchi P, Leoni M, Scardi P. Thermal diffusivity/microstructure relationship in Y-PSZ thermal barrier coatings. Journal of Thermal Spray Technology. 1999;**8**(1):102-109

[25] Vassen R, Tietz F, Kerkhoff G, Stoever D. New materials for advanced thermal barrier coatings. In: Lecomte-Beckers J, Schuber F, Ennis PJ, editors. Proceedings of the 6th Liége Conference on Materials for Advanced Power Engineering. Belgium: Universite de Liege; Nov 1998

[26] Cao XQ, Vassen R, Stoever D. Ceramic materials for termal barrier coatings. Sciencie Direct, Journal of European Ceramic Society. 2004;**24**:1-10

[27] Singletary BJ. Ceramics, Refractories and Glasses. In: Rittenhouse JB, Singletary JB. editors. Space Materials Handbook. 3rd ed. United States of America: NASA SP-30 51; 1969. pp. 120-130

[28] Nguyen CH, Chandrashekhara K, Birman V. Multifunctional thermal barrier coating in aerospace sándwich panels. Mechanics Research Communications. 2012;**39**:35-43

[29] Consejo Nacional de Ciencia y Tecnología. Comunicado 64/14, Invertirán CONACYT y Agencia Espacial Mexicana más de 30 MDP en investigación e innovación aeroespacial. Ciudad de México, Oct 8, 2014. Available from: https://www.conacyt.gob.mx/index.php/comunicacion/comunicados-prensa/381-invertiran-conacyt-y-agencia-espacial-mexicana-mas-de-30-mdp-en-investigacion-e-innovacion-aeroespacial [Accessed: Dec 16, 2017]

[30] Consejo Nacional de Ciencia y Tecnología. Agencia Informativa, Montserrat Muñoz. Misión suborbital ATON – Ulises I: Ciencia y arte en la FIL. Dec 7, 2015. Available from: http://www.conacytprensa.mx/index.php/ciencia/universo/4619-prioritaria-mision-suborbital-aton-ulises-i-ciencia-y-arte [Accessed: Apr 12, 2017]

5

Cassini Spacecraft-DSN Communications, Handling Anomalous Link Conditions, and Complete Loss-of-Spacecraft Signal

Paula S. Morgan

Abstract

Once spacecraft are launched, it is impossible for engineers to physically repair anything that breaks onboard the vehicle. Instead, remote solutions must be employed to address spacecraft anomalies and fault conditions. To achieve this goal, telemetered data from the spacecraft are collected and assess by ground personnel to resolve problems. However, if the ground-to-spacecraft communication system breaks down, or the vehicle delivers an anomalous signal, a rigorous protocol must be employed in order to re-establish or fix the telecommunications link. There are several factors that can contribute to link problems, such as malfunctions or mishandling of the ground station equipment, onboard failures of the spacecraft's flight software coding, or even mishaps caused by the space environment itself. This chapter details the anomaly recovery protocols developed for the Cassini Mission-to-Saturn project, to resolve anomalous link problems as well re-acquisition of the spacecraft should a complete Loss of Signal (LOS) condition occur.

Keywords: Cassini, spacecraft, Saturn, deep space network communications, fault protection, loss-of-spacecraft-signal, anomalous downlink

1. Introduction

Despite the vast distance between remote-controlled interplanetary spacecraft launched from earth and the Deep Space Network (DSN) ground stations that operate them, the communications link to the spacecraft is very reliable, thanks to the extraordinary telecommunication capabilities built into NASA's DSN antennas around the world and the spacecraft's own system design. For the Cassini Mission-to-Saturn spacecraft (**Figure 1**), it takes nearly an hour

Figure 1. The Cassini-Huygens spacecraft.

and a half for commands from the Spacecraft Operations Flight Team (SOFS) here on earth to reach Cassini, where the orbiter is touring the Saturnian system (~8.5 AU). Yet, an anomalous downlink (D/L) signal condition can occur (or complete LOS) from several sources: environmental effects such as bad weather conditions at the DSN station or station problems (broken equipment), erroneous ground commands uplinked (U/L) to the spacecraft by the SOFS team, errors in the onboard running sequence, spacecraft pointing errors, internal FSW errors, or computer platform failures can cause problems when attempting to acquire the spacecraft's D/L signal. The space environment itself can also contribute to an LOS condition, since cosmic ray bombardment on the spacecraft's systems can cause spurious Solid State Power Switch (SSPS) trip-off of the spacecraft's Radio Frequency System (RFS) units, as well activations of the onboard Fault Protection (FP) routines which will reconfigure to redundant backup RFS units, so that reconfiguration by the ground is required in order to lock-up on the spacecraft's D/L signal.

To safeguard against these DSN-spacecraft link problems, troubleshooting methods have been developed by the Cassini SOFS team to diagnose and resolve conditions that inhibit spacecraft signal acquisition. A "Loss of Downlink Signal Recovery" protocol was developed for the SOFS team to follow in the event of an anomalous D/L signal (or completed LOS), as well as special FP which is implemented into Cassini's onboard FSW. This algorithm will monitor for prolonged absence of ground commanding, eventually invoking a "Loss of Commandability" FP (FP which is typically implemented into most deep space missions to safeguard against these undetected, sometimes waived or ground-induced failure conditions). Called "Command Loss FP" (from the perspective of the spacecraft since it's no longer receiving ground commands), this "catch-all" type of autonomous monitor-response algorithm will observe the absence of ground commands for a predetermined (programmable) period of time, using a "countdown timer" which decrements until it is reset by a ground command or reaches "0" (which triggers the response). An extended series of actions are then commanded by FP to re-establish ground commandability by configuring various telecom arrangements and spacecraft attitudes in an attempt to find a viable U/L path. Each attempt by the response to command a new path is separated by an appropriate ground response interval for the SOFS team to re-acquire the spacecraft via U/L command.

In all anomalous spacecraft D/L cases, it is desirable to re-establish spacecraft communications before the Command Loss Response activates in order to avoid the autonomous commanded actions of the FP: termination of the onboard running sequence (lost science opportunities), device swaps, propellant consumption via commanded turns, etc. Therefore, an expedient method for identifying possible anomalous/LOS causes is highly desirable before the FP activates, if possible. To aid in this goal, an Excel tool was developed to supplement the LOS Recovery Protocol in "timeline" format. Described herein are the optimized solutions implemented on Cassini for re-acquisition of the spacecraft's signal during anomalous D/L and LOS events, as well as an expedient method for recovery from the actions of the Command Loss Response, if activated.

2. The Cassini mission

NASA's Cassini Mission-to-Saturn spacecraft is the first robotic mission ever to orbit the planet Saturn. Managed by the Jet Propulsion Laboratory (JPL) in Pasadena, this flagship-class mission is composed of 11 operating scientific instruments which study many intriguing features of Saturn, its moons, and ring system. The Cassini Program is an international cooperative effort involving primarily NASA, the European Space Agency (ESA), and the Italian Space Agency (Agenzia Spaziale Italiana, ASI). Cassini is the fourth spacecraft to visit the Saturnian system (but is the first vehicle to enter its orbit), and is composed of the NASA/ASI Cassini orbiter and the ESA-developed Huygens probe. Cassini launched on October 15, 1997, arriving at Saturn in 2004, after performing scientific observation of Earth's moon, Venus, and Jupiter (as well as participating in several scientific experiments) during its 6.7 year cruise period. Cassini's suite of (currently operating) science instruments consists of the following (**Figure 2**):

1. Composite Infrared Spectrometer (CIRS)
2. Ion & Neutral Mass Spectrometer (INMS)
3. Visible & Infrared Mapping Spectrometer (VIMS)
4. Ultraviolet Imaging Spectrograph (UVIS)
5. Imaging Science Subsystem (ISS)
6. Magnetospheric Imaging Instrument (MIMI)

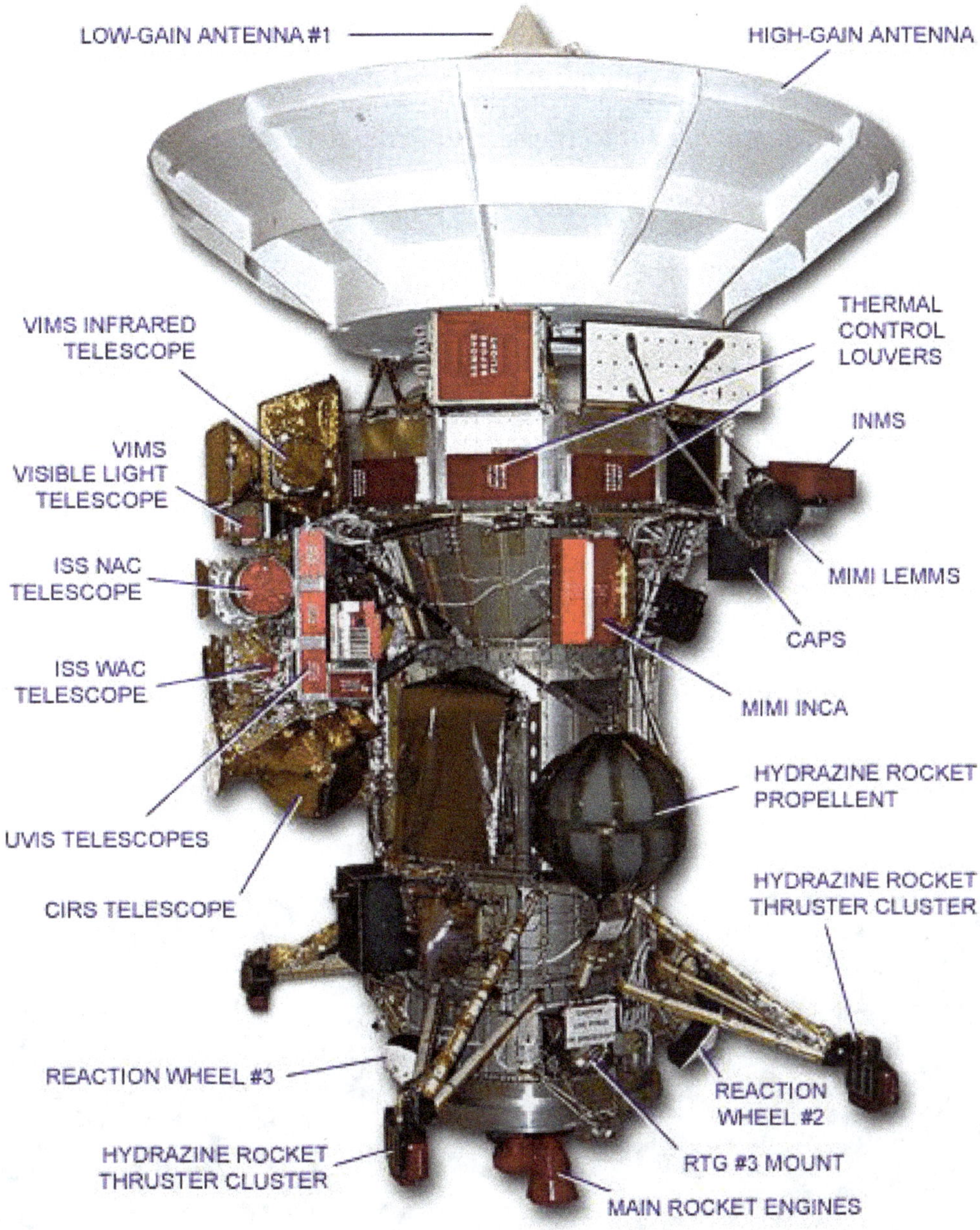

Figure 2. Cassini's instrument suite.

7. Dual Technique Magnetometer (MAG)
8. Cosmic Dust Analyzer (CDA)
9. Radio & Plasma Wave Science instrument (RPWS)
10. Radio Science Subsystem (RSS)
11. Radar

Also included onboard Cassini is the Huygens Probe; an atmospheric laboratory designed to collect data in the Titan Moon atmosphere and its surface. Deployed in January 2005, the probe consisted of six scientific instruments which performed experiments in aerosol collection, descent imaging & spectral radiometry, gas chromatography & mass spectrometry, atmospheric sampling, and surface science. The entire Cassini mission consists of seven phases:

- Launch and initial acquisition of the spacecraft (October 15, 1997)
- Inner cruise (beginning October 20, 1997)
- Outer cruise (beginning February 2000)
- Science cruise (starting July 2002)
- Saturn Orbit Insertion (SOI; July 2004)
- Huygens Probe Release (January 2005)
- Saturn Tour continues (2004–2017)

During the cruise portion of the journey to the Saturnian system, two gravity assist maneuvers were required from Venus, one from Earth, and one from Jupiter. Until Cassini reached 2.7 AU from the sun (during the inner cruise phase), communications between earth and the spacecraft were accomplished via the Low Gain Antenna (LGA), since the 4-m diameter High Gain Antenna (HGA) must be used to shield the spacecraft from the sun's heating (i.e. used as a sunshade). After reaching this distance (begin Outer cruise phase), communications begin on the earth-pointed HGA.

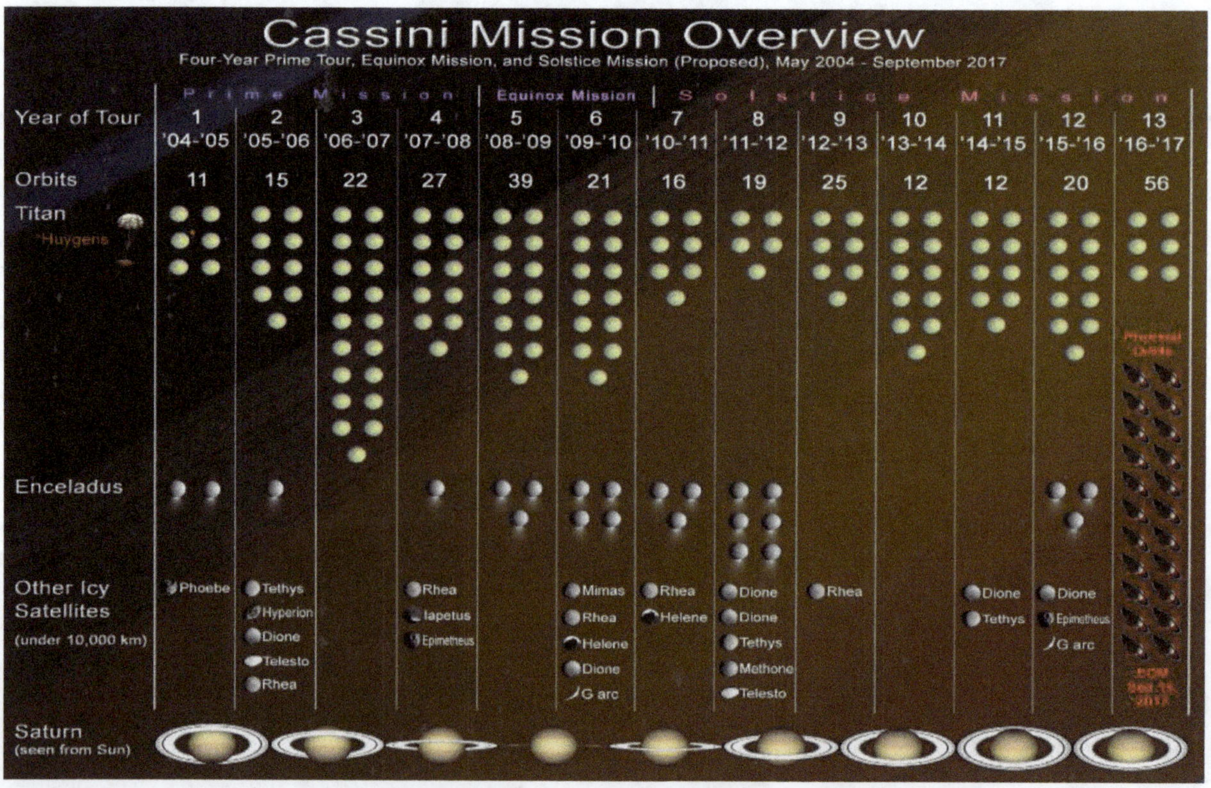

Figure 3. Cassini's prime, equinox XM, & solstice XXM tours.

Cassini's "Prime Tour Mission" began in 2004, where planet/moon science investigation activities continued until 2008. Two mission extensions were granted: the "Equinox Mission" from 2008 to 2010, and the "Solstice Mission" from 2010 to 2017 (**Figure 3**, [1]). The spacecraft's 20 year mission ends with 42 orbits around the main ring system (**Figure 4**, [2]). Beginning on November 30, 2016, Cassini's orbit reoriented the spacecraft to the outer edge of the main rings to perform a series of 20 F-Ring orbits; a region of Saturn's rings which look like an odd "interwoven" structure. The last time that Cassini observed these rings close-up was at Saturn arrival in 2004, which allowed observation of only the dim, backlit side. But in November of 2016, numerous opportunities became available to examine the F-Ring's structure, with high-resolution observation of both sides of the F-Ring. The final mission phase called "The Grand Finale" began in April 2017 with a close flyby of Saturn's giant moon Titan, which provided re-orientation of the spacecraft's trajectory, allowing it to pass through the gap between Saturn and the D-Ring; the closest ring to the planet. With only a 1500 mile-wide corridor to fly through, Cassini will investigate this unexplored region of the Saturnian system, making the closest observations of Saturn to date. During these last 22 (D-Ring) orbits of the Cassini mission, the planet's magnetic and gravity fields will be mapped with high precision, and extremely close views of the atmosphere will be observed. New insights into Saturn's interior structure, the precise length of a Saturnian day, and the age and total mass of the rings will also be evaluated. On September 15, 2017, Cassini will end its 20 year mission with a fiery plunge into Saturn, providing valuable data about the planet's chemical composition as the friction forces (from the atmospheric entry) cause the vehicle to burn up, thus satisfying Planetary Protection requirements [3].

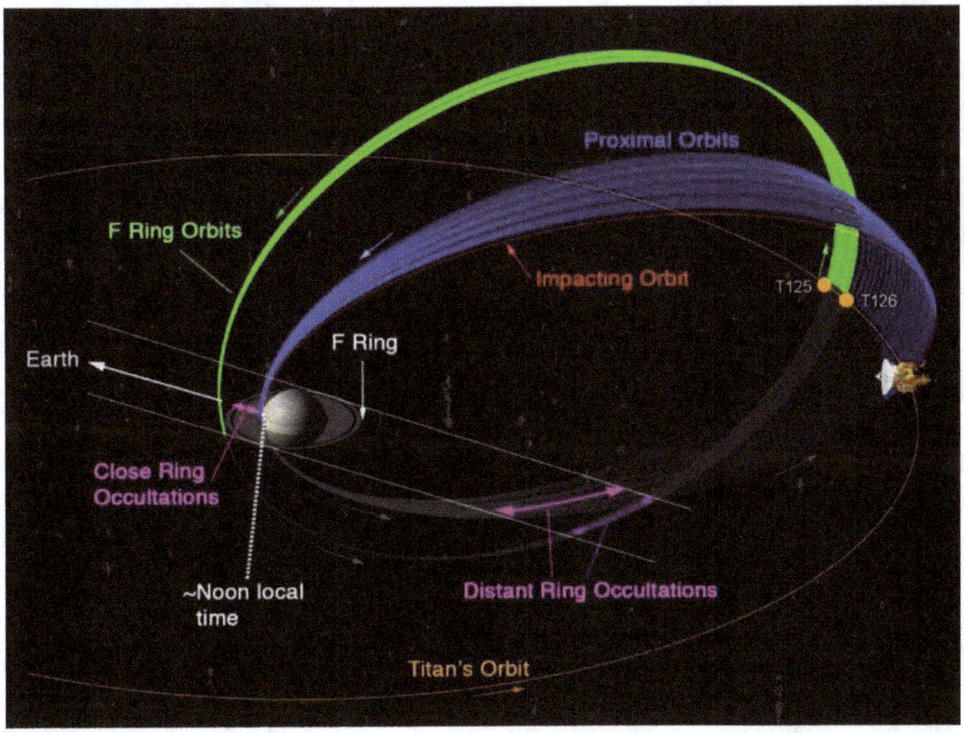

Figure 4. Cassini's end-of-Mission F & D Ring Orbits.

3. The Cassini radio communications system

Cassini's onboard telecommunications system consists of three antennas: a High-Gain Antenna and two Low-Gain Antennas (LGA-1 & LGA-2); all which interface with the RFS system (which performs command, telemetry, and radio-metric communications) and Radio Frequency Instrument Subsystem (RFIS); **Figure 5**. Cassini's 4-m Cassegrain HGA

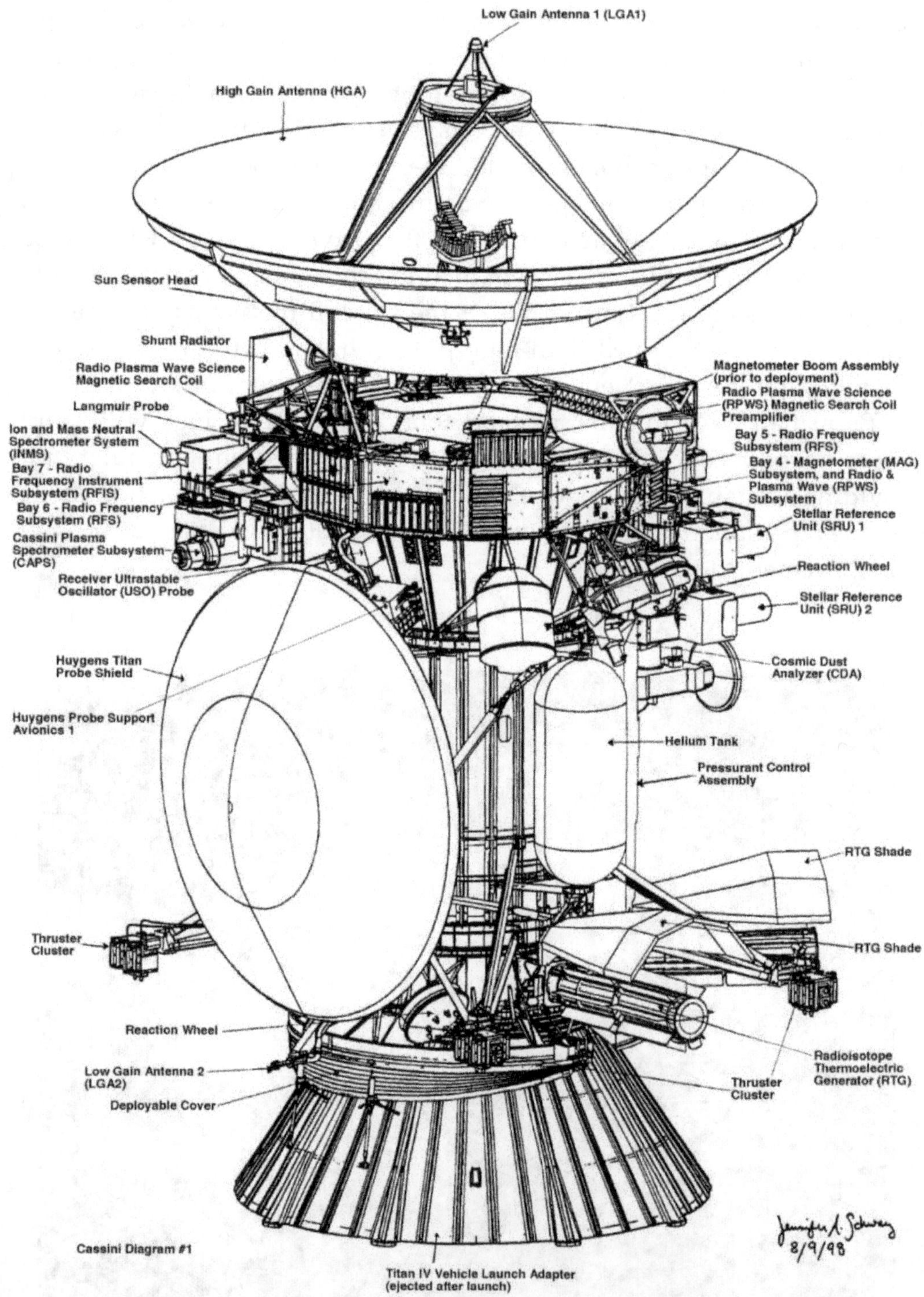

Figure 5. The Cassini spacecraft.

communicates with earth on X-band, and on S-Band with the Huygens probe (and radio-science). It also communicates on Ka-band to support radio science activities, and Ku-band for the imaging RADAR subsystem. The two LGA antennas operate on X-band only, with LGA-1 mounted on the top of the HGA (giving it an unobstructed field of view of 112°), and LGA-2 which is mounted on a boom below the Huygens probe near the bottom of the vehicle, yielding a 120° field of view. The LGA antennas were used for communication with the ground when the HGA could not be configured on earth-point due to thermal constraints (when in close proximity to the sun). In this case, the spacecraft had to be shielded by the HGA, leaving the LGA antennas to transmit and receive data at very low delivery rates. The LGA antennas are configured when FP executes.

Spacecraft are typically equipped with transmitters of relatively low radiating power for communication with earth (20 Watts for Cassini). This telecommunications link must bridge the distance of over a billion kilometers (earth-Saturn distance), which is achieved by employing frequencies in the microwave range using reflectors onboard the spacecraft to concentrate all available power into a narrow beam pointed precisely towards earth. Cassini's HGA is used to achieve this goal (as opposed to the LGA antennas which sacrifice gain but provide relatively uniform coverage over a wide range of spacecraft orientation angles). At the DSN station, large aperture Cassegrain reflectors are used to pick up the spacecraft's signal. These radio antennas use cryogenically cooled (low-noise) amplifiers to first amplify the faint spacecraft signal, followed by sophisticated receivers and decoders which can lock onto and extract the data with virtually with no errors at all.

The signal delivered from the spacecraft to earth's ground station is called a "downlink," and the transmission of commands and sequences from the ground to the spacecraft is called an "uplink." When a D/L signal is received from the spacecraft, the communication is called "one-way" (or if the D/L signal is generated onboard the spacecraft itself, the communication is also called "one-way"). When the U/L signal is being received by the spacecraft at the same time a D/L is being received by the ground station, the communication is called "two-way." Both U/L and D/L consist of a pure Radio Frequency (RF) tone which is called a "carrier." In order to carry information to or from the spacecraft, the carrier signal must be "modulated." A modulated signal may be sent from the ground station to transmit commands to the spacecraft. Likewise, the modulated signal is generated by the spacecraft to transmit science and engineering data to earth on its D/L carrier. The spacecraft's carrier signal is also used for tracking and navigation (as well as some types of science experiments such as radio science or gravity field mapping). Each DSN complex uses a hydrogen-maser-based frequency unit which is maintained in an environmentally controlled room (in the basement), sustained by an uninterruptable power supply. The maser serves as the reference for generating a precisely known U/L frequency. When an U/L signal is received by the spacecraft, it can choose to use the received U/L carrier to control its D/L carrier transmission (called 2-way coherent transmission). This ground-generated reference frequency is multiplied by a predetermined constant (1.1748999 for Cassini) and the transmitted D/L signal is phase coherent with the U/L signal (this multiplier prevents the D/L signal from interfering with the U/L signal which is being received from the ground). Precise tracking of the spacecraft is accomplished through this method, as well as the ability to carry out high precision science experiments onboard the orbiter.

Cassini carries its own Ultra-Stable Oscillator (USO). During the one-way phase when the spacecraft transmits its signal to the ground (before two-way communication is established), the spacecraft must generate its own D/L signal using the on-board USO. Once the ground's U/L signal is acquired by the vehicle, it will abandon its own D/L signal to regenerate the D/L, thus changing the frequency. During this time, the ground station will "lose lock" on the spacecraft and must tune in the new frequency. This "out-of-lock" condition is predetermined by the ground (on the order of a minute or two), so that data delivery to the ground is temporarily halted during this transition period, in order to preserve the precious science data. The USO device is quite reliable in generating a stable D/L signal, more so than the 2-way method with the ground, because the ground U/L signal phase is subject to corruption by atmospheric effects, solar wind, etc. Therefore, the USO is more desirable than the hydrogen maser. However, the USO frequency cannot be precisely known if the D/L frequency changes due to relative motion of the spacecraft (as well as vehicle drifting). Since ranging is fundamentally a phase measurement, the ground must use the hydrogen maser referenced U/L along with phase coherent receivers on the spacecraft and on the ground to determine the correct measurement.

4. Cassini mission telecommunications operations in flight

NASA's DSN is a part of JPL, consisting of a worldwide network of US spacecraft communication facilities. Placed approximately 120° apart around the Earth, three deep-space telecommunications stations are located in Goldstone, California (US), Madrid, Spain, and Canberra, Australia. The placement of these ground stations permits constant observation of spacecraft like Cassini as the Earth rotates. Unlike near-earth orbiters which move quickly round the earth, few ground stations are required to support deep space missions since they are visible for long periods of time. As mentioned before, these earth-based DSN ground stations contain steerable, high-gain, parabolic reflector antennas, providing a two-way communications link that tracks robotic interplanetary spacecraft like Cassini, acquiring telemetry data, transmitting commands, uploading software modifications, tracking spacecraft position and velocity, measuring variations in radio waves to support radio science experiments, and collecting science & engineering data. Interplanetary spacecraft such as Cassini, require huge DSN antennas with ultra-sensitive receivers and powerful transmitters in order to transmit/receive information over the vast earth-planet distances, with the largest antennas of the DSN often called upon during spacecraft emergencies. Nearly all spacecraft are designed to use the smaller DSN antennas (e.g. 34 m diameter) for nominal operations, but for a spacecraft emergency, the largest antennas are typically used (e.g. 70 m diameter) since the onboard FP typically configures low transmitter power, so that recovering any available telemetry is crucial to assessing the spacecraft's health in preparation for recovery actions. In the case of Cassini, the LGA is configured by FP with very low U/L & D/L rates.

Ground commands from earth travel at the speed of light (referred to as "One-Way Light Time;" OWLT), reaching Cassini from approximately 1 hour. 15 minute to 1 hour 30 minute, depending on the relative distance between earth and Saturn, given the change in relative distance due to the earth's rotation around the sun and the spacecraft's motion around the Saturnian system. Therefore, the majority of commands sent to the spacecraft for operations and science investigations must be uplinked to the Command & Data processing System (CDS) computers in large "command sequences," which consist of several weeks of planned commanding. These sequences typically consist of commanded turns to point Cassini's 11 operating instruments towards specific targets, providing high precision (down to the sub-milliradian) via two Attitude, Articulation, & Control System (AACS) computers. Captured science data is recorded on two Solid-State Recorders (SSR) during off-earth observation periods. These science activities (e.g. moon and ring encounters) are paused typically once each day (or two) for approximately 9 hours to establish communication with earth (via a scheduled DSN station) to downlink the science & engineering (housekeeping) data.

Once Cassini's earth-pointed attitude is stabilized, its D/L signal is received by the DSN station. Ten minutes later, the ACE initiates the U/L signal for commanding and navigational purposes. The data is transmitted from the spacecraft in the format of "symbols" which are "wiggles" in Cassini's radio signal's phase. The DSN receives the symbols and decodes it into "0" and "1 seconds" in order to reconstruct the telemetry data (engineering housekeeping data, science digital images, etc.). After the 9 hours of telemetry data have been downlinked to earth's DSN ground station, the spacecraft reduces its data rate, suspends its data playback (from the SSRs), and turns to the next science target via the onboard running sequence to collect new science data [4].

4.1. Nominal S/C acquisition

Prior to spacecraft acquisition at JPL's Space Flight Operations building in Pasadena, California, the "Cassini ACE" Real-time Operations Engineer must prepare to receive the data transmission stream from Cassini, and is in voice contact with the DSN station staff (in California, Australia, or Spain). The Cassini ACE provides their station operator with a 2 minute briefing to review the expected events for the day, before the DSN pass starts (any planned Reaction Control System (RCS) burns or Main Engine (ME) maneuvers, Flight Software (FSW) patches or uploads, etc.) and provides any pertinent updates. The DSN station operator, in turn, provides a weather report (clear skies or rain, plus wind conditions) and that all equipment is in working order (green), or has suffered a system breakdown (red). The designated (34 m or 70 m) antenna at the DSN station for the day's 9 hour pass has already been pointed precisely towards Saturn where Cassini's faint signal will be received. Once the spacecraft's signal has been acquired, the DSN station operator reports to the Cassini Ace that the station's receiver is "in lock." The Cassini Ace then acknowledges that the telemetry at his/her workstation is being received and looks nominal. From this point, the 9 hour DSN pass is in progress with Cassini transmitting its telemetry data. Thousands of engineering telemetry measurements (i.e. temperatures, voltages, pressures, computer statuses detailing the vehicle's health and status) are interleaved with the science data.

4.2. Anomalous D/L conditions

An "out of lock" condition can occur suddenly if Cassini's signal strength drops out (LOS condition). This can be caused by rain at the DSN station from too many water molecules in the vicinity of the antenna which give off an abundance of radio noise that can literally drown out the spacecraft's signal. In this case, the "DSN Receiver Status" on the Cassini ACE's console will light up with an "OUT OF LOCK" reading. The measured system operating noise temperature on the console should rise high enough to indicate that rain is the reason for the signal loss. But if bad weather is not the cause of the LOS condition, or caused by an unforeseen problem in the ground system equipment itself, the ACE will contact the Operations Chief (who is concurrently working with the Cassini ACE at JPL), to request that a second DSN antenna look for the spacecraft's signal, if available. If no signal is detected, the Cassini ACE will declare a "LOS condition" and proceed to follow the "LOS/Anomalous Downlink Contingency Plan" Procedure which requires that he/she contact the appropriate SOFS team members. These are spacecraft subsystem experts who must evaluate the situation and concur with the Cassini ACE that there is no earth-based problem causing the LOS condition (ground station or weather). In this case, the most likely explanation is that an onboard RFS-related FP routine has triggered. Numerous fault monitors are installed into Cassini's FSW that are constantly running to detect faults in spacecraft systems. Upon fault detection, a "canned" response routine(s) is executed autonomously to fix the problem, which is typically followed by an activation of the Safing Response. This response places the spacecraft in a predictable state, configuring lower power consumption with low U/L and D/L rates on LGA, commanding the HGA to sun-point (off earth-point). In the case of a RFS FP routine activation, the RFS device states might be altered, as a swap to a redundant RFS unit is commanded which changes the telecommunications configuration for D/L signal acquisition.

The ACE knows that Cassini will have transitioned from the HGA to the LGA antenna, should the FP activate. The LGA provides an extremely weak D/L signal since its beamwidth is much larger than the HGA beamwidth. At Saturn, the spacecraft's signal is so weak that telemetry delivery is only possible at 5 bps, requiring nearly 18 hours to receive all 30 decks of telemetry data that are needed for the SOFS team members to verify the spacecraft's health and determine its post-fault states. Recovery from any fault is extremely slow, but if no attitude control system problems are present and spacecraft attitude knowledge is preserved (no faults in the AACS computers), a second FP routine called the "High Gain Antenna Swap (HAS) Response" will automatically activate 1 hour after the Safing Response concludes. This FP will increase the U/L and D/L rates (D/L = 1896 bps), followed by a turn of the spacecraft's HGA to earth-point. In this configuration, all 30 decks of telemetry data are delivered to the ground in approximately 10 minutes, making recovery from the fault much more expedient. For typical FP activations, the SOFS team will examine the spacecraft telemetry for off-nominal conditions, sometimes reading out additional sections of Cassini's computer memory to confirm the diagnosis, and then prepare commands for the ACE to send which will recover the spacecraft from the FP activation, and restart the onboard running sequence once again.

In certain cases, complete LOS can occur. Resolution of a LOS fault may require extra DSN coverage, depending on the difficulty in determining the fault cause. As mentioned previously, the Cassini ACE also looks for other DSN tracks that can be borrowed from other flight projects or scheduled maintenance for the next few days. If the LOS condition persists, a "spacecraft emergency" will be declared to guarantee continuous DSN coverage to support spacecraft recovery efforts.

4.3. No spacecraft signal acquisition (LOS)

Unlike most faults that trigger the onboard FP, a fault causing total LOS means no acquisition of the spacecraft's signal at all (i.e. no lock-up on the expected or post-FP RFS configuration) by the DSN station. There are several reasons why a LOS condition can occur. These include DSN station breakdowns, misconfigured lock-up parameters, or even faults which are not detected by the FP design. Unfortunately, not every spacecraft fault case can be precluded by the onboard FP. In spite of the best efforts of pre-launch designers to identify all possible fault scenarios and produce a FP system to support them (detect, isolate, & resolve), certain failure modes are sometimes missed or are very difficult to avoid. Most JPL projects like Cassini strive to meet a "Single Point Failure" (SPF) policy [5], but certain failures cannot be easily detected, or are not identified during the design phase, and some failures can actually occur even though they have been exempted or waived [6]. Other LOS fault possibilities are problems that occur in devices which are intentionally not protected by the onboard FP. These devices include the HGA or LGA antennas, Waveguide Transfer Switches (WTS), and the USO on Cassini. Multiple faults are also a possibility, since they do not fall under FP design guidelines due to the SPF policy.

Hence, LOS can occur from several sources: erroneous ground-generated commands uplinked to the spacecraft, onboard sequence failures, multiple failures which are not typically required to be addressed by the onboard FP, spacecraft pointing errors, failed telecom configurations (via ground commanding), internal FSW errors, computer platform failures, bad weather, or DSN ground equipment failures. Also, not only can RFS FP swap to redundant units due to device faults and malfunctions, thus inhibiting the ground from locking up on Cassini's signal (since the RFS D/L signal path has changed), but environmental effects can also cause a LOS condition. SSPS trip-off of RFS units (caused by cosmic ray bombardment) can also cause temporary loss of the spacecraft's signal. To address this condition, the Cassini ACE must perform several "uplink sweeps" on different variations of the RFS units in an attempt to re-acquire the spacecraft's D/L signal. Once ground problems and weather are ruled out as an LOS cause, the assumption is that hopefully the onboard FP has executed and commanded a RFS device swap to a redundant unit. Otherwise, determination of the fault cause becomes increasingly difficult to diagnose.

5. Cassini LOS experiences

Cassini has experienced several LOS events during its mission lifetime. Some events have been caused by relatively minor problems, but two events are of significance. The first occurred on May 1, 2006. At the beginning of the DSN track, the DSN station was unable to acquire the spacecraft's "one-way" carrier signal (i.e. the ground-received spacecraft signal), which in turn, initiated the anomaly response process. However, after Round-Trip-Light-Time (RTLT; twice OWLT) had elapsed, the DSN station was able to lock up on the "two-way" carrier signal and the spacecraft's data. Telemetry indicated that Cassini's USO had suffered an SSPS trip event [7].

Cassini's power system consists of power control boards which contain 192 SSPS. SSPS trip events occur spuriously and without warning, on average 2–3 times per year due to the unforeseen environmental effects of galactic cosmic ray bombardment [8]. This condition is thought to be caused by one or more photon hits on the voltage comparator of the device, resulting in a false indication that the current load is anomalously high, thus tripping off the switch. Because of this phenomenon, a new "SSPS Trip FP" monitor & response algorithm was uploaded to the Cassini spacecraft's FSW. The monitor examines one SSPS switch state per second, (starting with switch number 1), and proceeds through all 192 SSPS switches. If a SSPS trip is detected, the response contains a table of appropriate actions for FP to act upon, based upon the specific SSPS switch and its function. The actions of the original SSPS FP response table for the USO (uplinked prior to 2006) only recorded the USO trip event (USO SSPS is #68) and cleared the tripped condition by commanding the unit OFF. However, after this USO trip event occurred, the response table was augmented (via uplink command) to command the device on (see **Table 1**).

Five years later on December 23, 2011 at the Beginning of the DSN Track (BOT), once again, no D/L signal was seen from the Cassini spacecraft. The DSN station at Canberra was supporting Cassini at the time. Following ACE direction, additional tracking was obtained using a Canberra station antenna, as well as a Goldstone station antenna, but without successful acquisition of the spacecraft's signal. The SOFS Anomaly Team was called together to diagnose the problem. At RTLT, Cassini was once again acquired in 2-way mode, confirming that the problem was with

Switch No.	SSPS Switch	Log Event	Cmd Switch "Off"?	Cmd Switch "On"?	Cmd Alt. Switch?	Alt. Switch SSPS No.	Switch State (On/Off)
68	USO Load Current	Y	Y	Y	N	-	-
				Updated via uplink command: N=>Y			

Table 1. SSPS trip FP for USO trip (post-2006).

the spacecraft's USO. Commands were sent on Christmas Day to inhibit the USO and swap to the Auxiliary Oscillator as the frequency source for the D/L signal until the fault within the USO device could be evaluated.

The next step for the SOFS team was to evaluate whether one-way operation of the USO was functioning properly (the two-way U/L must be halted in this case). Once configured, the DSN station was unable to lock onto Cassini's one-way signal which indicated that the USO was not operating properly. After a second attempt to establish the one-way link failed, a command was sent to inhibit the USO, allowing the Auxiliary Oscillator to take over again for spacecraft operations. Further tests conducted in January of 2012 confirmed that normal USO operation could not be re-established. After consulting with Radio Science and Applied Physics Laboratory (the builder of the USO), it was decided that the USO would be power cycled in an effort to "reset" the unit, although it was thought unlikely to work since the USO is an analog device. On January 9, 2013 the USO was powered OFF permanently and the Auxiliary Oscillator has been in operation ever since.

6. LOS protocol

For Cassini, addressing an "anomalous downlink" or LOS condition starts with the RFS Subsystem's "LOS/Anomalous Downlink Contingency Plan" Procedure to help identify possible reasons for the abnormal (or absence of) the spacecraft's D/L signal. This procedure describes possible troubleshooting methods and recovery actions needed for both off-nominal signal levels (e.g. carrier power is too low or too high) as well as partial lock-up conditions (e.g. no subcarrier, symbol, telemetry, or frame lock-up), and complete LOS. The procedure provides diagnoses & recovery actions in the form of flowcharts for the ACE and SOFS Anomaly team members to follow. Five partial signal loss/LOS candidate faults are considered when determining required anomaly resolution actions:

1. Spacecraft is not on earth-point when expected due to an incomplete turn, a fault in the AACS system, or FP activation.
2. DSN ground-station problem: station is not tracking the spacecraft properly, station receiver is down, breakdowns, weather, etc.
3. Spacecraft telecom problem: there is a problem in the telecommunications system (error caused by the onboard sequence commanding, ground U/L commanding, or the FP has executed)
4. Loss of the CDS (most likely a multi-fault condition)
5. Multiple faults or a catastrophic failure

RFS FP response actions are also noted in the recovery strategy flowcharts of the procedure and specify the expected post-fault RFS device states. Any attempt to re-acquire the spacecraft on the newly commanded RFS configuration is directly dependent on when the FP response has concluded. Attempted spacecraft recovery actions continue through each branch of the flowcharts until re-acquisition of the vehicle is successful (if possible).

For a complete LOS condition, the Cassini ACE must perform the "uplink sweep" on the correct RFS device configuration to re-acquire the spacecraft [9]. The assumption is that the activation of a RFS FP response will have swapped to its counterpart unit, possibly changing the polarity of the D/L signal. Depending on the failure (or number of failures), several RFS device combinations are possible with variations on the following components, depending on which FP has activated and what the current RFS prime units are:

- DST-A/CDU-A or DST-B/CDU-B (Deep Space Transponder; Command Detector Unit)
- TWTA-A or TWTA-B (Traveling Wave Tube Amplifire)
- TCU-A or TCU-B (Telemetry Control Unit)
- WTS-A or WTS-B (Waveguide Transfer Switch)
- LGA-1 (LGA-2 is no longer in use) or HGA antenna
- Auxiliary Oscillator or DST VCO (Voltage-Controlled Oscillator)

Figure 6 depicts the RFS Functional diagram for Cassini, whose prime RFS units are: DST-A/CDU-A, TCU-B, and TWTA-B; WTS-A used for U/L, WTS-B used for D/L. The use of these devices are listed below:

- DST: is used for both the U/L and D/L function
- CDU: is part of the DST and used for the U/L function
- TWTA: is an amplifier used in the D/L function
- TCU: controls the RFS system.
- WTS: provide switching capability for transmitting or receiving the signal through the HGA, LGA-1, or LGA-2 antennas.
- Auxiliary Oscillator: provides 1-way D/L carrier frequency reference.
- VCO: is part of the DST and provides 2-way D/L carrier frequency reference.

Also included in certain RFS FP response actions is a Power-on-Reset (POR) of the prime TCU and/or the Power subsystem where selected devices are turned off, reset, or reconfigured, which will select spacecraft components according to their own FP protocols. Further complicating the anomalous/LOS condition is the fact that RFS FP algorithms are multi-tiered (address several different fault types), and can activate at any time per their persistence counters (unique for each FP algorithm) which can range from seconds to minutes, further reconfiguring these device states after spacecraft re-acquisition is attempted, so that it is difficult to know which RFS combinations for the ACE to try (or which combinations should be re-tried or eliminated). Therefore, it is very important to keep track of when RFS related FP responses have timed out.

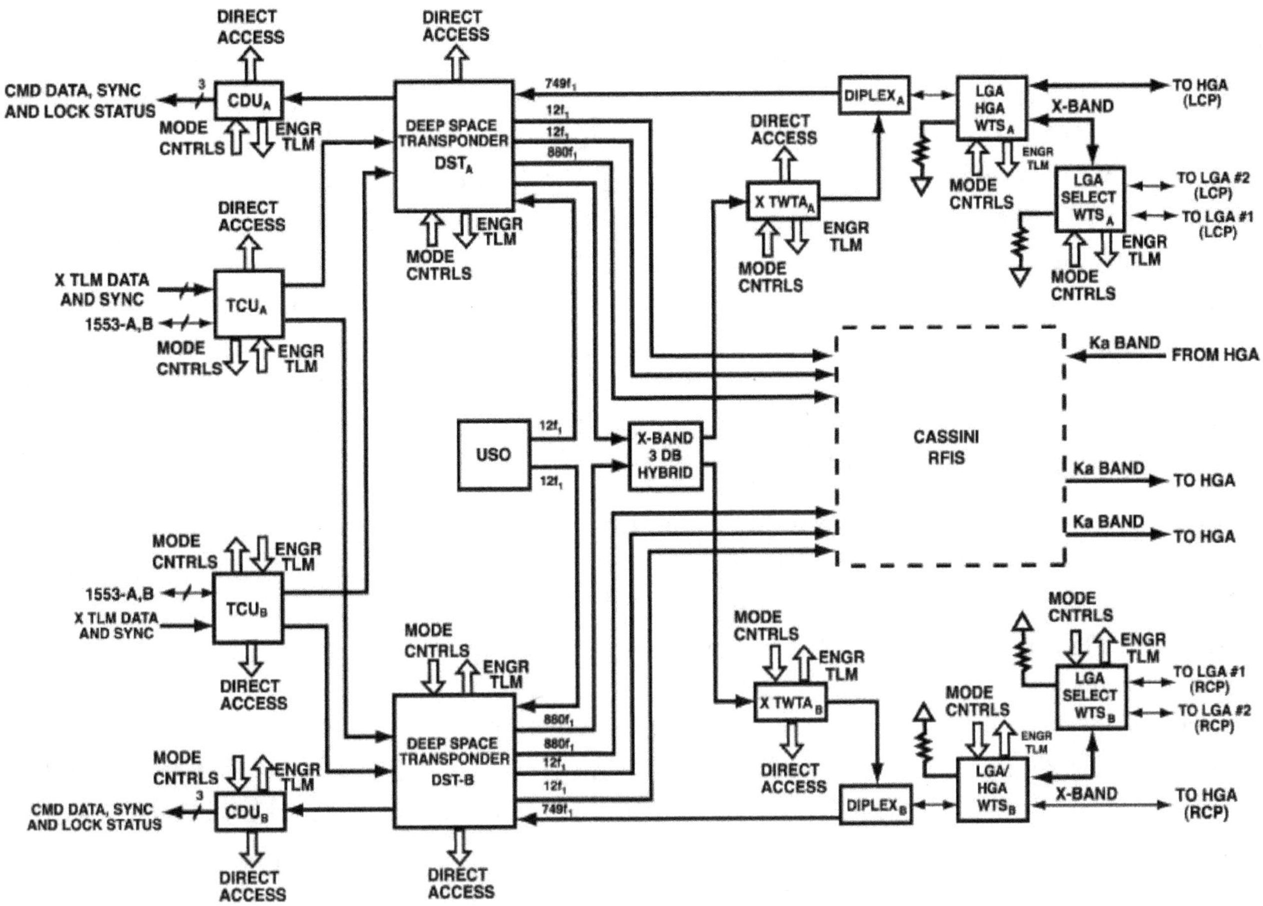

Figure 6. Cassini's RFS functional diagram.

7. Command loss FP

An unresolvable LOS condition where the ground is no longer able to deliver commands to the spacecraft will eventually lead to the activation of a LOS FP response. The actions of this response can help to re-establish the U/L. In Cassini's FP design, loss of D/L fault coverage is not protected in an "end-to-end" manner since the D/L is not considered to be a critical spacecraft function which requires autonomous restoration. But restoration of the U/L however, is considered crucial to mission success and is therefore allocated "end-to-end" protection through a "Loss of Commandability" algorithm [10]. Although several other (higher priority) FP routines are installed into Cassini's FP suite to protect against these same type of failures in the U/L path (which provide more timely action), the Loss of Commandability algorithm provides a "safety net" type of FP which has the potential to restore both U/L and D/L. With this scheme in place, multiple levels of FP defense are provided (covering up to 3 faults).

This catch-all type of FP is referred to as a "Command Loss FP" (from the perspective of the spacecraft since it is no longer receiving ground commands) and is typically an "endless-loop" response. The Command Loss Monitor aboard Cassini will detect an extended period of time during which no commands have been received by the spacecraft from the ground. The Command Loss Monitor is configured with a timer which counts down from a programmable value (usually days) until it reaches "0" seconds or is reset via ground command (on Cassini, this "Command Loss Timer" (CLT) is currently set to 115 hours). The receipt of a valid U/L command by the spacecraft will reset the timer to its original value and restart the countdown. This provides an end-to-end check on command functionality between the vehicle and the ground. If triggered (timer reaches "0"), the Command Loss Response will initiate an extended series of actions which are designed to re-enable ground commandability onboard the spacecraft. The response will attempt to command various telecom configurations and spacecraft attitudes in an attempt to find a viable uplink path. Each reconfiguration of a new uplink path is separated by an appropriate ground response interval for the SOFS team to re-acquire the spacecraft.

Figure 7 illustrates Cassini's Command Loss Response chain. Once triggered, it progresses through a series of "Command Groups" divided by multi-hour "Command Pauses" which allow the SOFS team to react by sending an U/L command to halt the response. The Command Groups consist of actions to reconfigure redundant hardware and re-command spacecraft attitude and antennas. Each Command Pause allow several hours for the SOFS team to attempt re-acquisition of the spacecraft upon the newly commanded spacecraft configuration (the pause durations are set to a minimum of two RTLT periods). As shown in the figure, the first Command Group will select the Auxiliary Oscillator and execute the Safing Response which will turn off non-essential spacecraft loads, place the spacecraft in a lower power state, and re-direct the spacecraft's High Gain Antenna to sun-point, placing the spacecraft in a low U/L & D/L state through the LGA-1 antenna. After the first Command Group has executed, a 15 hour wait period (Command Pause) allows sufficient time for the SOFS Anomaly team to assemble at JPL and attempt re-establishment of the U/L, if possible, before RFS hardware swaps begin in successive Command Groups. If the re-acquisition attempt fails after Command Group #1 execution, the response will proceed with the next

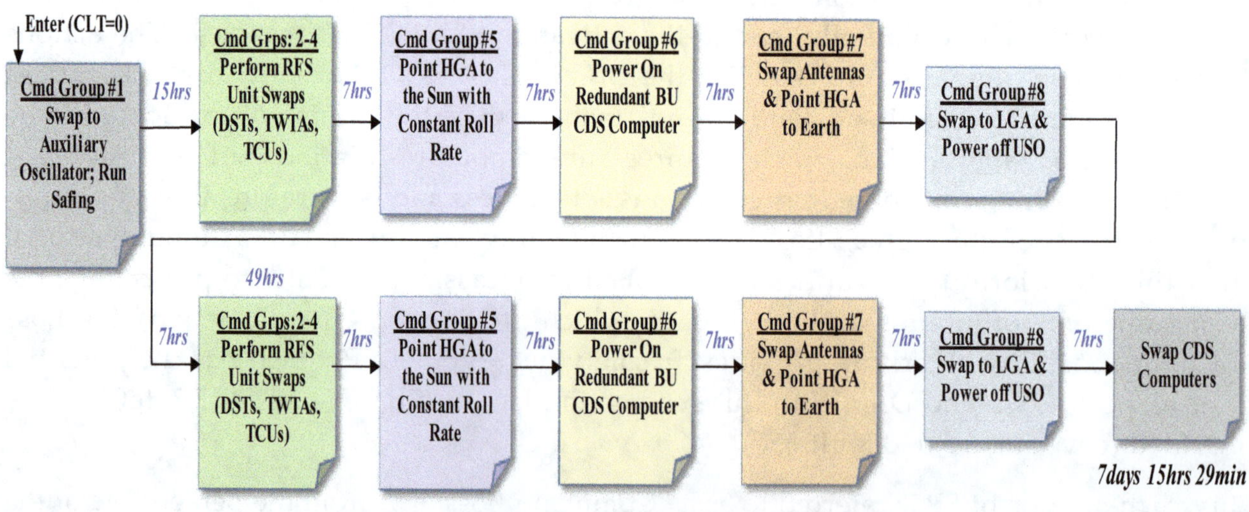

Figure 7. Cmdloss response actions.

course of actions specified in Command Group #2, which starts the series of RFS hardware unit swaps. Seven hour Command Pauses are installed between each subsequent Command Group to allow the SOFS team sufficient time to re-acquire the spacecraft on the newly commanded configuration. If the SOFS Anomaly team is able to re-acquire the vehicle within the first 71 hours (during the RFS unit swap phase), it is permissible for the HAS Response FP to execute (1 hour after the Command Loss Response has been terminated) via the selected (6NOP) U/L command which halts the response. Faults resolved during this first 71 hours are deemed to be "non-severe," since they are associated with RFS device failures. The HAS Response will increase the post-Safing U/L & D/L rates and swap from LGA-1 to the HGA antenna. However, if the Command Loss Response proceeds to Command Groups #5, it must be halted using the HAS FP "disable" command to keep the spacecraft on LGA-1 with the lower U/L & D/L rates, since the fault is considered to be too severe to transition to the higher rates. At the end of the Command Loss Response chain (approx. 7 days 15 hours), a swap to the redundant CDS is commanded and the Command Loss Response will start all over again on the redundant backup computer. The response will run endlessly until an U/L command is received by the ground. Once the spacecraft receives a ground command which restores the uplink successfully, the response will terminate and reset its Command Loss Timer, thus leaving the spacecraft on the last (successfully) commanded RFS/antenna configuration.

8. The LOS/Cmdloss timeline EXCEL tool

In all cases, it is desirable to re-acquire the spacecraft before the Command Loss algorithm times out and triggers its response, if at all possible, since this FP routine will configure the LGA antenna, which yields extremely slow data delivery. Should this response trigger, the Command Group actions (device swaps, etc.) most likely cannot be confirmed in telemetry with the very slow D/L rate of 5 bits per second. Therefore, it was determined that two timelines were needed to provide visibility into fault possibilities and to supplement the LOS/Anomalous Downlink Contingency Plan Procedure recovery efforts: 1) a pre-Command Loss Response "LOS Timeline" containing FP expiration times (and the corresponding RFS configurations) to eliminate the numerous fault possibilities, 2) a timeline to track the Command Loss Response actions if activated. This goal was accomplished through the development of an EXCEL tool which receives minimal user inputs, utilizing the Space Flight Operations Schedule (SFOS) file which is used daily by both the ACE and SOFS teams. The "LOS/Cmdloss EXCEL Tool" provides the following:

- Sheet #1: instructions for using the EXCEL Tool & required inputs taken from the SFOS file
- Sheet #2: Timeline #1 starting from LOS occurrence = > CLT = 0 seconds (Command Loss Response trigger time)
- Sheet #3: Timeline #2 detailing the Command Loss Response actions from CLT = 0 seconds through one entire CDS response cycle
- Sheet #4: all corresponding end conditions for each FP response activation in Timeline #1 with the required recovery actions

9. EXCEL tool example: 2011 USO Failure

Experience gained from the failed USO/LOS event on December 23, 2011 at BOT led to the development of this LOS/Cmdloss Timeline EXCEL Tool. To demonstrate its use, an example is provided here for this USO failure event.

Once no signal was detected from Cassini on Day of Year (DOY) 357 of 2011, the ACE proceeded to follow the "LOS/Anomalous Downlink Contingency Plan" Procedure, performing sweeps of the spacecraft on different RFS configurations to attempt re-acquisition of the vehicle. A second DSN station was requested and confirmed no acquisition of Cassini's signal (ruling out weather and station configuration problems). Had the EXCEL tool been available at the time, the following data would have been collected from the SFOS file as noted in **Figure 8**:

1. Time of LOS = > 17:15:00 UTC
2. OWLT = > 1 hour 23 minute 51 seconds
3. Year = > 2011
4. Last time CLT was reset = > DOY357 @ 02:15:00 UTC

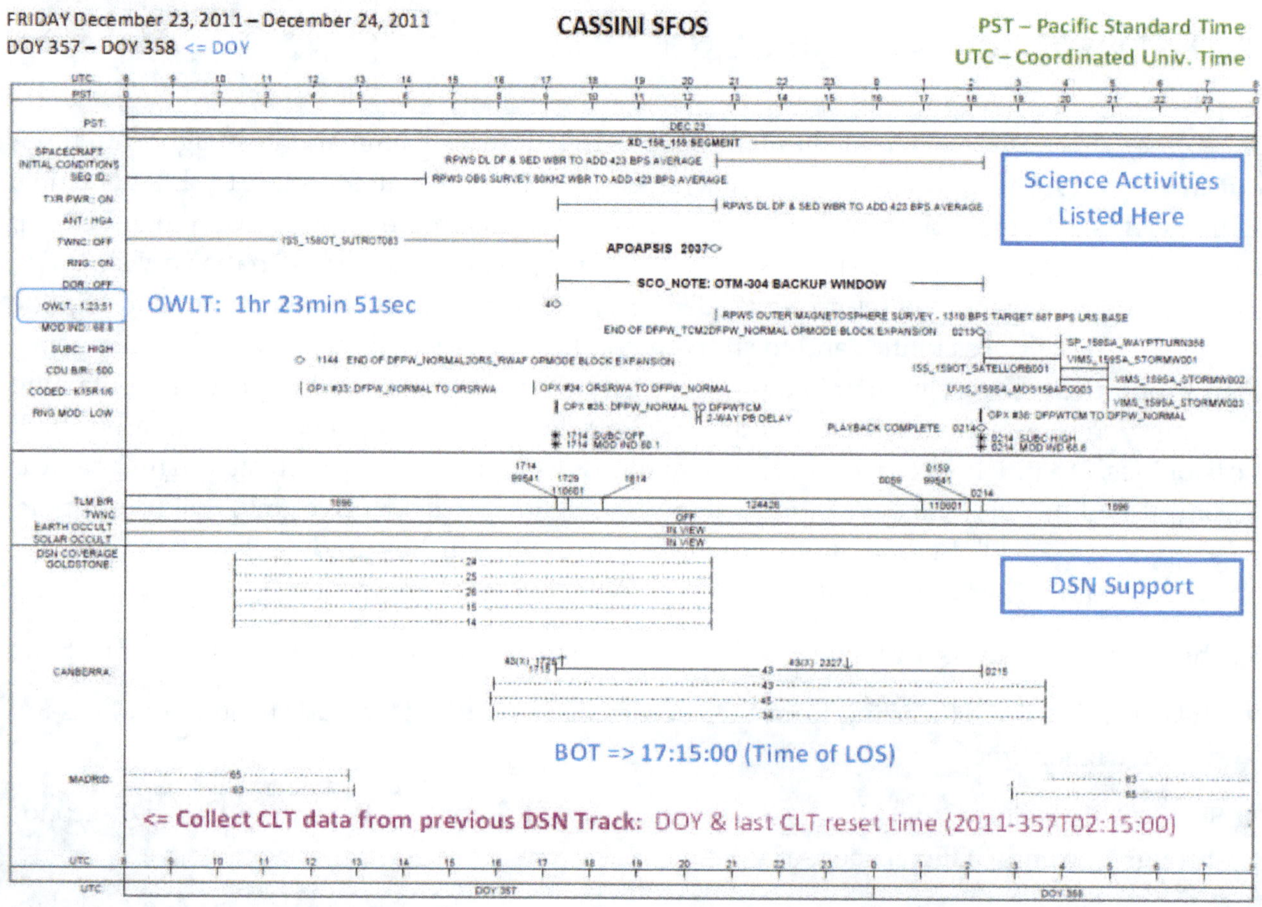

Figure 8. SFOS file for USO failure event.

5. Command Loss Defaults = > 115 hours

6. Prime RFS Devices set to: DST-A, TCU-B, TWTA-B

Once these data had been collected per the instructions listed in Sheet #1, EXCEL Sheet #2 inputs would be entered in the YELLOW spaces as shown in **Figure 9**, which in turn, will cause Sheet #2 through Sheet #4 to be populated with desired timing/post-fault configuration data. Copies of the SFOS and Sheet #2 though Sheet #4 would then be printed and distributed to each subsystem once the Anomaly team gathered to determine the cause and resolution of the LOS condition. As the group followed along with the SFOS file in LOS Timeline #1, spacecraft recovery efforts would have been coordinated with the Cassini ACE via telecom. All system-level FP responses are included in the LOS timeline for completeness (RFS-related responses are shown in red). These are the LATEST times that the FP responses would conclude, assuming that each activation started at BOT. Fault cases would be eliminated by the SOFS Anomaly team once re-acquisition for each completed FP response failed to re-establish the earth-spacecraft link.

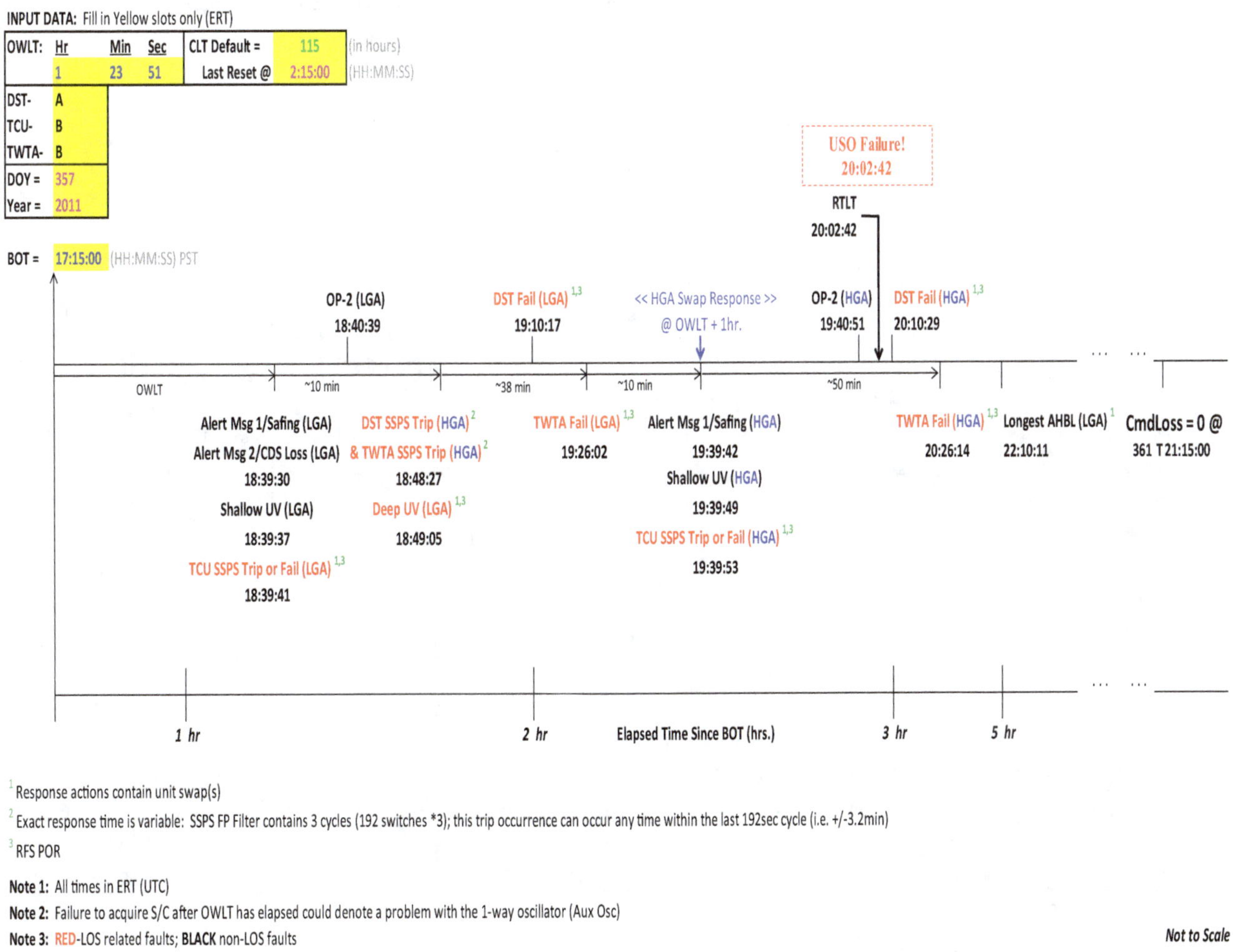

Figure 9. LOS timeline of SFP response expiration times.

In the figure, each completed response notes whether a RFS POR occurs, as well as RFS device swap occurrences. The end of the timeline calculates when the Command Loss Timer will decrement to "0" seconds. For each response case, the resulting antenna selected (LGA or HGA if the HAS response is executed for that particular response) is noted in the timeline. Corresponding RFS post-response states and end conditions of interest are listed in Sheet #4 (**Table 2**).

Although there are eight possible RFS combinations (see **Table 3**), there are only three DST/TCU/TWTA combinations of interest due to the selection of RFS prime units in the FP (i.e. the FP will never command the alternate combinations). Also, telemetry delivery on the post-Safing commanded LGA is minimal at best, so that the recommendation to the Cassini ACE would be to attempt re-acquisition with the FP commanded RFS combinations after the HAS response had concluded (since all RFS-related responses will execute the HAS response to swap to the HGA antenna and increase the D/L rate). According to the LOS timeline, no new RFS configurations will be commanded after 3 hours 20 minute (so that the nominal DST-A/TCU-B/TWTA-B arrangement is assumed), since all RFS-related FP responses will have executed. Problems to focus on from this point forward would be an onboard sequencing error, an activation of the AACS FP, undetected RFS failures not protected by FP, a LGA-1 or HGA antenna failure, WTS-B failure, multiple faults, or possibly a waived failure; all which will most likely leave the spacecraft on the LGA-1 antenna (note: for a USO failure, the DST's VCO will take over the D/L delivery once 2-way communication is established).

Elapsed Time is	Fault	Antenna	Osc	DST	TCU	TWTA	RFS POR?	Prime CDS	BU CDS	Prime AFC	BU AFC	U/L Rate (bps)	D/L Rate (bps)	Mod Index
18:39:30	Alert Msg 1 & 2	LGA	Aux Osc	DST-A	TCU-B	TWTA-B	N	CDS-B	CDS-A	AFC-A	AFC-B	7.8125	RTE-5	29
18:39:37	Shallow UV	LGA	Aux Osc	DST-A	TCU-B	TWTA-B	N	CDS-B	CDS-A	AFC-A	AFC-B	7.8125	RTE-5	29
18:39:41	TCU SSPS Trip	LGA	Aux Osc	DST-A	TCU-A	TWTA-B	Y	CDS-B	CDS-A	AFC-A	AFC-B	7.8125	RTE-5	29
18:39:41	TCU Failure	LGA	Aux Osc	DST-A	TCU-A	TWTA-B	Y	CDS-B	CDS-A	AFC-A	AFC-B	7.8125	RTE-5	29
18:40:39	OP-2	LGA	Aux Osc	DST-A	TCU-B	TWTA-B	N	CDS-B	CDS-A	AFC-A	AFC-B	7.8125	RTE-5	29
18:48:27	DST SSPS Trip	HGA	Aux Osc	DST-A	TCU-B	TWTA-B	N	CDS-B	CDS-A	AFC-A	AFC-B	Current Rate	Current Rate	Pre-Fault
18:48:27	TWTA SSPS Trip	HGA	Aux Osc	DST-A	TCU-B	TWTA-B	N	CDS-B	CDS-A	AFC-A	AFC-B	Current Rate	Current Rate	Pre-Fault
18:49:05	Deep UV	LGA	Aux Osc	DST-A	TCU-B	TWTA-B	Y	CDS-A	CDS-B	AFC-A	AFC-B/Sick	7.8125	RTE-5	29
19:10:17	DST Failure	LGA	Aux Osc	DST-B	TCU-A	TWTA-B	Y	CDS-B	CDS-A	AFC-A	AFC-B	7.8125	RTE-5	29
19:26:02	TWTA Failure	LGA	Aux Osc	DST-A	TCU-A	TWTA-A	Y	CDS-B	CDS-A	AFC-A	AFC-B	7.8125	RTE-5	29
19:39:42	Alert Msg 1 & 2	HGA	Aux Osc	DST-A	TCU-B	TWTA-B	N	CDS-B	CDS-A	AFC-A	AFC-B	250	RTE-1896	39
19:39:49	Shallow UV	HGA	Aux Osc	DST-A	TCU-B	TWTA-B	N	CDS-B	CDS-A	AFC-A	AFC-B	250	RTE-1896	39
19:39:53	TCU SSPS Trip	HGA	Aux Osc	DST-A	TCU-A	TWTA-B	Y	CDS-B	CDS-A	AFC-A	AFC-B	250	RTE-1896	39
19:39:53	TCU Failure	HGA	Aux Osc	DST-A	TCU-A	TWTA-B	Y	CDS-B	CDS-A	AFC-A	AFC-B	250	RTE-1896	39
19:40:51	OP-2	HGA	Aux Osc	DST-A	TCU-B	TWTA-B	N	CDS-B	CDS-A	AFC-A	AFC-B	250	RTE-1896	39
RTLT	Aux Osc Failure	HGA	DST VCO	DST-A	TCU-B	TWTA-B	N	CDS-B	CDS-A	AFC-A	AFC-B	Current Rate	Current Rate	Pre-Fault
20:10:29	DST Failure	HGA	Aux Osc	DST-B	TCU-A	TWTA-B	Y	CDS-B	CDS-A	AFC-A	AFC-B	250	RTE-1896	39
20:26:14	TWTA Failure	HGA	Aux Osc	DST-A	TCU-A	TWTA-A	Y	CDS-B	CDS-A	AFC-A	AFC-B	250	RTE-1896	39
22:10:11	AHBL (all tiers)	LGA	Aux Osc	DST-A	TCU-B	TWTA-B	N	CDS-B	CDS-A	AFC-A	AFC-B	7.8125	RTE-5	29
Anytime	WTS-A Failure (U/L path)	HGA	Aux Osc	DST-A	TCU-B	TWTA-B	N	CDS-B	CDS-A	AFC-A	AFC-B	Current Rate	Current Rate	Pre-Fault
Anytime	WTS-B Failure (D/L path)	HGA	Aux Osc	DST-A	TCU-B	TWTA-B	N	CDS-B	CDS-A	AFC-A	AFC-B	Current Rate	Current Rate	Pre-Fault
Anytime	Other (non FP) faults	HGA	Aux Osc	DST-A	TCU-B	TWTA-B	N	CDS-B	CDS-A	AFC-A	AFC-B	Current Rate	Current Rate	Pre-Fault

RED - LOS Related Fault
BLUE - Device swapped/POR case
PINK - Off-nominal post-fault configuration

Table 2. Post-response concluding end conditions (sheet #4).

No.	DST	TCU	TWTA
1	A	A	A
2	A	B	A
3	A	B	B
4	A	A	B
5	B	A	A
6	B	B	A
7	B	B	B
8	B	A	B

<= Nominal (expected) ABB Configuration (row 3)

Combination not commanded by FP (rows 2, 5, 6, 7)

Table 3. Possible RFS combinations.

9.1. Command loss response activation

If the SOFS Anomaly team was unable to re-acquire the spacecraft before the Command Loss Timer decremented to "0" seconds, the Command Loss Timeline in **Figure 10** would have been followed in synchrony with the SFOS file. In Sheet #3, the event times are listed in UTC (Universal

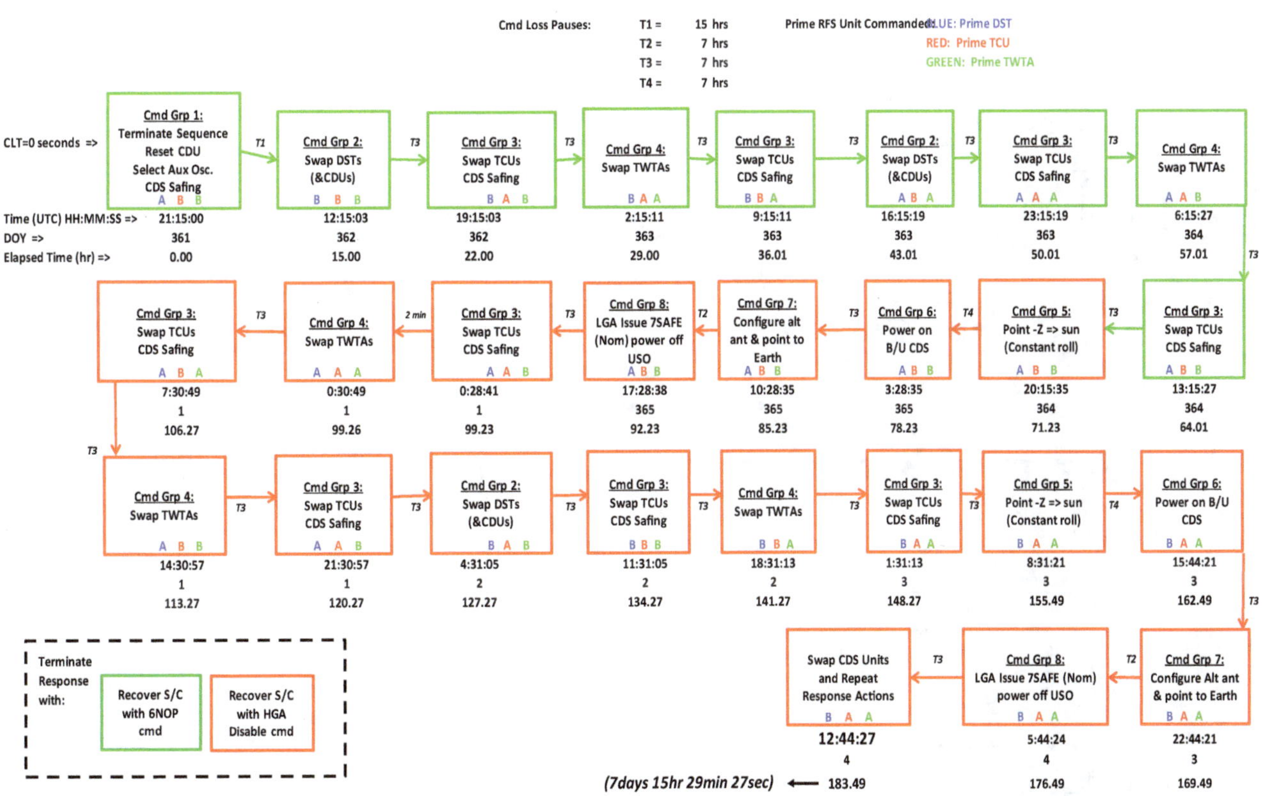

Figure 10. One command loss response cycle (sheet #3).

Time Coordinated), which is consistent with the SFOS file timeline (in successive pages to DOY357 which are not shown in this article). The timeline is also quoted in terms of DOY and elapsed time since the Command Loss Response triggered, showing each upcoming Command Group execution time. As mentioned before, the Command Groups consist of actions which reconfigure redundant hardware, eventually commanding spacecraft attitude and antennas in later Command Groups. Once a ground command is successfully received by the spacecraft, the response will be terminated, the CLT reset (to 115 hours), leaving the vehicle on the successfully commanded configuration.

The Command Loss Timeline is listed for one "CDS cycle" of the response. If all attempts to re-acquire the spacecraft have failed on the first response cycle of Command Groups on the prime CDS unit, the backup CDS computer will take over at the end of this response chain (after 7 days 15 hours 29 minute), so that the cycle is repeated on the redundant computer. As stated above, the Command Loss Response is an endless loop algorithm; below are the actions of the response cycles:

- 1st Response Cycle: The Prime CDS uses its RAM load; it is then re-booted with a FSW load stored on the SSR (at the end of the response cycle).
- 2nd Response Cycle: The BU CDS takes over immediately using its RAM load; it is re-booted with a FSW load stored on the SSR (at the end of the response cycle).

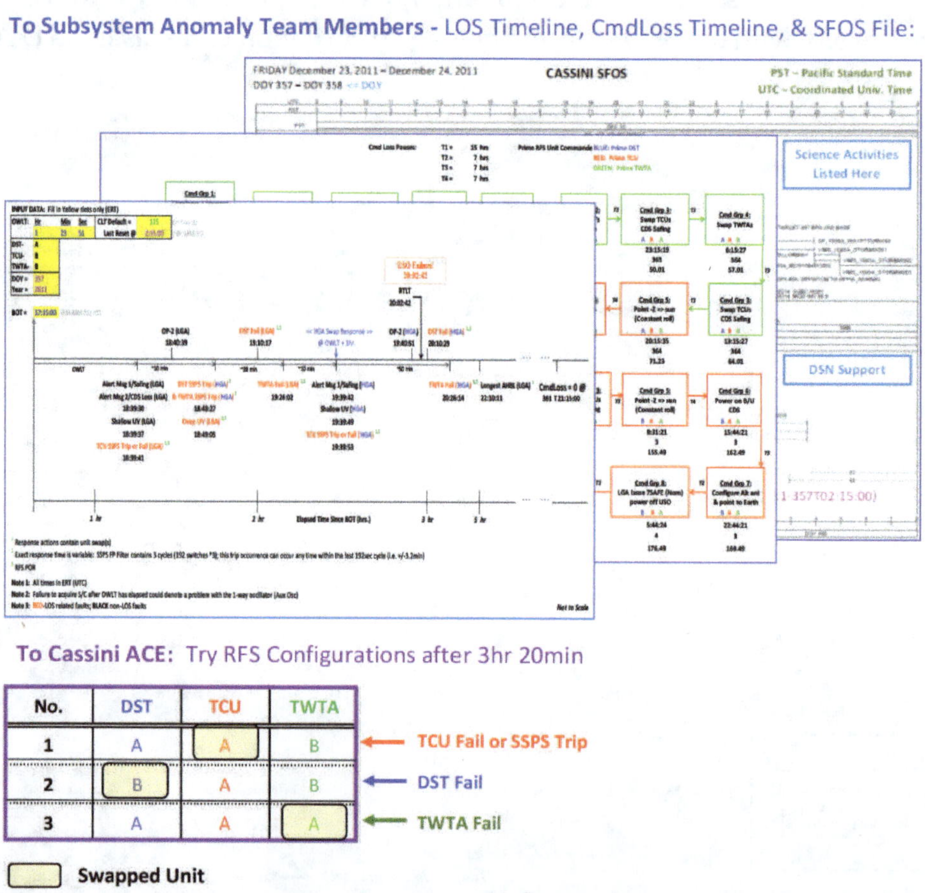

Figure 11. LOS/Cmdloss response info for SOFS team & ACE.

- 3rd Response Cycle: The Prime CDS uses the default SSR FSW load from the previous reset; the Command Loss Timer is set to the FSW default value of CLT = 5 days; at the end of this cycle, the CDS is re-booted with the same FSW load stored on the SSR (at the end of the response cycle), but must wait 5 days before continuing the response.

- 4th Response Cycle: The BU CDS uses the default SSR FSW load from the previous reset; the Command Loss Timer is set to the FSW default value of CLT = 5 days; at the end of this cycle, the CDS is re-booted with the same FSW load stored on the SSR (at the end of the response cycle), but must wait 5 days before continuing the response.

- 5th Response Cycle - ∞: Repeat cycles 3 & 4 above indefinitely.

For the 2011 USO failure event, the EXCEL LOS/Cmdloss Tool would have been used to generate the supporting data needed for trouble-shooting the anomaly for the SFOS Anomaly team, with recommendations included for the Cassini ACE as shown in **Figure 11**.

10. Other uses for the Excel tool

Cassini also relies upon the Command Loss Response to protect events of significant importance should a loss of U/L occur during science experiments and other selected spacecraft activities. **Figure 12** provides an example of this type of "Command Loss Response strategy" used to support the RSS LGA Gravity Experiment performed in 2015, where the HGA must be swapped to LGA-1 and then back again to HGA. The risk associated with this experiment was commanding the WTS switch during the HGA/LGA-1/HGA antenna swap series, where if a malfunction occurred on WTS-A, the U/L capability would

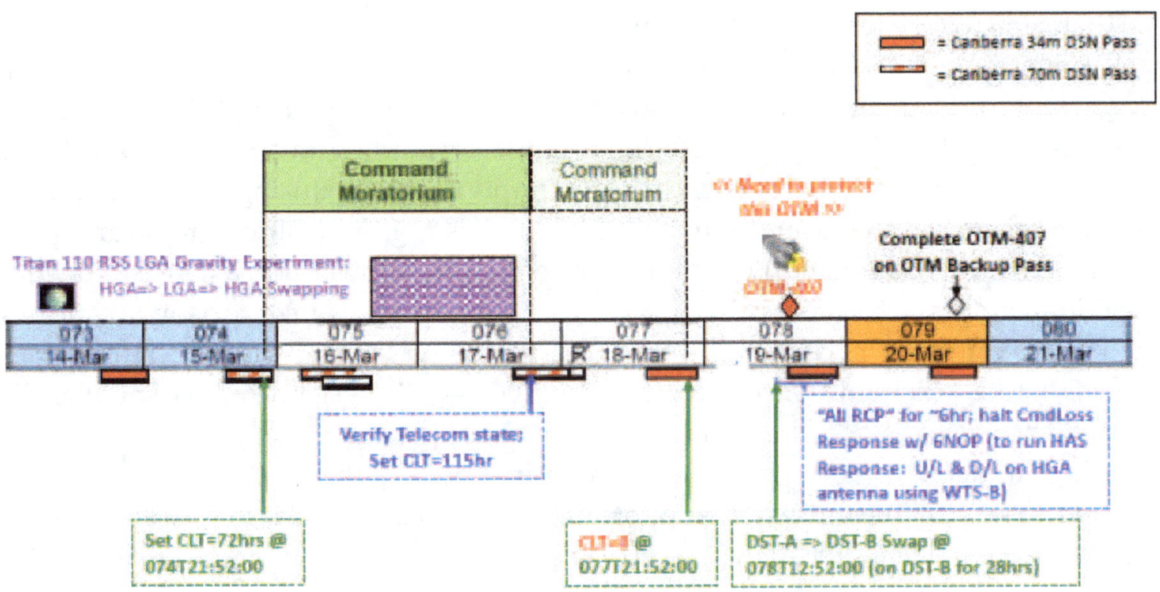

Figure 12. EXCEL tool support of 2015 RSS LGA gravity experiment.

be permanently lost (since there is no WTS FP on Cassini). In this case, the Command Loss Timer default of 115 hours would cause the Orbital Trim Maneuver (OTM) #407 to be missed on DOY078 should WTS-A fail (as well as the planned OTM backup opportunity on DOY079). To protect against loss of U/L after the WTS switch is commanded, the EXCEL tool was used to predict actions from the Command Loss Response which can provide a different U/L path through DST-B/WTS-B should WTS-A fail. The strategy shown in the figure depicts a reduced Command Loss Timer default of CLT = 72 hours with a "command moratorium" period implemented (no commanding allowed), which allows a controlled decrementation of the CLT timer during the RSS LGA Gravity Experiment. Once the test is complete on DOY075, an attempt to verify the telecom state by uplinking the original CLT default value of 115 hours is performed on DOY076. Should this U/L command fail to execute on the spacecraft, the command moratorium will continue until the CLT clocks down to "0" seconds, allowing the Command Loss Response to execute through to Command Group #2 which swaps DST-A= > DST-B, placing the U/L and D/L on WTS-B, just before the DSN track starts. The spacecraft would then be acquired on this new RFS configuration. The OTM would then proceed on the backup DSN pass. For Cassini, a failure of WTS-A would have meant that WTS-B must be used for the remainder of the mission, since the WTS-A switch is henceforth unusable. The actual execution of the RSS Gravity Experiment was successful without the need for FP intervention.

11. Conclusions & lessons learned

Overall, anomalous D/L and LOS occurrences are very challenging and can be difficult for the SOFS Anomaly team to diagnose and resolve. Once the spacecraft's D/L signal is lost, an expedient, accurate resolution process is needed for quick re-acquisition of the vehicle. Identification of FP responses, their conclusion times and corresponding end states, as well as plausible LOS causes, is extremely helpful in eliminating fault cases systematically, thus allowing the SOFS Anomaly team to focus on the actual cause of the LOS problem. Unfortunately, pre-launch FP analyses do not always protect against all LOS-related fault possibilities since design oversites, lack of schedule or funding in implementing FP algorithms, errors within the FSW, or even false assumptions made during the pre-launch testing phase (waived failures) can occur. In all cases, it is highly desirable to address a LOS condition before the Command Loss FP response activates. But if not, a concise timeline of this response and its actions is essential in order to coordinate team efforts in attempting to re-acquire the vehicle; especially since the LGA-1 antenna is commanded, configuring the very low D/L rate which must be delivered through Cassini's very noisy Auxiliary Oscillator (backup device used since the primary USO failed). Therefore, the "LOS/Anomalous Downlink Contingency Plan" Procedure in combination with "LOS/Cmdloss" EXCEL tool is expected to be very useful when supporting this challenging class of faults during the remainder of Cassini's highly successful 20 year mission, until its final plunge into Saturn's atmosphere on September 15, 2017.

Acknowledgements

This research was carried out at the Jet Propulsion Laboratory, California Institute of Technology, under a contract with the National Aeronautics and Space Administration.

Nomenclature

AACS	Attitude, articulation, & control system
ASI	Agenzia Spaziale Italiana (Italian space agency)
BOT	Beginning of (DSN) track
CDA	Cosmic dust analyzer
CDS	Command & data processing system
CIRS	Composite infrared spectrometer
D/L	Downlink
DOY	Day of year
DSN	Deep space network
ESA	European space agency
FP	Fault protection
FSW	Flight software
HAS	High gain antenna swap (algorithm)
HGA	High gain antenna
INMS	Ion & neutral mass spectrometer
JPL	Jet propulsion laboratory
LGA	Low gain antenna
LOS	Los of signal
MAG	Dual technique magnetometer
ME	Main engine
MIMI	Magnetospheric imaging instrument
OTM	Orbital trim maneuver
OWLT	One-way light time

RCS	Reaction control system
RF	Radio frequency
RFIS	Radio frequency instrument subsystem
RFS	Radio frequency system
RPWS	Radio & plasma wave science instrument
RSS	Radio science subsystem
RTLT	Round trip light time
SOFS	Spacecraft operations flight team
SOI	Saturn orbit insertion
SPF	Single point failure
SSPS	Solid state power switch
SSR	Solid-state recorder
U/L	Uplink
USO	Ultra-stable oscillator
UTC	Universal time coordinated
UVIS	Ultraviolet imaging spectrograph
VIMS	Visible & infrared mapping spectrometer
WTS	Waveguide transfer switch

Author details

Paula S. Morgan

Address all correspondence to: paula.s.morgan@jpl.nasa.gov

Jet Propulsion Laboratory, California Institute of Technology, Pasadena, California, United States of America

References

[1] Morgan P. Cassini. Mission-to-saturn spacecraft overview & cds preparations for end-of-mission proximal orbits. Jet Propulsion Laboratory/California Institute of Technology. In: Proceedings of the IEEE/AIAA Conference. Montana: Big Sky; March 2015

[2] Jet Propulsion Laboratory/California Institute of Technology Saturn Tour Highlights. In: Cassini-Huygens Website [Internet]. 2016. Available from: https://saturn.jpl.nasa.gov/news/2861/2016-saturn-tour-highlights/

[3] National Aeronautics and Space Administration's Office of Planetary Protection [Internet]. 2014. Available from: http://planetaryprotection.nasa.gov/

[4] Doody D. Deep Space Craft: An Overview of Interplanetary Flight. Chichester: Praxis; 2009

[5] Jones C. Cassini project pre-ship review/single point failures. Jet Propulsion Laboratory/California Institute of Technology. 1997

[6] Morgan P. Cassini spacecraft's in-flight fault protection redesign for unexpected regulator malfunction. Jet Propulsion Laboratory/California Institute of Technology. In: Proceedings of the IEEE/AIAA Conference. Montana: Big Sky; March 2010

[7] Jet Propulsion Laboratory/California Institute of Technology Cassini Significant Events, Cassini-Huygens News & Features. 2011. Available from: https://saturn.jpl.nasa.gov/news/1935/cassini-significant-events-122111-1312/

[8] Morgan P. Resolving the difficulties encountered by JPL interplanetary robotic spacecraft in flight. In: Ghadawala R, editor. Advances in Spacecraft Systems and Orbit Determination. 1st ed. Croatia: InTech Open Access; 2012. p. 235-264. ch11

[9] Taylor J, Sakamoto L, Wong C. Cassini Orbiter/Huygens Probe Telecommunications Deep Space Communications and Navigation Systems Center of Excellence (Descanso) Design and Performance Summary Series. 2002

[10] Morgan P. Robotic spacecraft health management. In: Johnson S, Gormley T, Kessler S, Mott C, Patterson-Hine A, Reichard K, Scandura P, editors. System Health Management: With Aerospace Applications. 1st ed. Wiley; 2011. p. 543-554. ch34

6

Low-Thrust Control Strategies for Earth-to-Mars Trajectories

Marco Gómez Jenkins and
Jose Antonio Castro Nieto

Abstract

Recent advances in electric propulsion systems have demonstrated that these engines have the potential to be used for long-duration travels, with applications such as cargo and human transportation for interplanetary voyages. The Variable Specific Impulse Magnetoplasma Rocket (VASIMR) is an example of this type of engine, possessing the ability to operate at a wide range of specific impulse levels. This chapter presents the results of a study comparing three different thrust control strategies for Earth-Mars trajectories, using the VASIMR engine at a power of 150 kW. These are constant thrust trajectories, trajectories with coasting periods, and trajectories with variable specific impulse, resulting in variable thrust. To achieve this, an optimization tool was created using spherical coordinates to model the dynamics of the spacecraft, optimal control theory to setup the optimization problem, and a differential evolution algorithm to minimize the cost function. A novel approach to model variable specific impulse and coast-arcs in the trajectories for spherical coordinates is presented as well. The optimization tool was utilized to find optimal trajectories from Earth to Mars orbit, and it was concluded that using variable thrust reduces propellant consumption for a variety of trajectories, when compared to the other two methods.

Keywords: low-thrust trajectories, high power electric propulsion, global optimization

1. Introduction

The National Aeronautics and Space Administration (NASA) announced in 2015 its partnership with commercial industry to develop 12 key technologies that will allow space and human exploration to deep-space destinations, such as the Moon and Mars [1]. The Next Space Technologies for Exploration Partnerships (NextSTEP) include concepts in advanced propulsion, habitation, and small satellites. Among these, three companies developing high power electric propulsion systems were selected to develop engines in the 50–300 kW range, with high specific impulse (2000–5000 s) and efficiency (greater than 60%). The purpose of the development of these engines is to obtain propulsion systems that can operate continuously for long periods, to enable deep space transportation using highly efficient propulsion.

The selected companies for NextSTEP are:

- Ad Astra Rocket Company of Webster, Texas
- Aerojet Rocketdyne Inc. of Redmond, Washington
- MSNW LLC of Redmond, Washington

Although all three companies are working on electric propulsion systems, these engines operate under different principles. Ad Astra Rocket Company's Variable Specific Impulse Magnetoplasma Rocket (VASIMR) uses radio waves to ionize and energize a propellant, converting it to a plasma state, and a magnetic field to guide and expel the plasma, producing thrust [2]. Aerojet Rocketdyne is working on a high power Hall thruster, which uses electrons trapped in a magnetic field to ionize propellant and accelerate the propellant to produce thrust, while neutralizing the plume to avoid the spacecraft from acquiring a charge [3]. The electrodeless Lorentz force (ELF) thruster developed by MSNW LLC, is a pulsed propulsion system that generates a high density and magnetized plasmoid, known as a field reversed configuration (FRC), using radio waves to produce a rotating magnetic field (RMF) [4]. These FRC sources are pulsed devices where the plasmoid evolves from neutral gas injection and ionization, to plasmoid growth and acceleration, and finally to plasmoid ejection.

If the parameters specified by NASA for engine performance are reached, these propulsions systems could be powered by solar energy for interplanetary flight. These type of systems are called solar electric propulsion (SEP) and would require approximately 10 times less propellant to operate than the typical chemical propellant that are currently operating [5]. Furthermore, SEP systems with thrust control could provide even more propellant savings compared to continuous thrust system. The main motivation of this study is to test whether this thrust strategy is indeed more efficient in terms of propellant consumed for interplanetary travel.

This chapter aims to find the optimal low thrust control strategy for transfers from Earth to Mars using three different thrust control strategies: (1) constant thrust trajectories, (2) trajectories with coasting periods, and (3) trajectories with variable specific impulse, resulting in variable thrust. To achieve this goal, an optimization tool was created to compute the optimal trajectory, given a fixed time of flight, for each thrust control strategy. The optimal trajectory was selected based on propellant consumption for each transfer. The engine used for the study is the VASIMR, given its ability to operate at a wide range of specific impulse values, and therefore thrust levels. Section 2 presents a description of this engine, while Section 3 presents the optimization tool created for this study. The results of the analysis are presented in Section 4, leading to the conclusions presented in Section 5.

2. Variable Specific Impulse Magnetoplasma Rocket

The VASIMR is an electric thruster of the electromagnetic kind. It uses magnetic fields to guide plasma through an exhaust, producing thrust in the process. The concept was created by Dr. Franklin Chang Díaz during his time as a graduate student at the Massachusetts Institute of Technology (MIT) and has been developed since the late 1970s [2]. During the 1990s, development of the engine took place in the Advanced Space Propulsion Laboratory (ASPL) at NASA's Johnson Space Center. The experimental engine tested at the laboratory operated at 10 kW and was later upgraded to a 50 kW version producing 0.5 N of thrust. Ad Astra Rocket Company was then created as a spin-off of the NASA laboratory and the engine has seen a significant development in technology during the company's lifespan. The most recent version of the engine (VX-200 or VASIMR eXperimental 200) runs at 200 kW and produces a maximum thrust of approximately 6 N at an specific impulse of 5000 s.

Currently, researchers are improving the engine to operate at steady state. In 2015, Ad Astra Rocket Company was awarded a 3-year, $9 million contract from NASA to develop the maturity of the VX-200 engine [6]. Specifically, by the end of the contract, company must demonstrate that the engine is able to operate at a power level of 100 kW for 100 h. Ad Astra is currently on schedule with this goal, and has successfully completed a NASA review after its second year of contract. Currently, the engine has operated for a total 10 h and there have been considerable changes to the vacuum chamber where the VX-200 operates. These modifications are necessary, so the engine can handle the thermal load produced by the engine. After demonstrating successful steady-state operations, a flight version of the engine called the VASIMR Flight 200 (VF-200) is planned to be constructed and tested in space.

Figure 1 presents a schematic of the VASIMR and its operating principles. The propellant (in gaseous form) enters the first stage of the engine and is converted to plasma by a helicon radio frequency (RF) generator. This was established in nuclear fusion experiments and consists of ionizing the gas. The plasma is guided forward using a magnetic field created by superconducting magnets. It then advances to the second stage where it is energized using ion cyclotron resonance heating (ICRH). The high-energy plasma is then exhausted using a magnetic nozzle, creating thrust. One unique feature of this engine is a technique

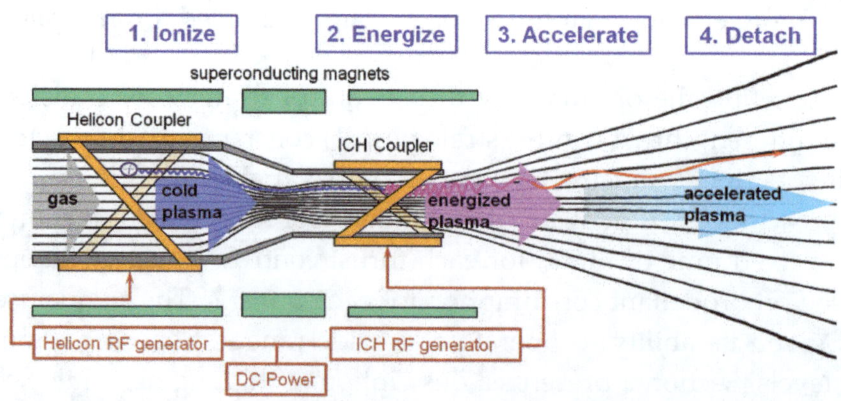

Figure 1. VASIMR operating principles.

called constant power throttling (CPW) [2]. This means that the engine can vary its thrust and specific impulse using constant power settings. The throttling is possible by controlling the amount of power that goes to each stage: if more power is directed to the first stage, more plasma is created generating more thrust, but at a lower specific impulse. If more power is directed to the second stage, less plasma is created but it will have a higher exhaust velocity (higher specific impulse), since it gets a greater energy boost from the ICRH. This variation in thrust and specific impulse is a great advantage since the engine can fit many mission profiles due to its flexibility. Additionally, the VASIMR can be scaled up in power (theoretically to MW capability), enabling crewed interplanetary flights using electric propulsion [7].

3. VASITOS

A low-thrust spacecraft trajectory optimization tool, called the Variable Specific Impulse Trajectory Optimization Software (VASITOS), was created to analyze the optimal thrust strategy. This section presents the software environment in which it was created, the propagation scheme used to model the dynamics of the spacecraft, and the global optimization algorithm incorporated to compute optimal low-thrust trajectories.

3.1. Software environment

The software environment in which VASITOS was developed consists of two sections: Spyder and PyGMO. The former was used to model the propagation of the orbit, while the latter was used for optimization. Spyder is an integrated development environment (IDE) that combines various open source packages written in Python [8]. These include some for scientific computing (NumPy and SciPy) and other for plotting (Matplotlib). It offers several advantages over other programs for scientific computing, mainly that it is an open source and that it is written in Python, a language, which is quite intuitive.

The Parallel Global Multiobjective Optimizer (PaGMO) is an optimization toolbox created by the Advanced Concepts Team at the European Space Agency (ESA) to solve complex optimization problems [9]. It is available for C++ and Python (the Python version is called PyGMO). The software features the generalized island model (GIM), which allows parallel computing in order to reduce computation time. PyGMO includes several optimization algorithms and global optimization problems, such as the genetic algorithm (GA), differential evolution (DE), particle swarm optimization (PSO), and adaptive simulated annealing (ASA), among others. The parallel computing scheme was implemented in the software and optimization simulations were performed in a Lenovo U410 with an Intel Core i5. This has multithreading, which means the operating system can identify up to four CPUs. Therefore, four islands were included in the parallel computing scheme.

To operate VASITOS, the user will input the initial and target orbit into the tool, along with the thruster specifications. VASITOS will run simulations until the end condition specified for the optimization algorithm is met. For example, for GA and DE, one must define the number of generations required in the simulation. The output will be the optimal path, propellant mass consumed, time of flight, and the offsets. These are defined as the difference between the target state and the final simulated state. If the results meet the mission requirements, then the user will process them further by creating plots and analyzing which trajectory is best based on mission needs.

3.2. Propagation

Spherical coordinates were the preferred method of modeling for this project since it has been successfully used for first-order mission analysis of interplanetary trajectories, resulting in an efficient computation time [10]. The position of the spacecraft in the two-dimensional Euclidean space is defined by the radius vector and the angle θ. The x–y coordinate system is centered at the main body (Sun for interplanetary trajectories). At the center of the satellite, there is another coordinate system defined, consisting of the radial axis and the θ axis. The velocity vector, originated at its center of mass, defines the velocity of the spacecraft. Another vector that starts at the same position is the thrust vector. The angle between the θ axis and the thrust vector is called the pitch angle (α). It is one of the control parameters in the optimization problem (further explained in the following chapter). The radial and tangential acceleration components due to thrust are defined as:

$$a_{r,T} = \frac{T}{m} \sin \propto \quad (1)$$

$$a_{\theta,T} = \frac{T}{m} \cos \alpha \quad (2)$$

where m is the mass of the spacecraft. The state can then be defined using four parameters: r, θ, v_r, and v_θ, where the last two parameters are the radial and tangential velocity, respectively. The mass of the spacecraft must be included as well, since it is using propellant to transfer from one orbit to the other. Therefore, the final state **X** is defined as:

$$\mathbf{X} = [r, \theta, v_r, v_\theta, m]^T \quad (3)$$

Once the state parameters were selected, the following step is to define their rate of change. This is essential to compute the future state. They are defined as [11]:

$$\dot{r} = v_r \quad (4)$$

$$\dot{\theta} = v_\theta \quad (5)$$

$$\dot{v}_r = \frac{\mu - r v_\theta^2}{r^2} + \frac{T}{m} \sin \alpha \quad (6)$$

$$\dot{v}_\theta = \frac{v_r v_\theta}{r} + \frac{T}{m} \cos \alpha \quad (7)$$

$$\dot{m} = -\frac{2\eta P}{(g_0 I_{sp})^2} \quad (8)$$

where μ is the gravitational parameter of the central body and T is the thrust of the low-thrust system. Most variables in equation $\dot{m}$ are engine specifications: η is its efficiency, P is the power, and I_{sp} is the specific impulse. The parameter g_0 is the standard acceleration due to gravity. The thrust magnitude is defined as:

$$T = \frac{2\eta P}{g_0 I_{sp}} \tag{9}$$

The equation shows that the thrust magnitude and specific impulse are inversely proportional, meaning that if one is increased, then the other is decreased. For this study, it is assumed that the engine efficiency and power are constant, so the specific impulse is an independent variable while the thrust is the dependent one. This will be important when selecting the former variable as a control parameter. Once the initial state of the system is defined, it can be combined with this system of equations to compute the state of the spacecraft at future times using an integrator.

3.3. Optimization

The rates of change of the state parameters are essential to form the Hamiltonian. In the context of optimal control theory, the Hamiltonian does not possess any physical meaning; it is a parameter derived from calculus of variation, which aids in finding the optimal trajectory. In a recent study, optimal control theory was applied to a spherical system, which only considered the radius, radial velocity, and tangential velocity [12]. Additionally, the only control parameter defined was the pitch angle. This chapter expands on previous work by including the position θ of the spacecraft within the trajectory and the mass of the vehicle. Furthermore, it includes the specific impulse as a control parameter. For the system defined in Section 3.2, the Hamiltonian is expressed mathematically as:

$$H = \lambda_r \frac{dr}{dt} + \lambda_\theta \frac{d\theta}{dt} + \lambda_{v_r} \frac{dv_r}{dt} + \lambda_{v_\theta} \frac{dv_\theta}{dt} + \lambda_m \frac{dm}{dt} \tag{10}$$

where λ's are the costates of each parameter that makes up the state. These costates represent the cost of changing one parameter relative to another. For example, if one simulates a transfer, where the change in radius is much greater than the change in angle θ, then the costates of the radius and radial velocity will be greater in magnitude than the ones associated with θ. The rate of change of the costates over time can be obtained by using the following property derived from optimal control theory:

$$\dot{\lambda}_i = -\frac{\delta H}{\delta i} \tag{11}$$

This results in the following expressions:

$$\dot{\lambda}_r = \frac{v_\theta^2 \lambda_{v_r} - v_r v_\theta \lambda_{v_\theta}}{r^2} - \frac{2\mu \lambda_{v_r}}{r^3} \tag{12}$$

$$\dot{\lambda}_{\theta} = 0 \tag{13}$$

$$\dot{\lambda}_{v_r} = -\lambda_r + \frac{v_\theta \lambda_{v\theta}}{r} \tag{14}$$

$$\dot{\lambda}_{v\theta} = \frac{-2v_\theta \lambda_{v_r} + v_r \lambda_{v\theta}}{r} \tag{15}$$

$$\dot{\lambda}_m = \frac{T}{m} \frac{\lambda_{v_r} + \lambda_{v\theta}}{\lambda_m \sqrt{\lambda_{v_r}^2 + \lambda_{v\theta}^2}} \tag{16}$$

Computing the costates is of the utmost importance in optimal control theory since the control parameters depend on them. For this study, there are two of them: the thrust direction and the thrust magnitude. The former is defined as the angle of attack α. The latter is inversely proportional to the specific impulse, meaning that if we control the specific impulse, we control the thrust magnitude. To obtain the profile of both control parameters, we need to use Pontryagin's Minimum Principle, which is expressed mathematically as:

$$\frac{\delta H}{\delta u} = 0 \tag{17}$$

where u is the control parameter. Since we have two control parameters, the resulting equations are:

$$\frac{\delta H}{\delta \alpha} = 0 \tag{18}$$

$$\frac{\delta H}{\delta I_{sp}} = 0 \tag{19}$$

By solving these two equations, we obtain the following control laws:

$$\sin \propto = -\frac{\lambda_{v_r}}{\sqrt{\lambda_{v_r}^2 + \lambda_{v\theta}^2}} \tag{20}$$

$$\cos \propto = -\frac{\lambda_{v\theta}}{\sqrt{\lambda_{v_r}^2 + \lambda_{v\theta}^2}} \tag{21}$$

$$I_{sp} = \frac{2m\lambda_m}{\sqrt{\lambda_{v_r}^2 + \lambda_{v\theta}^2}} \tag{22}$$

The angle of attack is divided into sine and cosine to ensure the right sign (+/−). It is important to use the atan2 function when computing the magnitude and direction of this angle. The optimal specific impulse I_{sp}^* defines the optimal thrust T^* in the following fashion:

$$T^* = \frac{2\eta P}{g_0 I_{sp}^*} \tag{23}$$

The value of the optimal specific impulse will depend on the boundaries defined by the engine specifications. This is expressed mathematically as:

$$\mathrm{I_{sp,L}} < \mathrm{I^*_{sp}} < \mathrm{I_{sp,U}} \tag{24}$$

where $I_{sp,L}$ and $I_{sp,U}$ are the lower and upper boundaries of the specific impulse, respectively. If the user wishes to introduce coast arcs (assuming that the specific impulse of the engine is constant), then the following bang-bang strategy is applied:

$$if\ I_{sp} > I^*_{sp}\ then\ T^* = T \tag{25}$$

$$else\ if\ I_{sp} < I^*_{sp}\ then\ T^* = 0 \tag{26}$$

Now, there are 10 equations for rate of change of the state and costate parameter (5 equations for states and 5 for costates). We also have the initial and final values for the states, which are defined by the users. The only thing we are missing is the initial values for the costates. These are called the design variables and are stored in the decision vector, which is defined as:

$$\xi = [\lambda_r(0), \lambda_\theta(0), \lambda_{v_r}(0), \lambda_{v_\theta}(0), \lambda_m(0)]^T \tag{27}$$

The goal of the optimization process is to find the decision vector that minimizes the following cost function:

$$J = W_r \Delta r + W_\theta \Delta\theta + W_{v_r} \Delta v_r + W_{v_\theta} \Delta v_\theta \tag{28}$$

where the Δ's are the offsets (defined as the absolute difference between the final simulated value and target value for selected state parameters) and the W's represent the weights assigned to each offset. The weights are selected by the user and are modified according to the mission needs. This optimization method is indirect since the function we are minimizing does not include the main parameter to minimize: the time of flight. By obtaining the optimal costate profiles and ensuring the final conditions are met, the time of flight is ensured to be minimized (which is why the method is called indirect). For this project, the optimal decision vector was obtained using a numerical method called differential evolution, which is part of the family of evolutionary algorithms. A detailed description of the algorithm can be found in [13].

4. Thrust control strategies

Electric propulsion systems have considerable potential for interplanetary travel, but to analyze its feasibility, one has to consider not only the spacecraft's optimal path, but thrust strategy. Three strategies are considered in this study:

- Continuous thrust
- Coasting
- Variable thrust

The first one consists of operating at a constant thrust throughout the trajectory, meaning that the engine is operating continuously. The second strategy consists of using "coast arcs," defined as periods where the engine is not producing thrust. Finally, variable thrust control will be tested given that the VASIMR has the ability to modify this parameter given that it features variable specific impulse.

4.1. Simulation parameters

Each thrust strategy was considered for a transfer from Earth's orbit to Mars' orbit in a two-dimensional heliocentric reference frame. Furthermore, it was assumed that the orbits of both planets are circular. The initial and final orbital parameters are displayed in **Table 1**. It can be observed that the final position in the target orbit is not specified, since the aim in these simulations is to reach the orbit, not the planet. The forces acting on the spacecraft are due to

	Initial orbit: Earth	Target orbit: Mars
Radius, r (km)	149.597×10^6	227.937×10^6
Velocity, v (km/s)	29.785	24.130
Position, θ (deg)	0.0	—

Table 1. Initial (Earth) and target (Mars) orbits to test control strategies.

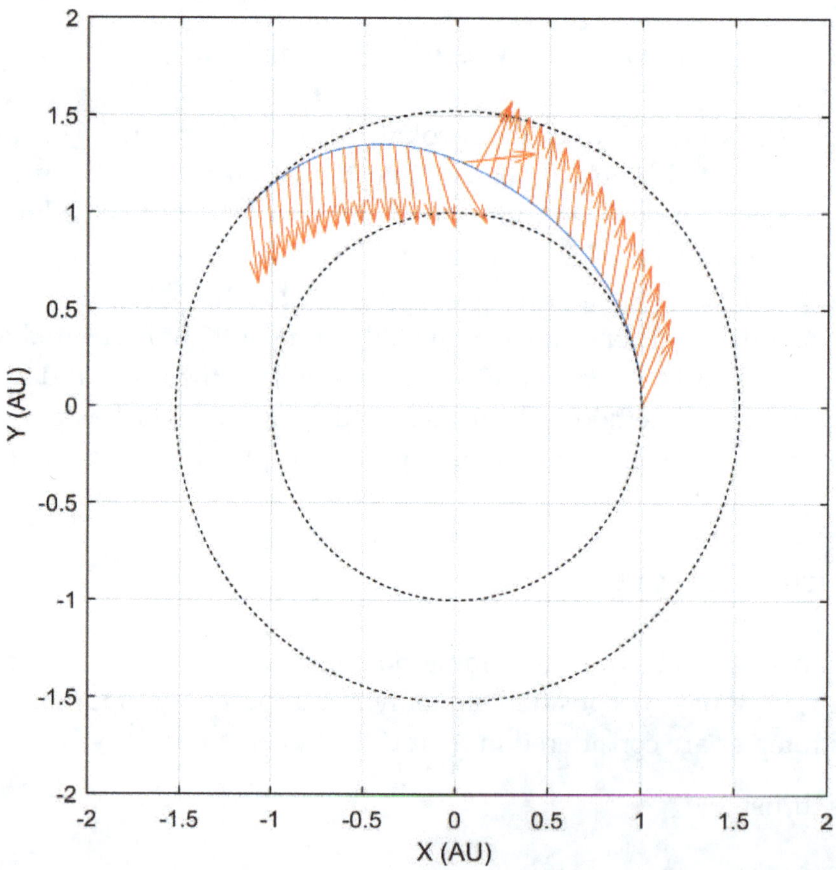

Figure 2. Transfer from Earth to Mars orbit using continuous thrust.

the Sun's gravity and the thrust produced by the engine. Third body perturbations from the planets on the spacecraft are not considered, nor the position of the planets on arrival and departure of the spacecraft.

The spacecraft was assumed to have a wet mass of 4500 kg, with a propellant mass of 1500 kg, and a VASIMR engine with 150 W of power and 65% efficiency. The specific impulse ranges from 5000 to 30,000 s, which are the theoretical limits of the engine [14]. When operating at a constant specific impulse, it was assumed that the specific impulse is equal to the lower boundary. The step size defined in the simulation was 24 hours, while the integrator used for propagation was the fourth order Runge-Kutta method. The differential evolution algorithm was set to a population size of 20, running for 500 generations.

Continuous thrust	Coasting	Variable thrust
1303 kg	1267 kg	1267 kg

Table 2. Propellant consumption for Earth to Mars transfer for three different thrust strategies.

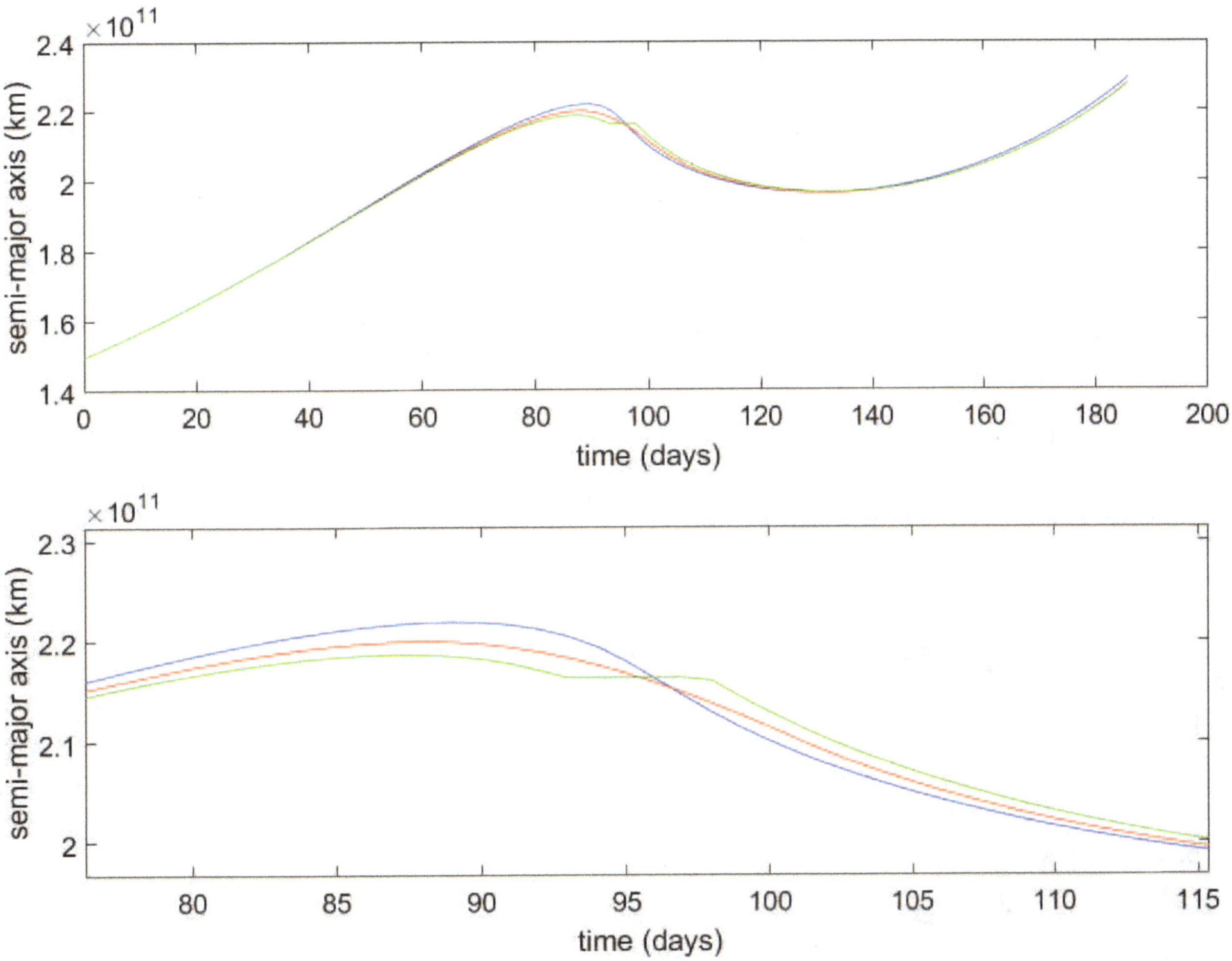

Figure 3. Semi-major axis profile for Earth-Mars trajectories using different thrust control methods. Top image displays the complete profile, while the bottom figure displays the profile at mid-flight (blue line = continuous thrust, green line = coasting, and red line = variable thrust).

4.2. Results

Figure 2 presents the results of a transfer from Earth to Mars orbit in a heliocentric reference frame in astronomical units (AU). The dashed inner circle represents Earth's orbit, while the dashed outer circle represents Mars' orbit. The curve represents the spacecraft's trajectory, while the arrows represents the thrust magnitude and direction. This last parameter demonstrates how the thrust direction was controlled to obtain the optimal trajectory. The spacecraft starts thrusting almost normal to the velocity vector and reverses direction at approximately mid-flight until reaching the final orbit. The thrust magnitude is not considered as a control parameter for this simulation since the thrust is assumed to be continuous. The final orbital trajectory results in a time of flight of 185.78 days and a propellant consumption of 1303 kg.

The same transfer was computed for the coasting and variable thrust case. For both of these cases, the time of flight was set to 185.78 days, which was the optimal time for the continuous thrust case. The propellant consumed to achieve the transfer for each case is presented in **Table 2**.

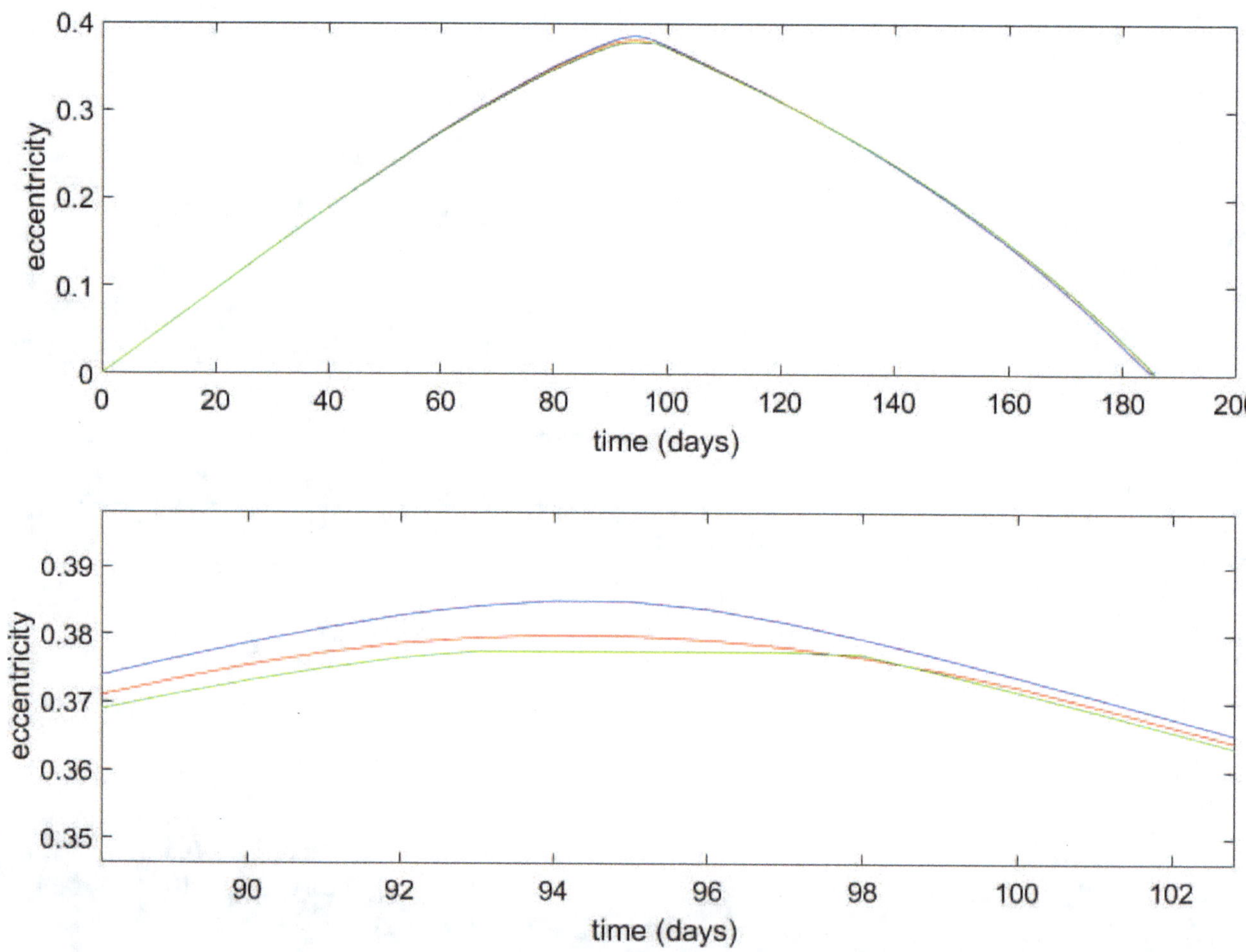

Figure 4. Eccentricity profile for Earth-Mars trajectories using different thrust control methods. Top image displays the complete profile, while the bottom figure displays the profile at mid-flight (blue line = continuous thrust, green line = coasting, and red line = variable thrust).

From **Table 2**, one can observe that using coasting and variable thrust results in a 2.7% reduction in propellant consumption relative to the continuous thrust strategy. To properly understand why this reduction occurs, one must analyze the in-plane orbital elements, as well as the thrust profile for each control method.

Figures 3 and **4** present the semi-major axis and eccentricity profile for the three thrust control methods, respectively. Additionally, **Figure 5** presents the specific impulse for each case. With this figure, the thrust profile can be deduced, given that the specific impulse is inversely proportional to the thrust of the engine. Presenting the specific impulse was favorable to ensure that the engine is operating within its limits. For the coasting case, the specific impulse was set to infinity during periods when the spacecraft is required to coast as dictated by the control law, resulting in zero thrust.

From **Figure 3**, it can be observed that the overall trend in the semi-major axis is an increase throughout the trajectory, except at approximately the halfway point. Here, there exists a considerable decrease in this parameter because the spacecraft performs a radical change in thrust direction: nearly 180°. With the use of a coast arc, the majority of the change of direction is performed without thrust (see **Figure 5**), meaning that there is a smaller change in the semi-major axis during this period, resulting in a lower loss of orbital energy. The strategy is more

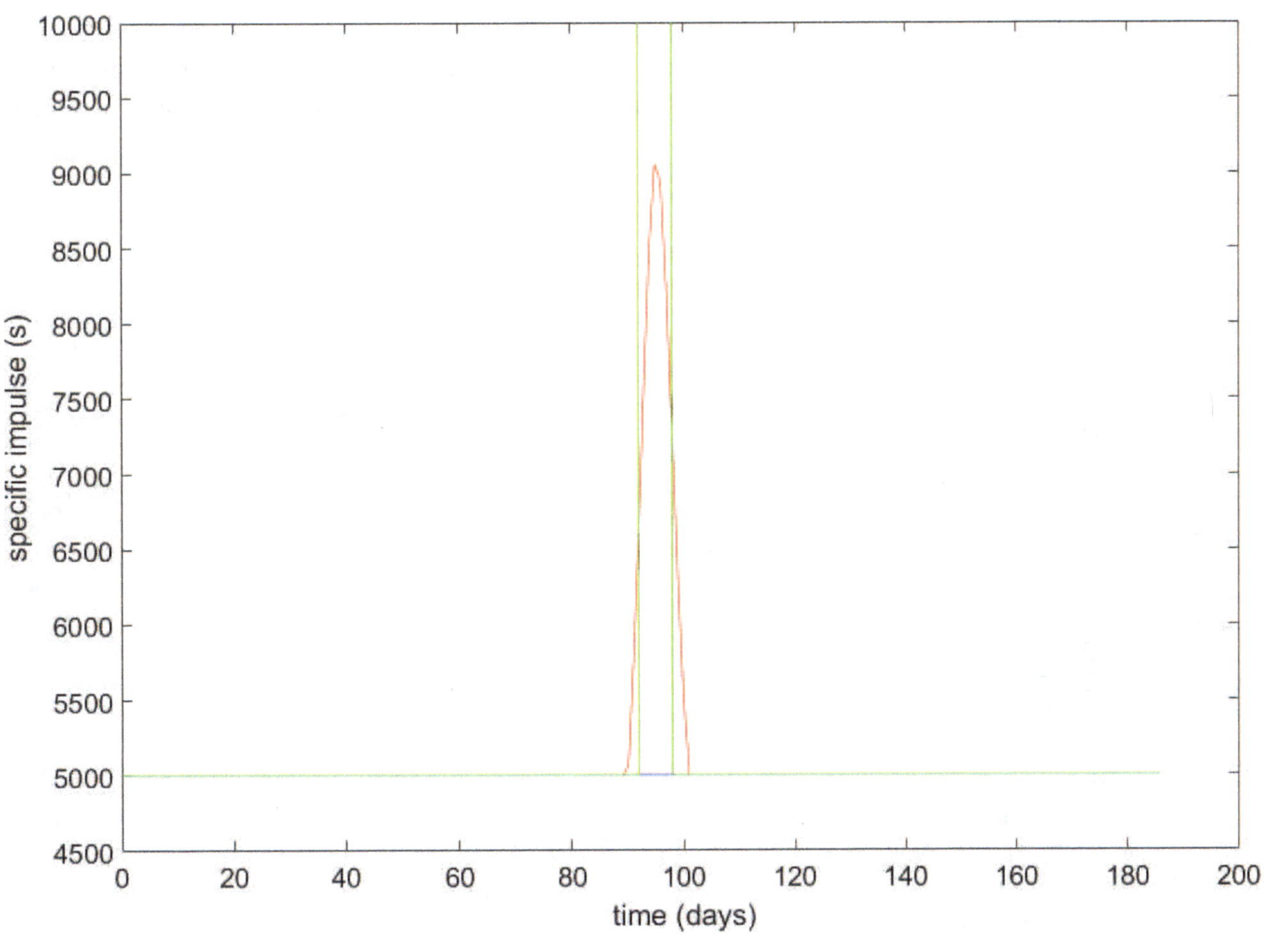

Figure 5. Specific impulse profile for Earth-Mars trajectories using different thrust control methods at mid-flight (blue line = continuous thrust, green line = coasting, and red line = variable thrust).

efficient compared to the constant thrust case, given that it requires less propellant. A similar phenomenon is observed when using variable thrust, where the thrust is lowered when performing the change of direction (see **Figure 5**).

Figure 4 explains why the spacecraft performs the rapid change in thrust direction. It is observed that the change in the eccentricity can be divided into two segments: the first one is a uniform increase while the second one is a uniform decrease. The change occurs at the halfway point, where the spacecraft performs the turn. The eccentricity profile is similar to the Hohmann transfer, considered an optimal transfer strategy for chemical rockets. In this type of transfer, the semi-major axis and eccentricity are increased instantly (modeled as an impulsive burn) when the spacecraft enters the transfer orbit and then the former is further increased but the latter return to zero.

4.3. Variable time of flight

In Section 4.2, it was observed that using coasting or variable thrust resulted in a more efficient transfer than using continuous thrust. Another advantage when using these two control strategies is that the mission designer can vary the time of flight to transfer from the initial to the target orbit, to account for the position of the target planet when the spacecraft arrives at its orbit. By varying the time of flight for this transfer, one could also analyze which control strategy would be best for different flight periods. To achieve this goal, both coasting and variable thrust methods were tested for Earth to Mars transfers using fixed time of flights of 195, 205, and 215 days. The results for the propellant consumption for each case are presented in **Figure 6**, along with the case presented in Section 4.2.

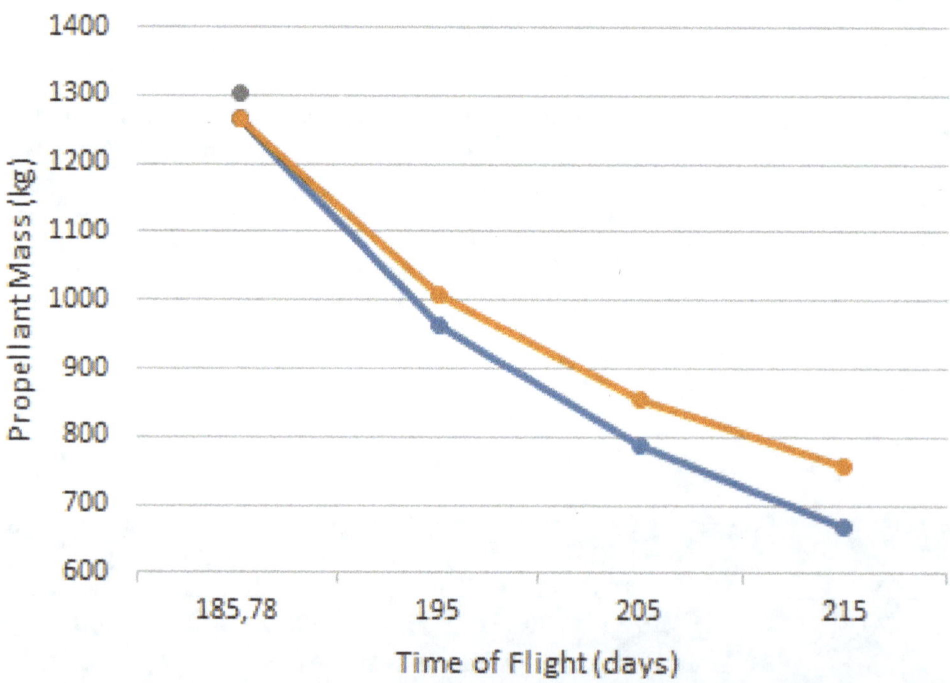

Figure 6. Propellant consumption for Earth to Mars transfers for different cases of time of flight using three thrust control strategies (gray dot = continuous thrust, orange line/dot = coasting, and blue line/dot = variable thrust).

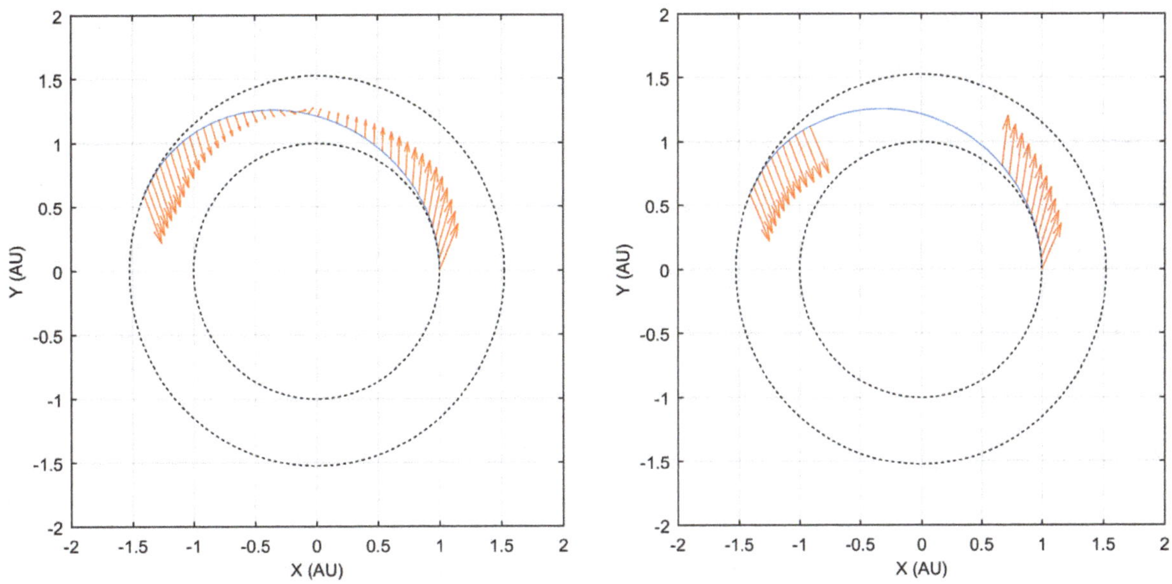

Figure 7. Transfer from Earth to Mars orbit using variable thrust (left) and coasting (right).

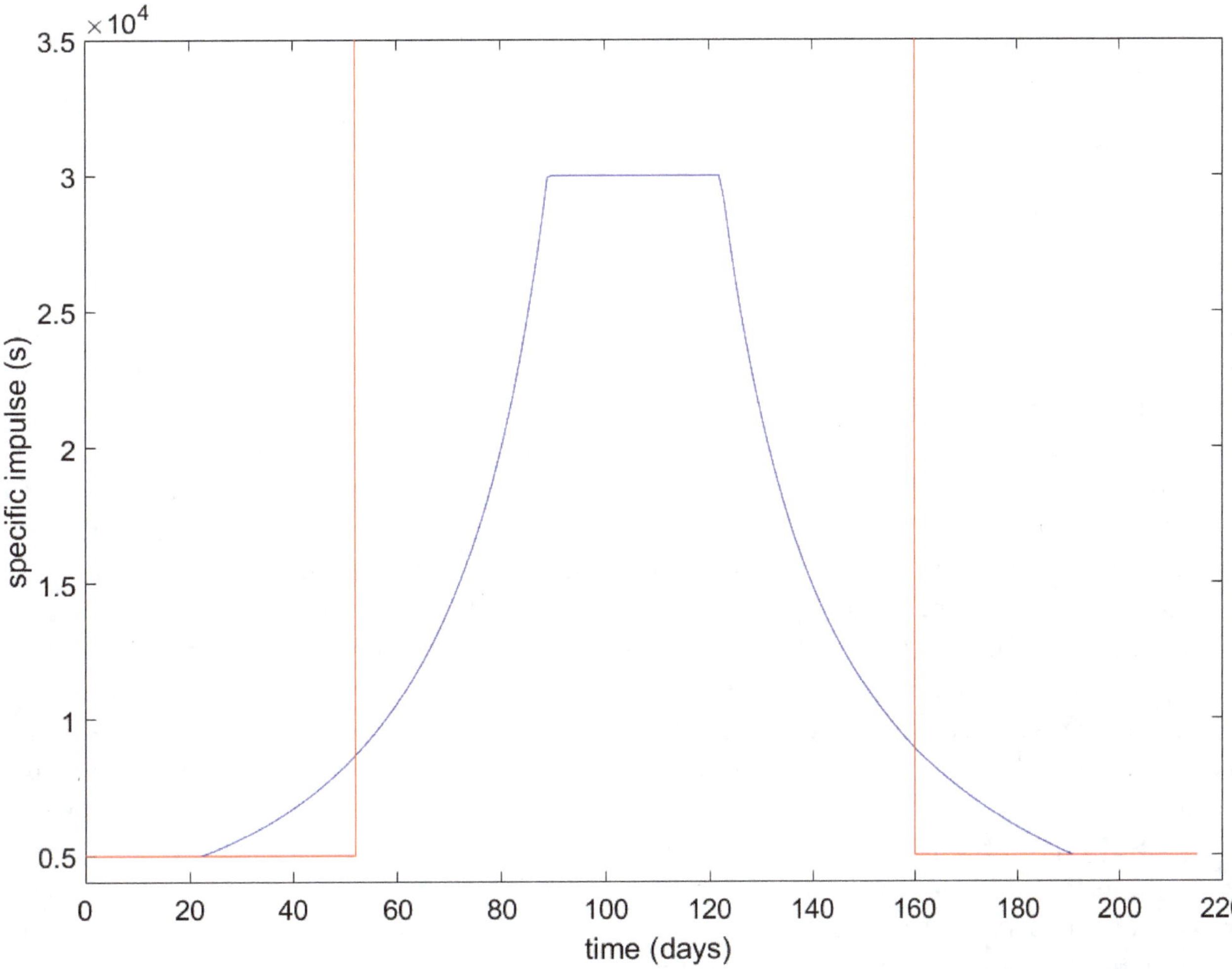

Figure 8. Specific impulse profile for Earth-Mars trajectories using different thrust control methods (blue line = variable thrust, and red line = coasting).

Figure 6 displays that as the time of flight increases, the variable thrust control strategy is more efficient than coasting in terms of propellant. To properly understand this phenomenon, the trajectory for both strategies was plotted for the case where time of flight was equal to 215 days. For this case, the propellant mass was reduced by 12% when using variable thrust compared to coasting.

Figure 7 shows the trajectory in blue, with the red arrows representing the thrust magnitude and direction. The thrust direction profile is similar to what was computed in Section 4.2, with the main difference being that the trajectory is longer, since the time of flight defined is approximately 30 days greater. The thrust magnitude for the variable thrust control strategy is constant at the beginning, but decreases as the spacecraft starts to change direction. At mid-flight, this parameter reaches its minimum but then starts increasing until it reaches at maximum at the end of the trajectory. For the coasting strategy, it is observed that the thrust is constant until approximately a quarter of the time of flight, when the coasting period begins. The thrust resumes in the opposite direction when there is a quarter of the time of flight remaining. This can also be observed in **Figure 8**, where the profile of the specific impulse is plotted. It is seen that using variable specific impulse creates a more gradual change in the orbit, when compared to the coasting mechanism, resulting in a more efficient transfer with a lower propellant consumption. Additionally, it is observed that the engine operates at its highest specific impulse for approximately 35 days, demonstrating the importance of achieving these high levels of specific impulse for interplanetary orbits.

5. Conclusion

Growing interest in high-power electric propulsion systems motivated the analysis of their performance when used to transfer from Earth to Mars orbits. VASITOS was created to study not only the optimal thrust direction, but the optimal thrust magnitude as well. Three thrust control laws were studied: continuous thrust, coasting, and variable thrust. By using a 150 kW thruster with a specific impulse of 5000 s and an efficiency of 0.65 on a 4500 kg spacecraft, it was computed that the optimal time of flight for the transfer using constant thrust was 185.78 days. Additionally, it was observed that there was a loss in orbital energy mid-way through the transfer. By using a variable specific impulse system (with boundaries of 5000–30,000 s), the propellant consumption was reduced by 2.7% due to the system's ability to throttle down at the point where the energy loss occurred. The coasting strategy resulted in a 2.7% propellant reduction as well since the engine stopped thrusting at the point of energy loss. Further results include the comparison of the coasting and variable thrust strategies for fixed time of flights. As the time of flight was increased, it was observed that the propellant consumption of the former strategy was less than the latter. For example, for a fixed time of flight of 215 days, the propellant consumption of the variable thrust strategy was 12% less. From these simulations, it was concluded that the best thrust control law for Earth to Mars transfers was variable thrust, due to its ability to gradually change the orbit relative to the other methods studied, resulting in a lower propellant consumption.

Author details

Marco Gómez Jenkins[1]* and Jose Antonio Castro Nieto[2]

*Address all correspondence to: marco.gomez@itcr.ac.cr

1 Space Systems Laboratory, Costa Rica Institute of Technology, Cartago, Costa Rica

2 Ad Astra Rocket Company, Costa Rica

References

[1] National Aeronautics and Space Administration. Advanced Electric Propulsion NextSTEP BAA Activity Project [Internet]. 2015. Available from: techport.nasa.gov [Accessed: November 16, 2017]

[2] Chang-Diaz F. The VASIMR rocket. Scientific American. 2000;**283**:89-97

[3] Reilly S, Hofer R. Thermal analysis of the 100-kW class X3 hall thruster. In: 47th International Conference on Environmental Systems; 16–20 July 2017; Charleston, South Carolina, USA. 2017

[4] Kirtley D, Pihl J, Pihl C. Development of a steady operating pulsed power system for FRC and inductive thrusters. In: International Electric Propulsion Conference; 2013

[5] Satellite Today. All Electric Satellites: Revolution or Evolution? [Internet]. May 2013. Available from: http://www.satellitetoday.com/publications/2013/05/01/ [Accessed: November 2017]

[6] Ars Technica. NASA's Plasma Rocket Making Progress Toward a 100-hour Firing [Internet]. 2017. Available from: https://arstechnica.com/science/2017/08/nasas-plasma-rocket-making-progress-toward-a-100-hour-firing/ [Accessed: February 11, 2017]

[7] Longmier B, Squire J, Olsen C, Cassady L, Ballenger M, Carter M, Ilin A, Glover T, McCaskill G, Chang-Diaz F, Bering E. VASIMR VX-200 improved throttling range. In: 48th AIAA/ASME/SAE/ASEE Joint Propulsion Conference & Exhibit; 30 July–01 August 2012; Atlanta, Georgia, USA; 2012

[8] Hoekstra J. Programming and Scientific Computing in Python for Aerospace Engineers. 3rd ed. Delft, Netherlands: Delft University of Technology; 2015

[9] Izzo D, Rucinski M, Biscani F. The generalized island model. Parallel Architectures and Bio-Inspired Algorithms Studies in Computational Intelligence. 2012;**415**:151-169

[10] Spangelo S, Dalle D, Longmier B. Integrated vehicle and trajectory design of small spacecraft with electric propulsion for earth and interplanetary missions. In: 29th Annual AIAA/USU Conference on Small Satellites; Utah, USA; 2015

[11] Vallado DA. Fundamentals of Astrodynamics and Applications. 1st ed. Portland: Microcosm Press/Springer; 2007

[12] Conway B, editor. Spacecraft Trajectory Optimization. 1st ed. Cambridge: Cambridge University Press; 2010

[13] Storn R, Price K. Differential evolution: A simple and efficient heuristic for global optimization over continuous space. Journal of Global Optimization. 1997;**11**:341-359

[14] Ilin A, Cassady L, Glover T, Carter M, Chang Diaz F. A Survey of Missions using VASIMR for Flexible Space Exploration. Ad Astra Rocket Company Technical Report; 2010

7

Cost-Effective Platforms for Near-Space Research and Experiments

Kobi Gozlan, Yuval Reuveni, Kfir Cohen,
Boaz Ben-Moshe and Eyal Berliner

Abstract

High-altitude balloons (HABs) are commonly used for atmospheric research. In recent years, newly developed platforms and instruments allow to measure position, temperature, radiation, humidity and gas profile in the troposphere and stratosphere. However, current platforms, such as radiosonde, have limited bandwidth and relatively small number of possible sensors on board. Furthermore, all the measuring instruments carried on board the balloon cannot be reused since most of the times the radiosonde cannot be retrieved. In this chapter, we present a generic near-space research platform based on an improved radio frequency (RF) communication, an advanced set of sensors that might also include a return-to-home (RTH) micro-UAV. We present the overall structure of an advanced HAB payload, which is equipped with a low-cost sophisticated set of sensors along with HD camera system, which weight less than 300 g. The payload is tied to a weather balloon with a smart autonomous release mechanism and two-way RF telemetry channel (LoRa or Iridium communication). The payload can be released from the balloon at any given time or position, allowing it to fall at a predicted area. In case the payload is attached to a micro UAV, it can return autonomously by multioptional smart decline to a predefined location using a built-in autopilot. The suggested new strategy is presented using several case studies and field experiments.

Keywords: atmospheric and climate research, testing space components, near-space experiments, autonomous near-space flights, high altitude weather balloons, long-range RF communication

1. Introduction

1.1. Motivation

Traditionally, the space industry was mainly founded by governmental or military organizations. Yet, in recent years, the "new-space" environment attracts several major private companies such as Google, Facebook and OneWeb, each having a large-scale communication project involving global coverage using low earth orbit (LEO) nano-satellite swarm.

The vision of having a reliable and affordable global network, which can be accessed from any location on Earth at any given time, is a challenging scientific task, which attracts both industrial and academic efforts during the last few decades. Currently, the majority of all proposed solutions are based on a network of numerous LEO nano-satellites, which will establish a global network using radio frequency (RF) communication data received on Earth. Major companies such as Google, Qualcomm, Facebook and SpaceX have each invested in similar projects, commonly referred as new-space and near-space projects. OneWeb is one example for such initiative project involving a large constellation of LEO satellites. Other projects such as Google's Loon or Facebook's Aquila Drone are not directly focused on satellite constellations but on near-space massive constellation of drones or balloons. The new-space industry includes various small to medium size companies, which are currently developing products for the near-space environment, e.g., Planet Labs and Spire companies are two examples for such effort, which is focused on global imaging and IoT.

Constructing a cost-effective global network requires the use of low-cost electronics, unlike the traditional space industry, which uses dedicated expensive hardware. In order to perform a "space-qualified" testing platform on such components, a flexible modulated testing platform is needed. In this work, we present a new generic methodology for performing near-space experiments based on advanced low-cost payload, which is tied to a weather balloon. The suggested strategy is based on more than dozen balloon-launch experiments encompassing a large number of components (electronics and mechanics), which were tested at 10–30 km heights.

1.2. Related scientific work

High-altitude balloon (HAB) platforms have been used for direct atmospheric measurements for more than a century [1]. Measuring devices, which send data from HAB to a base-station located on the ground, using pocket-sized radio frequency (RF) transmitters and are widely known as radiosondes, were first invented by the French scientist Robert Bureau in 1929 [2]. Recently, HAB platforms have started to gain the ability of measuring, recording and transmitting other sources of data from a vast variety of instruments, substantially increasing HAB payload capabilities [3]. Furthermore, the increasing supporting evidence for climate change along with the understanding of real-time atmospheric composition measurements, both in the upper troposphere and lower stratosphere, is a key feature for studying radiative effects in our plant's climate system [2, 4], emphasizing the need for developing upper-air climate observation platforms [5, 6].

Although the main objective for HAB measurements is to monitor changes in temperature and water vapor vertical profiles in the troposphere and stratosphere, several new upper-air radiation profile measurements indicate supplemental valuable information regarding atmospheric absorption and emission of radiation with respect to the recently revealed interesting insight regarding the radiation obsorbtion/emission dependency with altitude [2]. In addition, due to the fact that our plant's atmosphere is continuously bombarded by energetic particles, mainly galactic cosmic rays (GCR), along with sporadic space weather events, additional particles are introduced into the stratosphere and troposphere [7]. Regardless of this recurring impact, the effects of energetic particles in the troposphere and lower stratosphere are still inadequately understood. There are numerous mechanisms for explaining how weather and climate could potentially be modulated [8], but the majority of energetic particle effects in the lower atmosphere are linked to their potential for ionizing the surrounding air. The formed ions can accumulate on cloud tops, contributing to the microphysics [9], may play a key role in the formation of aerosol (e.g., [10]). In addition, atmospheric ions can absorb directly infrared radiation (IR) [11, 12], and high-energy particles are also presumed to impact lightning rates [13].

Above land and within the boundary layer (~few hundred meters), the atmosphere is mainly ionized by the radiation emitted from radioactive isotopes decay in the Earth's crust [3]. Hess [14] postulated that the ionization profile in the atmosphere should decrease with altitude due to the fact that the radioactive element source is located near the surface. However, after conducting balloon measurements, Hess discovered that the ionization increased at altitudes above 10 km and claimed that it is caused by GCR source. He also determined that penetration depth of these particles depends on the energy spectrum of the incoming radiation [14]. Two decades later, Regener extended Hess' experiments using HAB, measuring ionization rates up to altitudes of 20 km [15]. They discovered that cosmic ray ionization reaches its maximum value between altitudes of 17–24 km and is known as the Regener-Pfotzer maximum (RP max). The Pfotzer Maximum, which is also geomagnetic-latitude dependent [16], formed within the tropopause layer below the stratosphere where primary particles (pions and hadrons) decrease and secondary particles (muons) increase [17]. This is a major source of ionization in the Earth's atmosphere. The establishment of an electromagnetic-muon stream results in ambient air ionization during the release of primary energies by the excitation of air molecules deeper in the atmosphere [17]. During this stream, a portion of the primary particles reach the ground as high-energy secondary particles [18]. The electromagnetic field also interacts with incoming particles, as the sun's solar radiation penetrates the atmosphere. This mixing is directly associated with the pressure decrease as the differential absorption rate within tropopause heights varies [17].

The necessity for developing new techniques and platforms for measuring and identifying energetic ionizing radiation in the atmosphere becomes vital. However, despite numerical model simulations for estimating flight trajectories, high-precision global positioning system (GPS) technology and the relatively slow balloon descent, recovering high-cost payload yet remains challenging, difficult and time-consuming, specifically around mountains or coastal areas [2].

Retrieving the payload enables us to acquire all the recorded data during the flight and that we were not been able to send using wireless communications. This is easier said than done and in practice, HAB payloads are not expected to be retrieved. For retrieving the payload, one should know the exact landing location of the payload, and more important, one must have access to that location. Thus, knowing the payload's exact landing location is not enough as it can "land" in the middle of the ocean or in a high peak of a mountain.

The rest of the paper is structured as follows: in Section 2, we survey the basic principle of flying a high-altitude balloon (HAB). In Section 3, we present design for a disposable cost-effective payload for low-bandwidth applications, which provide the base platform for our experiments. In Section 4, we cover the HAB payload components, power supply behavior, thermal design and pre-flight tests. In Section 5, we present our investigations of long-range communications for low- and high-bandwidth applications. In Section 5, we present our own setup of a near-space return-to-home (RTH) micro-UAV for retrieving the payload with its recorded data. Finally, we discuss our efforts and future work.

2. Preliminaries: basic principle of high-altitude balloon

In general, high-altitude balloon (HAB) is composed of the following components:

- A latex balloon—comes in a wide range of weights, which basically reflects its ability to be inflated with helium, common use for HAB may weight 100–1200 g.
- A payload—which includes all the necessary components for conducting the experiment and retrieving the data. In **Figure 1**, two payloads are connected (black and white ice-cream boxes).
- A ground station (GS)—commonly includes an RF receiver. In **Figure 1**, the GS also includes a robotic telescope and transmitter to control the payload detaching process.

Consider a balloon with a self-weight of 1000 g, about 1000 l of helium is needed in order to allow the balloon to start floating (for each m^3 of helium—one can expect a lift of 1000 g—1 kg).

Figure 1. Launching a HAB—yet another day at the office.

Assuming 2000 l of helium were used, the 1000-g balloon should gain about 1000 g lift at 1 atm. Assuming a 500 g payload is attached to the balloon, one can expect an overall lift of 500 g. In **Figure 2**, a basic calculation of the expected balloon parameters is presented.

Given the desired requirements for the experiment, e.g., max altitude, payload weight and required floating duration, one can adjust the amount of helium in the balloon accordingly. **Table 1** presents few examples for such adjustment.

As the balloon inclines, its surrounding air pressure decreases. **Table 2** presents the expected air pressure with respect to the balloon height.

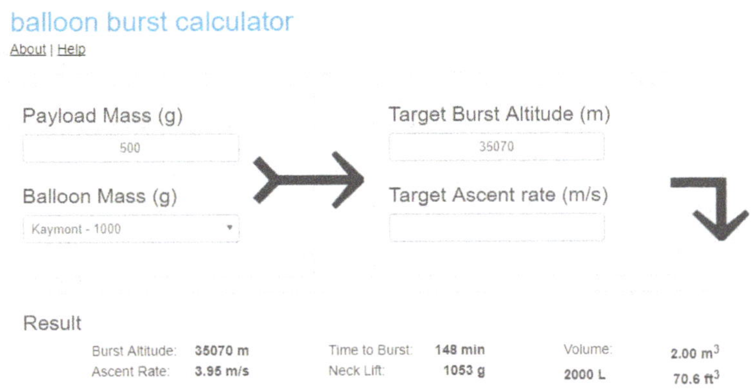

Figure 2. An HAB's lift and burst calculator, from: http://habhub.org/calc/.

Balloon type	Volume (L)	Payload (g)	Neck (g)	Burst altitude (m)	Ascent rate (m/s)	Duration (m)
300 Kaymont	600	200	316	27,890	2.96	157
600 Kaymont	1500	200	1453	29,280	5.95	82
1000 Kaymont	2000	500	1053	35,070	3.95	148
1000 Kaymont	3000	1000	2080	32,135	4.84	111
1000 Kaymont	3000	1500	2080	32,135	3.54	151
1000 Kaymont	4000	2000	3106	30,053	4.44	113

Table 1. HAB parameters: Few examples of the lift, duration and burst altitude with respect to the balloon type, payload mass and amount of helium.

Altitude (m, 15 cel)	Air pressure (atm)	Balloon volume (L)	Balloon diameter (m)
0	1.0	1000	1.24
2361	0.75	1333	1.36
5477	0.5	2000	1.56
10,278	0.25	4000	1.97
16,096	0.1	10,000	2.67
32,230	0.01	100,000	5.76
48,330	0.001	1,000,000	12.41

Table 2. Expected air pressure at a given altitude.

3. Long-lasting "floating" balloon

In a typical HAB configuration using latex balloon, the balloon will ascend and expand as the air pressure decreases with height due to the thin atmosphere. At a certain point, it will inflate up to its elastic point, explode and fall. This means that if we can make the balloon float in a relatively constant altitude, we can extend its lifespan and endurance. Moreover, fixing the balloon at high altitudes could also enable to test any desired hardware under near-space conditions.

Google's "loon project" is a good example for an HAB setting that is capable of floating up in the atmosphere for a long duration. However, such settings are expensive and complicated, thus they are not a practical solution for scientific researchers.

Our approach for "fixing" the balloon's altitude was directed toward a simple constriction, i.e., a main latex balloon and a cluster of foil balloons. Foil balloons are not elastic and cannot expand, thus their volume can be approximated as constant. This means that as the outer pressure drops due to the thin atmosphere, its upthrust force will weaken and might even change its direction as dictated by the buoyancy force equation:

$$F_B = \left(\rho_{air} - \rho_{gas}\right) gV \tag{1}$$

where ρ_{air} is the surrounding air density, ρ_{gas} is the helium density, V is the balloon volume and g is the gravitational force. We present a basic example of our current test setting design (**Figure 3**):

- A single 1000 g main latex balloon with a capability of 1300 g neck-lift.
- Five non-lasting foil balloons with a fixed volume of about 110 l each, with a self-weight of approximately 90 g. Combining these balloons implies a weight variance of about 500 g.
- A two-parted payload:

Figure 3. A long-lasting HAB experiment. This setting retained its floating state for about 2.5 h before the main balloon exploded.

Figure 4. The retrieved payloads from the above experiment were found on a distant field, about 100 km from the point of launch. The crashing location was transmitted by the iridium communication system after the crash. In the picture, the upper box is the secondary payload and the main is the lower box.

- Main payload (above) 500-g: Iridium transceiver, GPS, Solar panel, battery, Geiger counter and an autonomous release mechanism for a secondary payload.
- Secondary payload (lower) 400-g: A long-range HD video streaming system based on Wi-Fi and a directional antenna pointing down (14 dBi flat panel antenna).

In this test setting, each foil balloon has a net weight of approximately 0 g on ground level, while the expected weight at an altitude of 10 km is about 70 g. The overall setting provides a lift force of about 400 g on ground level. When reaching to 9–10 km height the system's net lift force should be about 0, making the system relatively altitude-stationary (**Figure 4**).

4. Sensors, energy and thermal design

In this subsection, we cover the HAB payload components. First, we present the common needed and used sensors in "near-space" experiments. Then, a brief discussion on energy and thermal design is presented—followed by a discussion of how to test a potential payload (on the ground) for its ability to operate under near-space conditions.

4.1. Sensors

- GNSS (e.g., GPS): Global Navigation Satellite Systems refer to a positioning sensor commonly used for computing the 3D position in a typical horizontal accuracy of 2–3 m in the open sky (the vertical accuracy is often not as accurate as the horizontal – errors of 10–20 m are common even in the open sky). We have mostly used U-blox GNSS relievers which are becoming the industry standard for most COTS (Commercial Off-The-Shelf) drones. Remark: one should configure the GNSS receiver to a "balloon-mode" (Airborne) else the positioning might be limited to a low altitude of 12 km or less. Due to the nature of the balloon "Airborne < 1 g" is the preferred model.

Modern GPS can support 10 Hz position sampling rate—yet for most coming measurements, such sampling rate is not needed—as the dynamics of the balloon is very low. Lowering the positioning rate may also help reducing the energy consumption of the GNSS receiver.

- **9DoF**: is basically a set of MEMS sensors: three axis magnetic field, three axis accelerometers, three axis gyroscopes. Combined they can be used to compute orientation. We have used Bosch BNO055 sensor which also has a true orientation filter—and found it both affordable and robust.
- **Barometer**: this sensor measures the atmospheric pressure and temperature. This combination enables us to compute a naïve estimation of elevation in submeter accuracy. However, in our experiments, we noticed that in altitudes higher than 10 km, the barometer's altitude estimation started to slow its elevation in a certain pattern, which repeated itself. Using the GPS sensor measurements, we were able to estimate its true altitude, which consists with the expected altitudes. It should be denoted that although barometers mostly have an elevation accuracy of submeter (in some models subfeet), in high elevation the accuracy gets worse, below is an example of real data of "faulty" barometer (**Figure 5**).
- **Temperature:** thermocouple sensors are simple and robust sensors and being used to measure the inner and outer temperature of the payload. These values are significant for the proper operation of the electronic components and the batteries.

4.2. Energy and thermal design

It should be noted that performance of all batteries drops drastically at low temperatures starting −10°C. At high elevation such as 10–30 km the outer temperature is expected to be

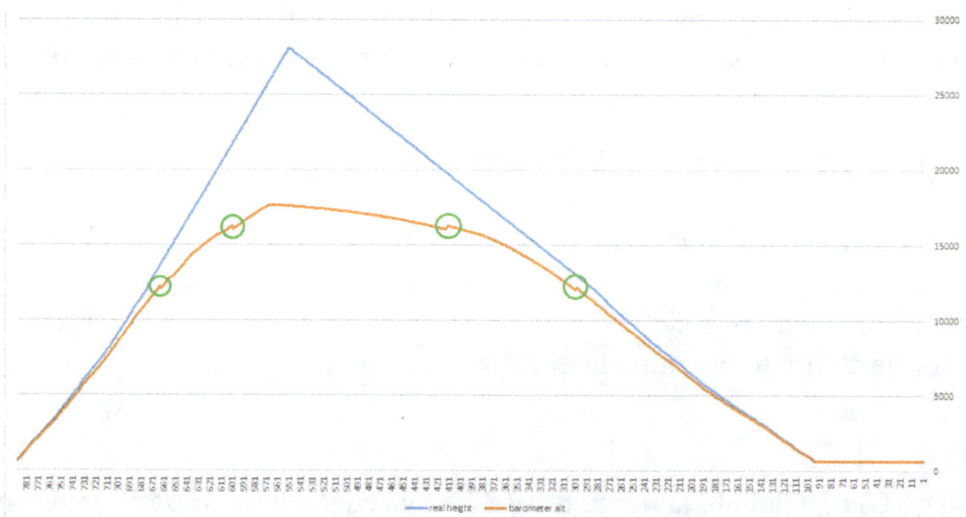

Figure 5. The balloon altitude over time, as recorded by the barometer sensor and our true altitude estimation.

[−60–45]°C, respectively. This makes the task of keeping the payload at "room level" temperature (i.e., [0, 45]°C) vital. Packaging the payload with COTS boxes made from materials that provide proper thermal insulation such as expanded polystyrene (EPS) is sufficient for such need. Recall that in height of 16 km, the expected air pressure is 0.1 atm, while at 31 km, it is about 0.01 atm, so air-based passive cooling is significantly less efficient than on the ground.

In practice: taking into an account the above considerations and the fact that most IoT components are suited for operating in near-space conditions it is easy to construct a thermal-balanced payload. Most of the required tests for the payload performances under near-space conditions can be performed with a simple setup, which consists of a vacuum chamber, a home freezer and a simple thermal camera (**Figure 6**). **Table 3** depicts a thermal analysis of a Samsung Galaxy S6 mainboard under low ventilation conditions.

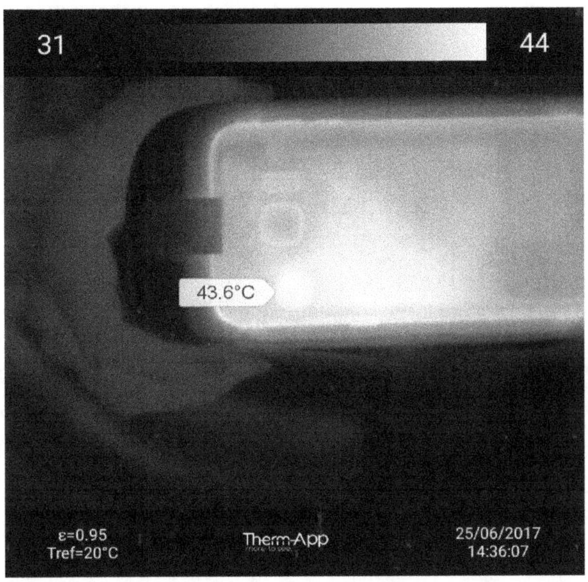

Figure 6. Thermal analysis of an android phone using a thermal camera (Op-gal's Therm-app).

Platform	Max altitude [m]	Max horizontal velocity [m/s]	Max vertical velocity [m/s]	Sanity check type	Max position deviation
Portable	12,000	310	50	Altitude and velocity	Medium
Stationary	9000	10	6	Altitude and velocity	Small
Pedestrian	9000	30	20	Altitude and velocity	Small
Automotive	6000	100	15	Altitude and velocity	Medium
At sea	500	25	5	Altitude and velocity	Medium
Airborne <1 g	50,000	100	100	Altitude	Large
Airborne <2 g	50,000	250	100	Altitude	Large
Airborne <4 g	50,000	500	100	Altitude	Large
Wrist	9000	30	20	Altitude and velocity	Medium

Table 3. Form: U-blox M8 N manual—make sure you use airborne mod (the default is portable—so the GPS will not work above 12 km).

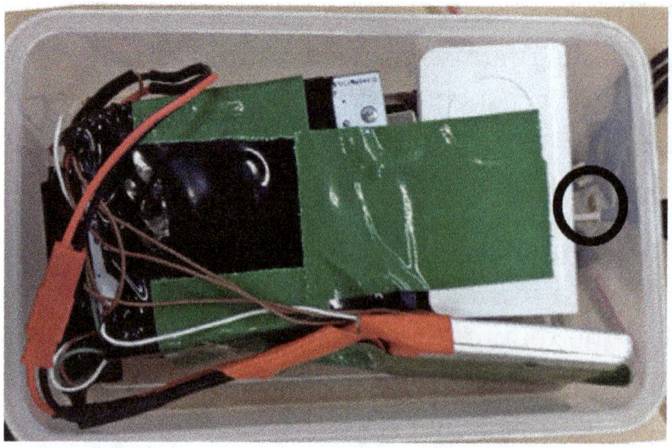

Figure 7. A typical thermal-balanced-payload, notice the ventilation hole marked with a circle.

A typical payload will include a GPS, microcontroller, LoRa modem, Geiger counter, barometer and humidity sensor (**Figure 5**). The total energy consumption is about 250 mW (**Figure 7**).

5. Disposable cost-effective payload for low-bandwidth sensor data applications

In most cases, we usually direct our efforts toward recording and transmitting low-bandwidth sensor data or testing electronics at near-space conditions. As retrieving the payload with its data is not always certain, we designed the payload to be cost-effective and disposable and yet capable of long-range low-bandwidth communications.

Our basic HAB payload setup typically includes the following components:

- Arduino MCU.
- 433 MHz LoRa radio transceiver.
- Versatile GNSS module capable of GPS, GLONASS.
- Environment conditions sensors (barometric pressure/altitude/temperature/humidity/Dewpoint).
- An actuator for releasing the payload on command.
- Geiger counter-based on the new solid-state technology (which reduces the weight and price of Geiger counter).

Such payload's BOM (Bill of Material) will cost about 100–120$. The weight of the payload can be reduced to a sum of 150 g, making it suitable for 300 g HAB. The total cost including the cost of the launch will cost less than 200$. In case there is no need for a Geiger counter, the overall BOM of the payload and balloon can be below 100$.

Using this affordable payload design, we were able to perform several experiments in which we measured Gamma counts with respect to altitude and location in relative high accuracy. **Figure 8** shows the real-time Geiger count as received at the GS from the payload (over 120 km range).

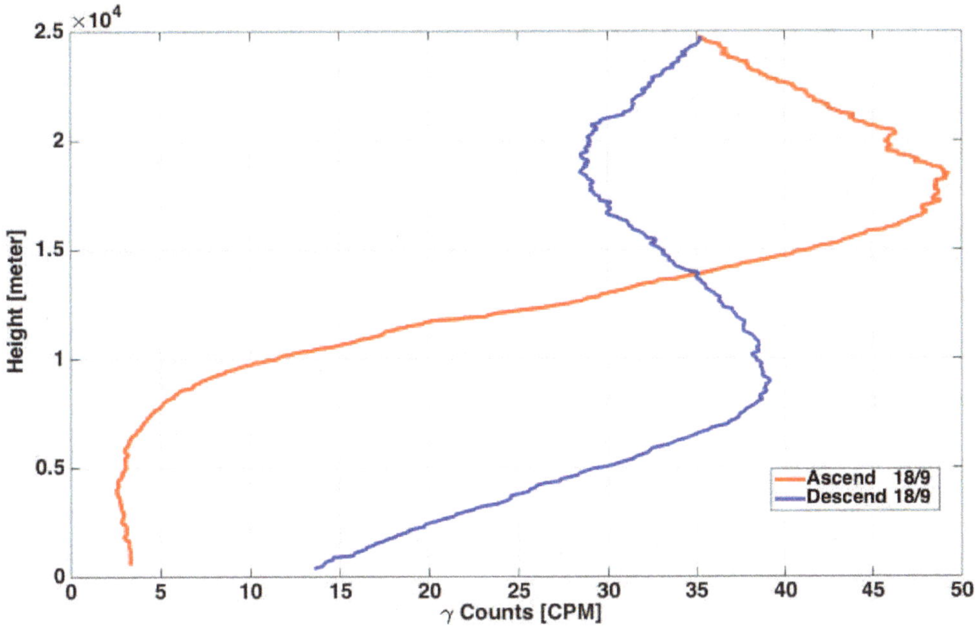

Figure 8. Gamma count vs. altitude.

6. Long-range communications

6.1. Current state

Radiosonde is the most common type of payload which is capable of long-range communications suitable for HAB. A radiosonde can be regarded as a black-box which includes a variety of sensors and a radio transmitter. Radiosondes may come in various shapes and technologies, but in general, they measure: position (GPS), barometric pressure, humidity and temperature. They also may include some other related sensors such as Ozone meter. These data are transmitted to the GS using RF communications, commonly—UHF 400–406 MHz, and 1675–1700 MHz. This solution's range is typically between 50 and 200 km that depends on environmental conditions.

As mentioned above, radiosonde payloads are closed systems that limit the user's ability to customize them. This means that in order to transmit additional sensors data, an additional communication device is required as well. Moreover, they provide low-bandwidth and half-duplex (download only) communications. It should be noted that the RF, which is used by a radiosonde, is not an ISM band, and therefore, it might require RF approval by local authorities. Other concerns about using radiosonde communication abilities include the lack of frequency reusability and security in most Radiosonde payloads.

Providing high-bandwidth communications enable us to obtain real-time measurements such as multi-spectral images and conducting high-resolution gamma-ray spectrometer measurements. Full-duplex communications enable us to interact with the payload, so we can remotely control the payload or the balloon motion. The ability to adapt the modem communications setting, i.e., reprogramming it in real-time makes it a more flexible solution that provides bandwidth and range according to the user's needs or environmental conditions. In our experiments, we consider the minimally required coverage range to be about 40–50 km that is required for conducting HAB missions.

Cellular 3G/LTE communications are intuitively a natural solution for full-duplex and high-bandwidth communications that is communally used by "makers". However, a cellular device that has been used at high altitudes can be easily detected by multiple base-stations simultaneously. Generally, such device will be blocked by the cellular providers thus making its 3G/LTE communications inoperable till the device returns to ground level.

This makes smartphones not suitable as a real-time communications solution for high altitudes. On the other hand, for low altitude applications or when it is known that the payload will fall in a cellular covered area, smartphones might be considered as a suitable communication solution. Denote that in many countries (e.g., USA) mobile phones are required to operate in "flight mode" while "in-air".

UHF RF communication such as 433, 866, and 915 MHz which are ISM RF bands can provide low-bandwidth and long-range communications solution. We have investigated many drones remote control (RC) two-way communication solutions, and we found that while they can provide long-range communications their high-energy consumption and cost make them less appealing for day-to-day HAB missions. As an example, the DragonLink RC technology, which is the gold standard for flying long-range drone's communications, required in our experiments 1.5 W transmitter for achieving the range of 40 km and a data rate of 19.2 kbps.

In our experiments, we found that LoRa technology-based devices are the most suitable and preferable solution for HAB missions' requirements. Meaning, they are robust, programmable, with a very low-energy consumption and affordable. With the right setting, we were able to achieve full-duplex communications with a 25 mW transmitter more than 120 km range and a data rate of 0.4 kbps.

Wi-Fi technology can provide high-bandwidth communications, however it was designed for as Wireless Local Area Network (WLAN). This means that with COTS hardware in a direct line of sight communications, its expected range is limited to hundreds of meters. We have designed a long-range Wi-Fi setting based on EZ-WiFiBroadcast settings. EZ-WiFiBroadcast is a special DIY design of Wi-Fi communications which is commonly used as a poor man's long-range HD FPV solution. With our current long-range Wi-Fi setting, we have been able to capture 720P video from a HAB at 9.8 km height and located about 15 km from the GS.

Free-Space Optical (FSO) also known as laser communications are a less common high-bandwidth communication solution which can be achieved by the use of a robotic telescope which tracks in real-time the HAB. In a clear day, such device can track an HAB for over 50 km. In our experiments, we successfully tracked HABs for more than 70 km using low-cost Celestron StarBright XLT telescope with 127 mm aperture Schmidt-Cassegrain lens. As shown recently by Google in their Loon project "Demonstration of free-space optical communication for long-range data links between balloons on Project Loon". This kind of high-bandwidth communications is still extremely complicated and requires technical skills and efforts which are not common in most research groups.

Another commercial solution is Global satellite communications (we have used Iridium's two-way Short Burst Data—SBD), this kind of solution requires a "pay per message" data plan

(~10 cents per 50 bytes)—so it is applicable for low-bandwidth missions. Yet it allows full control, two-way communication. Another satellite-related solution named "SPOT" is commonly used to track HABs. This one-way (transmission only) solution uses the "Global-Star" satellite network for near global coverage. Interestingly, we have found that the use of short message service (SMS) in cellular communications was relatively efficient and we were able to send and receive text messages from about 5000 m height when the expected height is about 2500 m.

6.2. Cellular 3G/LTE communications

In this section, we present methods for constructing simple (DIY) payloads based on COTS devices. We start by presenting a naive attempt to shoot high-resolution images from high altitude—as part of a class challenge in the undergraduate "Autonomous Robotics" course during the year 2017 (given in the Computer Science Department at Ariel University). All suggested solutions included an Android phone with an international sim card and an app which captures time-laps photos with position while attempting to upload them using existing cloud uploader tools (**Figure 9**). The balloon launches included the following setting:

- A regular latex 600, 1000 g balloon.
- Smartphone-based payload—100–200 g. Android phones with the needed apps for time-lapse camera (such as OpenCamera) and a cloud-based uploader app (such as Dropbox or Google drive). The phone was equipped with a sim card which can be used for uploading the data—using a prepaid data plan.
- Thermal Box: the most common is polystyrene (ice-cream box)—which is needed to maintain a controlled temperature for the phone electronics and batteries.

Five different solutions were implemented (see **Figure 10**) mainly using the OpenCamera android open source. None of the payloads could capture reasonable images from high altitude

Figure 9. Three payloads ready to be launched—each with a smartphone and software for uploading the gathered data. As part of the navigation graduated course in Ariel University (Israel). None of the payloads could actually transmit good and clear images from high altitude. All payloads eventually fall in Suraya. Over 200 km from launch.

—although at least three (out of five) phones made it back safely to the ground and two of them even sent few images—until it was discovered by a "lucky founder" or simply run out of power.

Although the suggested concept failed the overall solution of using a software-only solution based on affordable smartphones seems to be a feasible cost-effective solution to many near-space applications.

6.3. Long-range Wi-Fi communications

Long-range and high-bandwidth communication solutions suitable for HAB missions are not common, especially not as COTS hardware. High-bandwidth data applications such as multi-spectral imagery or high-resolution measurements have a great value for exploring various electrical phenomena such as lightening discharges, sprites or blue-jets in the atmosphere and other aspects of this environment.

For providing high-bandwidth communication capabilities, we are directing our efforts on utilizing COTS communications hardware based on IEEE 802.11 standard WLAN which is also known as Wi-Fi. Wi-Fi networks can easily provide high-bandwidth communications but with COTS hardware they have a very limited range. Using a much more sophisticated hardware can extend its range dramatically to a few kilometers, but such systems are costly, with high-power demands and with a form factor that is not suited for a typical HAB's payload (**Figure 11**).

In theory, the use of a high gain directional antenna about 18–24 dBi for the receiver at the ground station and a directional antenna with a gain of about 10–14 dBi should provide us a link budget greater than 150 dB. Such link budget should enable communications for long ranges estimated at 10–30 km on regular conditions. In optimal conditions and a Forward Error Correction (FEC) mechanism, the range can be extended to about 50 km (**Figure 12**).

This lead us to investigate a different approach that uses COTS Wi-Fi hardware but in a non-traditional way. Some of the IEEE 802.11 network interfaces can operate in a special debug

Figure 10. An image of the sky that was made by an android smartphone (Xiaomi Redmi 4A). In this experiment, the phone's camera was out of focus. It might be due to ice on the camera lens. The images were successfully uploaded after the payload has made it to the ground.

Figure 11. HAB ground station (GS): Left: The GS in general: two robotic telescopes with (auto-track) and a high gain 24dBi Wi-Fi antenna. Right: the robotic telescope: (a) A view-finder webcam. (b) A Wi-Fi + 3G router. So the telescope can be controlled globally. (c) A Pi-camera mounted to the telescope eye-view. (d) A Raspberry Pi which controls the telescope using either visual tracking and GPS coordinates.

Figure 12. Keep it simple: launching two simple payloads: Raspberry-Pi (upper) and an android smartphone (taking this image).

mode that allows them to transmit and receive Wi-Fi communications with no regards to the IEEE 802.11 standard itself. As such, "makers" have used this feature for creating a "poor man's" long-range HD FPV solution. We based our system on "bortek"'s version of EZ-WiFiBroadcast (**Figure 13**).

Typically, the Wi-Fi long-range system includes:

- Raspberry-Pi (RPi) device usually Raspberry Pi 0, with RPi-Cam camera and a Wi-Fi Network Interface Card (NIC) with an external directional 14 dBi flat panel antenna with its face directed down.
- Ground station based on another RPi device with an external 20 dBi directional antenna.

Even though that our research on this approach is at early stages we have been able already to capture 720P video from a HAB at 9.8 km height and in an estimated distance of about 15 km with our long-range Wi-Fi communication system. **Figure 13** is an image captured in this particular HAB mission. We found that flat directional antennas perform quite well as long as the angle between the balloon and the GS was not too wide.

As the GS design is compact it can be used as a mobile ground station located on top of a car which "chases" the balloon.

6.4. Long-range communication LoRa vs. iridium

In Europe, there is a well-established RF solution for tracking on HABs led by the UK High Altitude Society (https://ukhas.org.uk). This cooperative solution allows an online tracking mechanism based on COTS Software Defined Radio (SDR). This system is based on a fixed low-bandwidth protocol—mostly at the 434.075 MHz frequency and has been successfully in use for hundreds of launches annually. Yet, in many cases the UKHAS system is not suitable due to geo-location, bandwidth or even security reasons. In this work, we mainly focus on such cases in which a "real" Ground Station is needed. The iridium modem allows true global coverage and two-way communication, yet it is relatively expensive (300$) and requires a data plan which cost about 2$ for kB (a compressed single JPG image of 100 kB—will cost about 200$). This kind of pricing makes it applicable mainly for strictly low bit rate application. The LoRa modem is an affordable (10–20$) system with adaptive bit rate and works in unlessens band. The expected range for LoRa communications is over 120 km, while in a few places around the world, LoRa gateways are started to be deployed so that the expected route can be covered. But in general even with a single LoRa gateway it is expected to cover the balloon route (50–200 km)—using a standard UHF Yagi antenna in the expected range. We conclude that the LoRa solution can be an affordable complementary communication solution. It can be connected to a smartphone allowing long-range communications coverage and with actively connected to the Iridium system it can benefit the most to the satellite communication (**Figure 14**).

Figure 13. The smartphone payload shot from an upper payload based on a Raspberry-Pi camera equipped with a long-range Wi-Fi transmitter. The picture was taken at about 9.7 km above ground.

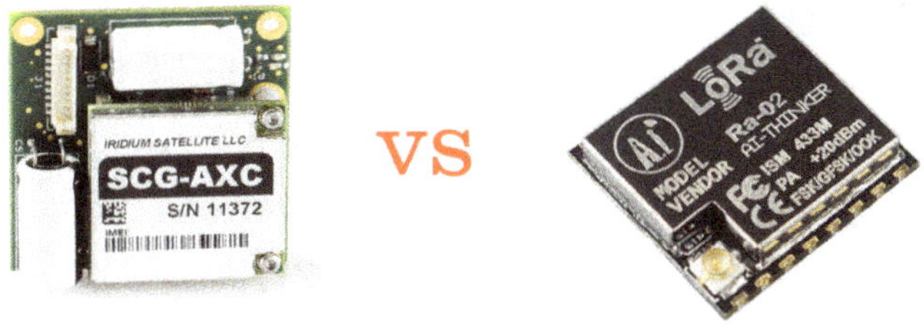

Figure 14. LoRa module vs. iridium module.

7. Near-space return to home micro drone

7.1. Drone structure

We present a near-space drone, which is affordable, robust and may weight below the FAA regulations (300 g). The micro-UAV has a unique RTH control algorithm adjust to near-space conditions and on board black box for storing a wide range of sensor measurements (**Figure 15**). The proposed platform has the following properties:

1. Low-cost, lightweight electric UAV which was equipped with: multiple real-time sensors, HD cameras, a Pixhawk flight controller, GNSS receiver and long-range RF communication system for RC & telemetry data.
2. Smart release mechanism with several parameters for autonomous operation.
3. Near-space flight mode for smart decline.

Figure 15. Four different models of RTH micro drones. Each of them was tested for autonomous flight launched from a balloon.

The basic requirement of the UAV is the ability of autonomous RTH or any other Geo location. The UAV needs to be lightweight, aerodynamic wing structure for fast and smooth flight and at list extended range of 50 km for RTH. Denote that in most cases flying back home will require flying against the wind (**Figure 16**).

7.2. Smart release mechanism

The smart release mechanism is established from two main elements: mechanical mechanism and autonomous smart release software. The mechanical mechanism has two construction sets: Servo or Fuse wire. The servo is operated with PWM signal, and the fuse wire burns from relay. One of the most important things is the way the balloon attached the release mechanism to the UAV without affecting the UAV fly ability and minimal change of the aerodynamic, because of that the release mechanism mounted on the balloon payload. The autonomous smart release software is an algorithm that gets a several sensor parameters and decides if to release the UAV. The algorithm has the next prioritization: balloon burst, RC signal, altitude, battery, and geo fence. The RC signal is the only parameter that comes from the ground, the rest calculated on the MCU (**Figure 17**).

7.3. Near-space flight mode

This mode has few parameters for controlling on smart decline. After the UAV release from the balloon, it will open parachute to altitude that set on the algorithm, the next step is to release

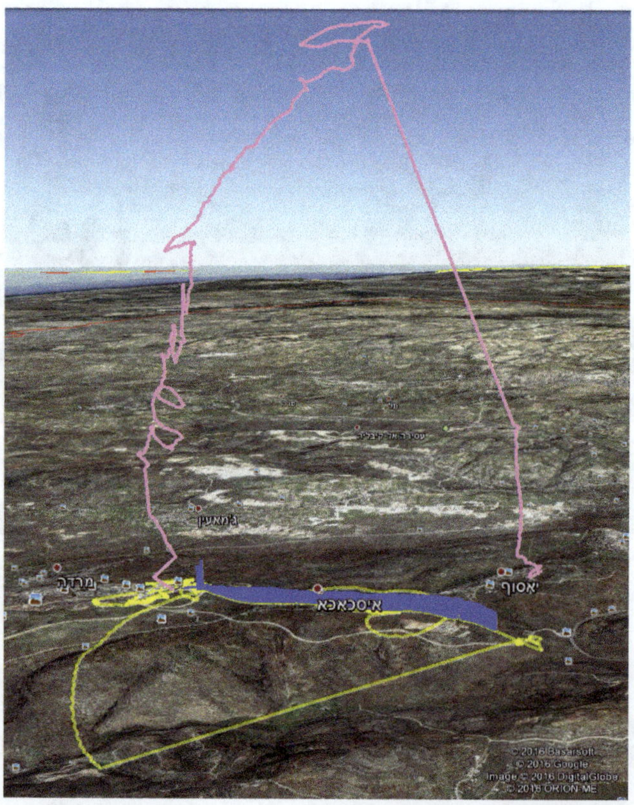

Figure 16. Full flight path of an HAB and RTH payload by a micro drone.

Figure 17. The RTH payload is going up.

Figure 18. Getting back home: A massive UHF transition caused the drone to get into the "fail-safe" state, releasing the drone which in turn flyied back to "home" autonomously.

the parachute and glide with a constant decline rate to altitude that set on the algorithm and then open the motor and fly back home (**Figure 18**).

Currently, we are constructing a micro wing-shape UAV with solar panels for energy harvesting; this will allow us to perform a much longer time and range experiment using super-pressure balloons. Release the drone on a "sunny morning"—allowing it to fly for up to 6 h during day-time covering 100–200 km. Such distance should be sufficient for finding a proper landing region (**Figures 19** and **20**).

7.4. Regulation and safety

Launching a HAB requires authorization and following local regulations. We present here some of the US Federal Aviation Administration (FAA) regulations. Please note that even though many countries tend to adapt these regulations, local regulations might differ from the following (FAA Part 101 and 14 CFR Part 48):

Figure 19. 290 g RTH micro-UAV, with a release carbon strip on its backend.

Figure 20. RTH micro drone lunching.

1. Any cellular phones must be turned off (airplane mode enabled) for any aircraft and/or balloon as soon as it leaves the ground.
2. Any individual payload must weight less than 4 pounds and have a weight-to-size ratio of less than 3.0 ounces/square inch (total weight of the payload only divided by its smallest face).
3. Total payload of two or more packages carried by one balloon must be less than 12 pounds total.

4. The balloon cannot use a rope or other device for suspension of the payload that requires an impact force of more than 50 pounds to separate the suspended payload from the balloon.

5. No person may operate any balloon in a manner that creates a hazard to other persons, or their property.

6. No person operating any balloon may allow an object to be dropped therefrom, if such action creates a hazard to other persons or their property.

7. The owner must register their HAB as part of the FAA's new Unmanned Aircraft System (UAS) laws. The registration number must be marked on each HAB flight.

Here are the main rules of thumb we have used in our HAB launches (on top of the local aviation regulations):

1. It is highly recommended to update the related FAA authorities and get a permission in advance.

2. Validate in real-time the conformation for the launch, a few minutes prior to the lunch.

3. Make sure you are not launching the HAB nearby airports or other no-flight-zones.

4. The overall weight of all payloads should not be more than 1 kg, "Return to Launch" UAVs should weigh less than 500 g—preferable below 300 g (FAA regulations).

5. The maximal declining speed of the falling payload (below 5000 m) should not exceed some velocity (say e.g., 12 m/s).

6. The usage of a parachute cannot guarantee declining speed or velocity. As in this method the overall max weight per square cm should be below some value, we strongly recommend a weight-to-size ratio of no more than 2.5 g per cm square, e.g., a cube payload of 1 l should not weight more than 250 g.

7. Secure each payload's component to prevent its fall.

8. If there are still some safety issues with the HAB, make sure its planned route is not above populated areas—preferably above the sea. Aborting a HAB-UAV mission into the sea is a safe backup plan—and in HAB lots can go wrong.

9. Only launch at a safe zone—where there are no power-lines or buildings.

8. Discussion and conclusion

In the last decade, HAB experiments, which were considered esoteric and rare, have become more applicable for scientific researchers and near-space experiments. Today, the overall cost of an HAB experiment can reach up to $500. Radiosondes are commonly used for transmitting the sensory data in real-time. However, using this technology has a limited communication capability and is very hard to customize. New long-range wireless communication technologies such as LoRa allow us to transmit a wide range of sensory data with both substantial low-cost and light weight setup. The maximum data rate provided by LoRa technology is 37.5 kbps, which is sufficient for two-way telemetry along with a wide range of sensory data but is not suitable for high-data-rate applications such as real-time video data. For that we found long-range Wi-Fi techniques to be a prominent strategy: allowing us transmission of live video data up to ranges of about 15–30 km. For long duration application in which the balloon may circle the world, we also present a global two-way communication solution based on Iridium modem.

As the state-of-the-art of communications is still limited, we presented a whole different approach which focused on retrieving the payload in a safe and secure way. Such solution overcomes the need for transmitting the measured data wirelessly—as all the needed information are stored on board of the UAV.

Moreover, this approach highly reduces the risk of losing precious equipment and enables reusing the experiment platform over and over again. In the past, developing and operating an autonomous UAV system was a complicated and costly project. However, in recent years the successful efforts of the toy and hobbies industries to make UAVs accessible and simple to operate provided the opportunity for using UAVs as a common research tool. As such it can be used as a practical and cost-effective solution for returning the payload home with a relatively simple release mechanism and auto-pilot controller.

Based on six different experiments performed during 2016–2017, we conclude that the suggested strategy of using an autonomous UAV as a generic multi-parametric near-space platform is suitable for tropospheric remote sensing and for testing electronic components in near-space conditions.

Current research focuses on exceeding the operational capabilities of long-range Wi-Fi to a full-duplex communication channel and extending its range even further with the development of a high-gain antenna tracker. The deployment of LoRa WAN infrastructure can extend the HAB's communication service over huge areas.

Finally, the current range of RTF autonomous micro UAV is about 30 km. We expect that after optimizing the algorithm for the decline mode (from near space to ground), such range may be extended to 50–100 km with a relatively high probability of success.

Author details

Kobi Gozlan[1], Yuval Reuveni[1,2,3,4], Kfir Cohen[1], Boaz Ben-Moshe[1]* and Eyal Berliner[1,2,5]

*Address all correspondence to: benmo@g.ariel.ac.il

1 K&CG lab, Ariel University, Israel

2 Department of Management, Bar-Ilan University, Israel

3 Department of Physics, Ariel University, Israel

4 Eastern R&D Center, Ariel, Israel

5 School of Sustainability, Interdisciplinary Center (IDC) Herzliya, Israel

References

[1] Hoinka KP. The Tropopause: Discovery, definition and demarcation. Meteorologische Zeitschrift (Sonderheft). 1997;**6**:281-303

[2] Kräuchi A, Philipona R. Return glider radiosonde for in situ upper-air research measurements. Atmospheric Measurement Techniques. 2016;**9**(6):2535-2544

[3] Yaniv R, Yair Y, Price C, Nicoll K, Harrison G, Artamonov A, et al. Balloon measurements of the vertical ionization profile over southern Israel and comparison to mid-latitude observations. Journal of Atmospheric and Solar-Terrestrial Physics. 2016;**149**(Suppl. C):87-92

[4] Solomon S, Rosenlof KH, Portmann RW, Daniel JS, Davis SM, Sanford TJ, et al. Contributions of stratospheric water vapor to decadal changes in the rate of global warming. Science. 2010;**327**(5970):1219-1223

[5] Seidel DJ, Berger FH, Immler F, Sommer M, Vömel H, Diamond HJ, et al. Reference upper-air observations for climate: Rationale, progress, and plans. Bulletin of the American Meteorological Society. 2009;**90**(3):361-369

[6] Bodeker GE, Bojinski S, Cimini D, Dirksen RJ, Haeffelin M, Hannigan JW, et al. Reference upper-air observations for climate: From concept to reality. Bulletin of the American Meteorological Society. 2016;**97**(1):123-135

[7] Bazilevskaya GA. Solar cosmic rays in the near earth space and the atmosphere. Advances in Space Research. 2005;**35**(3):458-464

[8] Mironova IA, Aplin KL, Arnold F, Bazilevskaya GA, Harrison RG, Krivolutsky AA, et al. Energetic particle influence on the Earth's atmosphere. Space Science Reviews. 2015;**194**(1): 1-96

[9] Harrison RG, Nicoll KA, Ambaum MHP. On the microphysical effects of observed cloud edge charging. Quarterly Journal of the Royal Meteorological Society. 2015;**141**(692):2690-2699

[10] Duplissy J, Enghoff MB, Aplin KL, Arnold F, Aufmhoff H, Avngaard M, et al. Results from the CERN pilot CLOUD experiment. Atmospheric Chemistry and Physics. 2010;**10**(4): 1635-1647

[11] Aplin KL, Lockwood M. Cosmic ray modulation of infra-red radiation in the atmosphere. Environmental Research Letters. 2013;**8**(1):015026

[12] Aplin KL, McPheat RA. Absorption of infra-red radiation by atmospheric molecular cluster-ions. Journal of Atmospheric and Solar-Terrestrial Physics. 2005;**67**(8):775-783

[13] Scott CJ, Harrison RG, Owens MJ, Lockwood M, Barnard L. Evidence for solar wind modulation of lightning. Environmental Research Letters. 2014;**9**(5):055004

[14] Hess VF. Über Beobachtungen der durchdringenden Strahlung bei sieben Freiballonfahrten. Physikalische Zeitschrift. 1912;**13**:1084-1091

[15] Regener E. New results in cosmic ray measurements. Nature. 1933;**132**:696-698

[16] Carlson P, Watson AA. Erich Regener and the ionisation maximum of the atmosphere. History of geo - and space. Sciences. 2014;**5**(2):175

[17] Carmichael-Coker MK. Increase of ionizing radiation at the Pfotzer maximum over the southern Appalachians. 2014 NCUR; 2015

[18] Mishev A. Short- and medium-term induced ionization in the earth atmosphere by galactic and solar cosmic rays. International Journal of Atmospheric Sciences. 2013;**2013**:9

8

Space Flight

George Dekoulis

"Πώς Γαία και Ήλιος ή δε Σελήνη αιθήρ τε ξυνός γάλα τ'ουράνιον και Όλυμπος ἐσχατος ἠ δ' ἀστρων θερμὸν μὲνος ωρμὴθησαν γὶγνεσθαι".

"How did heaven's Earth and the Sun, or the Moon, the Solar Wind, and, the Milky Way Galaxy and ultimate Olympus (Dias/Jupiter), or the astral thermo-stability, were generated?" (Parmenides, on Nature, 500 BC).

1. Introduction

Would you answer to this 2500 year old question? Based on the archaeological findings, prehistoric human societies had similar cosmogonical and cosmological wonderings, verified to a minimum of 12,000 years ago. Cosmogony refers to the creation of cosmos, while cosmology to its structure. Cosmos symbolises the actual decoration of the universe with its various structures, including humans on Earth, via the eternal flow of the vital divine energies, as demonstrated by this 7500 year old randomly picked artefact of **Figure 1**.

An interesting highlight about our prehistory is that all the documented major civilisations around the globe shared similar memories, moral, and mental values, no matter the physical distances amongst them. These fundamental philosophical concepts are still in use, some with the same names and some with adjusted ones. However, all these core ideologies, that include a lot of superstitions too, lack the support of scientific data, especially when it comes to beliefs regarding space.

As a giant step ahead for the human civilisation and space science itself, major space centres have been established globally over the last century. Space centre scientific observations are performed using three types of instrumentation, namely, ground-based, suborbital, and spaceborne [1]. All three types are scientifically competing with each other, and, more importantly they couple each other by extending the frequency ranges outside the spectrum of the spaceborne instruments [2]. In this way, scientists obtain a richer range of scientific observations. The demands for more capable ground-based and suborbital facilities have been

Figure 1. Sample artefact, 5500 BC, archaeological museum of Za-dar/Dia-dora, Croatia.

increased over the recent years [3]. Expenses related to cleanroom procedures, space qualification, launch, and operation have been kept to a minimum [4–8].

On the other side, the more costly near-Earth orbit and, especially, deep space missions are totally justified by their qualitative basis of technological capabilities that they offer [9, 10]. High-resolution magnetometry, UV, X-ray, stray light imaging power, etc. are simply samples of the superior in-situ measurement data that these missions have been providing to the scientific community [11].

The winner in the race for deep-space was stemmed early in the race and as early as in 1964. The immense joint effort of regular people, U.S. Government, the outstanding work of the scientists at Jet Propulsion Laboratory (JPL), the technological miracles achieved at National Aeronautics and Space Administration (NASA), in just 6 years after NASA's establishment, lead to the success of Mariner IV [12]. The success is even greater if it is taken into consideration that the 4th of July Independence Day was in the not so distant 1776.

Mariner in 1967 carried a slightly modified instrumentation [13], which was further adjusted to meet the expectations of Pioneer 10/11 to Jupiter and Saturn (1972/1973) [14]. A Pioneer 10/11-based flight-spare instrumentation was modified for ISEE-C [15], outperforming the FGM-based ISEE-A/B spacecrafts [16], in return-science [17]. The successful ISEE-C and Pioneer 10/11 designs lead to highly stable and low-noise instrumentation designs for Ulysses in 1990. Until 2008, Ulysses studied solar space physics [18] and performed accurate in-orbit observations [19].

The Cassini-Huygens mission to Saturn, Titan, and Saturn's moons was launched on 15 October 1997 and ended gloriously on 15 October 2017. Some flight-spare instrumentation from Ulysees was modified and added to Cassini to support the first-time in space S/VHM [20]. Cassini applied the dual-magnetometer (DM) technique [21]. DM accelerates the pre-launch magnetic cleanliness and calibration program, records the post-launch field variation, and controls the redundancy in interplanetary missions [22]. It remains the most innovative interplanetary mission ever achieved [23]. It is also the topic of the next section.

2. The Cassini-Huygens mission

The Cassini-Huygens mission exceeded all expectations and explored a planetary system that is different from ours. 635 GB of science data were collected and 453,048 fantastic images were transmitted back to Earth, as shown in **Figure 2**. This enhanced our knowledge regarding the solar system. The spacecraft travelled in total 4.9 billion miles (7.9 billion kilometres). Eighteen scientific instruments were onboard Cassini, and, a probe that landed on Saturn's moon, Titan. Titan is larger than planet Mercury. Scientists from 27 countries participated to the project. The mission assisted in verifying new remote sensing techniques and flight-proving this unique spacecraft design.

It took 7 years for the spacecraft to reach Saturn. In order to gain the required gravitational force to perform this journey, Casssini flew twice by Venus, by Earth and, then, by Jupiter, before reaching the Saturnian system. The mission was also supported by the Italian Space Centre (ASI), the European Space Agency (ESA), and the U.S. Congress. The Cassini-Huygens interplanetary spacecraft holds a record weight in its category of 6.1 tons, when fully fuelled. Cassini proved that Saturn produces lightning bolts ten thousand times more powerful than the strongest on Earth and equatorial winds in the range of 1100 mph. It also proved that the Titan has similarities with early Earth, due to its nitrogen-enhanced atmosphere. The complex organic structures in its atmosphere will eventually fall to its surface. This will be an equivalent point similar to the one when life is initiated on Earth. Further analysis of Titan's collected data will enhance our knowledge of how life was enabled on Earth. Subsequent study of the data collected by Cassini will assist in understanding how the universe itself and our solar system were created.

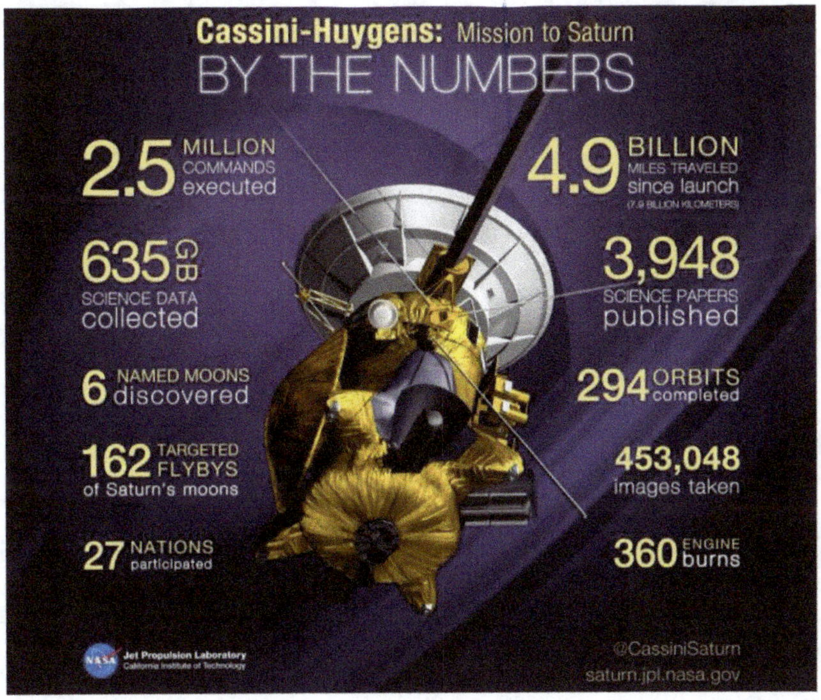

Figure 2. Cassini-Huygens by the numbers. Courtesy of JPL, NASA.

3. Juno-teaser for the space fans

Following the success of the Cassini-Huygens mission to Saturn (Chronos = time), Titan (King) and Saturn's moons, the Juno (Hera) mission to Jupiter (Dias) was the first competed mission selected for NASA's New Frontiers program to perform first-time in-depth observations of Jupiter's structure, atmosphere, and polar magnetosphere. The spacecraft was launched from Cape Canaveral Air Force Station on the 5th of August, 2011. Juno entered a polar orbit of Jupiter on the 5th of July, 2016.

JPL released the following composite image on the 7th of March, 2018, as shown in **Figure 3**. It consists of data collected by the Jovian Infrared Auroral Mapper (JIRAM).

- Do you think there is a connection between **Figures 1** and **3**?
- If yes, what do you think that this might be?

Please, visit NASA's JPL website to find the solution and more information regarding Space, Cassini-Huygens, and Juno. Additionally, valid educational material on Physics and the flow of liquid or air masses will assist you in solving the puzzle.

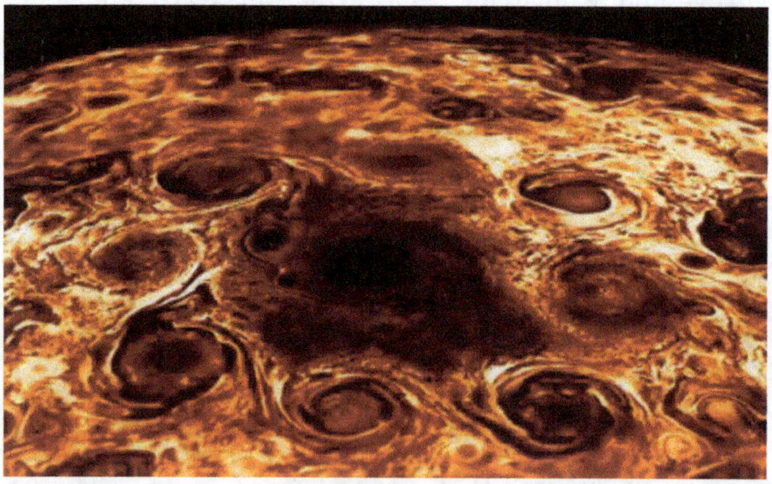

Figure 3. Cyclones encircle Jupiter's North Pole. Courtesy of JPL, NASA.

4. Conclusion

Space has always been intriguing people's imagination. However, space flight has only been feasible over the last 60 years. In this book, recent research results are presented in the areas of simulation, spacecraft navigation, propulsion, suborbital flight and seep-space operations. We hope this book will be advantageous to researchers and to also inspire the younger generations into pursuing studies and careers within the space industry.

Author details

George Dekoulis

Address all correspondence to: dekoulis@aeispace.org

Aeronautical and Space Engineering Department, Aerospace Engineering Institute (AEI), Nicosia, Cyprus

References

[1] Dekoulis G. Novel space exploration technique for analysing planetary atmospheres. In: Air Pollution Vanda Villanyi. London, UK: IntechOpen; 2010. DOI: 10.5772/10053

[2] Dekoulis G. Novel digital magnetometer for atmospheric and space studies (DIMAGORAS). In: Aeronautics and Astronautics Max Mulder. London, UK: IntechOpen; 2011. DOI: 10.5772/17326

[3] Dekoulis G. Novel digital systems designs for space Physics instrumentation [Ph.D. thesis]. Lancaster University; 2007

[4] Dekoulis G, Honary F. Novel low-power fluxgate sensor using a macroscale optimisation technique for space physics instrumentation. SPIE, Smart Sensors, Actuators, and MEMS III. 2007;**6589**:65890G-1-65890G-8

[5] Dekoulis G, Honary F. Novel sensor design methodology for measurements of the complex solar wind—Magnetospheric—Ionospheric system. Microsystem Technologies. 2008;**14**(4-5):475-482

[6] Dekoulis G. Field Programmable Gate Array. London, UK: InTech; 2017. ISBN 978-953-51-3208-0

[7] Dekoulis G. Robotics. London, UK: InTech; 2018. ISBN 978-953-51-3636-1

[8] Dekoulis G. Drones – Applications. London, UK: InTech; 2018. ISBN 978-953-51-5948-3

[9] Dekoulis G, Murphy N. New Digital Systems Designs for Validating the JPL Scalar Helium Magnetometer for the Juno Mission. NASA JPL Research Report; 2008

[10] Dekoulis G. 3D reconfigurable NoC multiprocessor imaging interferometer for space climate. In: COSPAR, Space Climate, Space Plasmas in the Solar System, Including Planetary Magnetospheres. NASA ADS: Istanbul, Turkey; 30 July-7 August 2016;**41**(1):D2.5-25-16

[11] Dekoulis G. 3D reconfigurable MPSoC for unmanned spacecraft navigation. In: COSPAR, Spacecraft Instruments and their Use, Space Studies of the Earth's Surface, Meteorology and Climate. NASA ADS: Istanbul, Turkey; 30 July-7 August 2016;**41**(1):A0.2-24-16

[12] Slocum RE, Reilly FN. Low field helium magnetometer for space applications. IEEE Transactions on Nuclear Science. 1963;**10**(1):165-171

[13] Connor B. Space magnetics: The Mariner V magnetometer experiment. IEEE Transactions on Magnetics. 1968;**4**(3):391-397

[14] Smith E, Connor B, Foster GJ. Measuring the magnetic fields of Jupiter and the outer solar system. IEEE Transactions on Magnetics. 1975;**11**(4):962-980

[15] Frandsen AMA, Connor BV, Van Amersfoort J, Smith EJ. The ISEE-C vector helium magnetometer. IEEE Transactions on Geoscience Electronics. 1978;**16**(3):195-198

[16] Ogilvie KW et al. Descriptions of experimental investigations and instruments for the ISEE spacecraft. IEEE Transactions on Geoscience Electronics. 1978;**16**(3):151-153

[17] Tsurutani BT, Von Rosenvinge TT. ISEE-3 distant Geotail results. Geophysical Research Letters. 1984;**11**(10):1027-1029

[18] Balogh A, Lanzerotti A, Louis J, Suess ST. The Heliosphere through the Solar Activity Cycle. UK: Praxis Publishing; 2008. ISBN: 978-3-540-74301-9

[19] Balogh A et al. The magnetic field investigation on the Ulysses mission: Instrumentation and preliminary scientific results. Astronomy and Astrophysics Supplement Series. 1992;**92**(2):221-236

[20] Kellock S, Austin P, Balogh A, et al. Cassini dual technique magnetometer instrument (MAG). Proceedings of SPIE. 1996;**2803**:141-152

[21] Russell CT. The Cassini-Huygens Mission Orbiter In Situ Investigations. The Netherlands: Kluwer Academic Publishers; 2005. ISBN: 978-1-4020-2773-4

[22] Voorhies CV et al. Preliminary Calibration Plan for the Advanced Particles and Field Observatory (APAFO) Magnetometer Experiment. NASA-TM-104545; 1991

[23] Doody D. Cassini/Huygens: Heavily instrumented flight systems approaching Saturn and Titan. IEEE Aerospace Conference Proceedings. 2003;**8**:3637-3646

9

Fractal Pyramid: A New Math Tool to Reorient and Accelerate a Spacecraft

Alexander P. Yefremov

Abstract

An original mathematical instrument matching two different operational procedures aimed to change orientation and velocity of a spacecraft is suggested and described in detail. The tool's basements, quaternion algebra with its square-root (pregeometric) image, and fractal surface are represented in a parenthetical but in a sufficient format, indicating their principle properties providing solution to the operational task. A supplementary notion of vector-quaternion version of relativity theory is introduced since the spacecraft-observer mechanical system appears congenitally relativistic. The new tool is shown to have a simple pregeometric image of a fractal pyramid whose tilt and distortion evoke needed changes in the spacecraft's motion parameters, and the respective math procedures proved to be simplified compared with the traditionally used math methods.

Keywords: spacecraft motion, operation, quaternion, fractal surface

1. Introduction

In classical mechanics, rotation of a rigid body (in particular, a spacecraft) and its translational motion are normally regarded as drastically different actions leading to changes in its position and are respectively described by different groups. Relativistic mechanics, in its turn, deals with these two types of motions "more homogeneously" since rotation and linear motion are described in this case by 4×4 matrices from the Lorentz group $SO(1,3)$. However, it is well known that the special relativity limits itself by inertial motions of the involved frames of reference while use of general relativity comprising any types of motion but demanding math methods of tensor calculus seems unapproved sophisticated. Happily, there exists a simpler vector version of the relativity theory admitting arbitrary accelerated motion of the frames. A brief formulation of the theory is made with the help of quaternion vector units, each set of the units representing a Cartesian-type frame of reference. In this case, the rotation-and-translation operator is given by 3×3 matrix belonging to the group $SO(3, C)$ known to be 1:1 isomorphic to the group $SO(1, 3)$. However, the calculations of the body's complex motions even within the framework of the vector-quaternion relativity remain prolonged and cumbersome, a simpler method is desired. Such a method is found due to existence of 1:2 isomorphism of the groups $SO(3, C)$ and $SL(2, C)$, the last being a spinor group operating in fractal two-dimensional complex-number valued space (a fractal surface). It is necessary to mention that the subgroup of $SL(2, C)$, rotational group $SU(2)$, is normally used in space-flight practice, providing comparatively simple mathematical computations for a spacecraft reorientation tasks [1, 2]. This method is based upon similarity-type transformations of the initial quaternion triad, in fact assuming nontrivial multiplication of at least three different quaternions, though it straightforwardly gives the data describing the axis of single rotation and value of the respective angle. However, this method provides no translational motion.

In this study, we suggest an essential development of the last (single rotation) method leading, first, to noticeable simplification of computations, and second, to possibility of introduction of additional parameters responsible for the spacecraft acceleration. This development is fully based on fundamental properties of subgeometric dyad forming the fractal space in a way underlying the 3D physical space. Moreover, we suggest subgeometric images (fractal joystick and fractal pyramid) of the math tools realizing the spacecraft's reorientation and acceleration tasks. As well, we give a brief comparative analysis of simplicity (or complexity) of conventional and new methods.

The study is composed as following. In Sections 2–4 we offer a detailed mathematical introduction. In Section 2, we renew our knowledge of quaternion algebra giving traditional (Hamiltonian) and more compact (tensor) notions and correlations. In Section 3, we briefly reproduce the quaternion version of the relativity theory. In Section 4, we consider main notions and properties of the 2D fractal space and show how to build a 3D frame out of a dyad element.

Sections 5–7 are devoted to new math methods making operations of a spacecraft simpler and more functional. Section 5 is devoted to presentation of three methods to reorient a spacecraft with accent on convenience of the single rotation method involving a fractal joystick model. In Section 6, we suggest a very simple way to introduce (apart from space rotation) an acceleration of the spacecraft and demonstrate a subgeometric image of the respective math tool having a shape of fractal pyramid. Finally, in Section 7, we give a sketch of a technological map previewing necessary steps to simultaneously reorient and accelerate the spacecraft followed by a series of relevant pictures.

2. Basic notions and relations of quaternion algebra

Quaternion (Q-) numbers were discovered by Hamilton in 1843 [3]. A quaternion is a math object of the type $q = a1 + b\mathbf{i} + c\mathbf{j} + d\mathbf{k}$ (in Hamilton's notation), where a, b, c, d are real coefficients at the real unit 1 (the symbol is normally omitted in the number) and at three imaginary units $\mathbf{i}, \mathbf{j}, \mathbf{k}$ forming the postulated multiplication table (16 equalities).

$$\begin{aligned}&1^2 = 1, \mathbf{i}^2 = \mathbf{j}^2 = \mathbf{k}^2 = -1, 1\mathbf{i} = \mathbf{i}1 = \mathbf{i}, 1\mathbf{j} = \mathbf{j}1 = \mathbf{j}, 1\mathbf{k} = \mathbf{k}1 = \mathbf{k},\\ &\mathbf{ij} = -\mathbf{ji} = \mathbf{k}, \mathbf{jk} = -\mathbf{kj} = \mathbf{i}, \mathbf{ki} = -\mathbf{ik} = \mathbf{j},\end{aligned} \quad (1)$$

Q-numbers and the multiplication law (1) can be more compactly rewritten in the vector (and tensor) notations $\mathbf{i}, \mathbf{j}, \mathbf{k} \to \mathbf{q}_1, \ \mathbf{q}_2, \ \mathbf{q}_3 \to \mathbf{q}_k, j, k, l, m, n \ldots = 1, 2, 3$; then, a quaternion is a sum of scalar (a) and vector $(b_k \mathbf{q}_k)$ parts $q \equiv a + b_k \mathbf{q}_k$, where $a, \ b_k \in \mathbf{R}$, and the multiplication table (1) has the form

$$1\mathbf{q}_k = \mathbf{q}_k 1 = \mathbf{q}_k, \ \mathbf{q}_k \ \mathbf{q}_l = -\delta_{kl} + \varepsilon_{klj} \ \mathbf{q}_j, \quad (2)$$

Summation in repeated indices is implied, and δ_{kl} and ε_{klj} are the 3D Kronecker and Levi-Chivita symbols (see e.g., [4]).

Quaternions admit the same operations as real and complex numbers. Comparison of Q-numbers is reduced to their equality: two Q-numbers are equal if coefficients at respective units are equal. Commutative addition (subtraction) of Q-numbers is made by components. Q-numbers are multiplied as polynomials; the rules (1, 2) state that multiplication is noncommutative (left and right products are defined), but still associative. A quaternion $q \equiv a + b_k \mathbf{q}_k$ has its conjugate $\bar{q} \equiv a - b_k \mathbf{q}_k$, the norm $|q^2| \equiv q\bar{q} = \bar{q}q$, and the modulus (positive square root from the norm) $|q| \equiv \sqrt{q\bar{q}} = \sqrt{a^2 + b_k b_k}$. Inverse number is $q^{-1} = \bar{q}/|q|^2$; so, for two quaternions q_1 and q_2, division (left and right) is defined as $(q_1/q_2)_{left} = q_2 q_1/|q_2|^2$ and $(q_1/q_2)_{right} = q_1 q_2/|q_2|^2$. If q is a product of two multipliers $q_1 = a + b_k \mathbf{q}_k$ and $q_2 = c + d_n \mathbf{q}_n$, then from definition of the norm one finds

$$|q|^2 = |q_1 q_2|^2 = (q_1 q_2)(\overline{q_1 q_2}) = q_1 q_2 \bar{q}_2 \bar{q}_1 = q_1 \bar{q}_1 q_2 \bar{q}_2 = |q_1|^2 |q_2|^2. \quad (3)$$

Written in components, Eq. (3) becomes the famous identity of four squares

$$\begin{aligned}&(ac - b_1 d_1 - b_2 d_2 - b_3 d_3)^2 + (ad_1 + cb_1 + b_2 d_2 - b_3 d_2)^2 + (ad_2 + cb_2 + b_3 d_1 - b_1 d_3)^2 +\\ &+(ad_3 + cb_3 + b_1 d_2 - b_2 d_1)^2 = (a^2 + b_1^2 + b_2^2 + b_3^2)(c^2 + d_1^2 + d_2^2 + d_3^2).\end{aligned} \quad (4)$$

Identities of the type (4) exist only in four algebras: of real numbers (trivial identity), of complex numbers (two squares), of quaternions (four squares), and of octonions (the last exclusive algebra with one real and seven imaginary units admits identity of eight squares; multiplication in this algebra is no more associative).

Geometrically, the imaginary Q-units are associated with three unit vectors initiating a Cartesian coordinate system (Q-triad, Q-frame). This image, in particular, follows from the fact that, according to Eq. (2), each imaginary unit appears as ordered product of the two others: $\mathbf{q}_1 = \mathbf{q}_2 \mathbf{q}_3, \mathbf{q}_2 = \mathbf{q}_3 \mathbf{q}_1, \mathbf{q}_3 = \mathbf{q}_1 \mathbf{q}_2$ (vector products in Gibbs-Heaviside algebra). One can easily construct a set of such units. To demonstrate this, we consider a couple of 2×2-matrices, $A = \begin{pmatrix} a & b \\ c & -a \end{pmatrix}, \ B = \begin{pmatrix} d & e \\ f & -d \end{pmatrix}$, traceless: $TrA = TrB = 0$, and not degenerate: $\det A \neq 0$, $\det B \neq 0$. We use the matrices to build two different imaginary units as

$$\mathbf{q}_1 = \frac{A}{\sqrt{\det A}}, \quad \mathbf{q}_2 = \frac{B}{\sqrt{\det B}}. \tag{5}$$

We form the product of the two units and demand that its trace vanishes that is given as

$$\mathbf{q}_1\mathbf{q}_2 = \frac{AB}{\sqrt{\det A \det B}}, \quad Tr(AB) = 0; \tag{6}$$

then Eq. (6) gives expression for the third imaginary Q-unit $\mathbf{q}_1\mathbf{q}_2 = \mathbf{q}_3$, and as a whole, we get the Q-triad $\mathbf{q}_k$, the real unit always remaining the unit matrix $1 \equiv \begin{pmatrix} 1 & 0 \\ 0 & 1 \end{pmatrix}$. One readily checks up that the triad given by Eqs. (5) and (6) identically satisfies the multiplication law (2). Built in a similar way, the simplest representation of Q-units $\mathbf{q}\tilde{k}$ is given by the Pauli matrices $\mathbf{p}\tilde{k}$ with factor –i: $\mathbf{q}\tilde{k} = -i\mathbf{p}\tilde{k}$

$$1 = \begin{pmatrix} 1 & 0 \\ 0 & 1 \end{pmatrix}, \quad \mathbf{q}_{\tilde{1}} = -i\begin{pmatrix} 0 & 1 \\ 1 & 0 \end{pmatrix}, \quad \mathbf{q}_{\tilde{2}} = -i\begin{pmatrix} 0 & -i \\ i & 0 \end{pmatrix}, \quad \mathbf{q}_{\tilde{3}} = -i\begin{pmatrix} 1 & 0 \\ 0 & 1 \end{pmatrix}, \tag{7}$$

and the imaginary Q-triad given as Eq. (7) describes a constant Q-vector frame.

However, a Q-frame may be variable, rotating, and moving. There are two types of transformations changing the frame but retaining the form of the multiplication law (2). The first is rotational-type transformation

$$\mathbf{q}_{k'} = O_{k'n}\mathbf{q}_n \tag{8}$$

where $O_{k'n}$ is a 3 × 3-matrix (its components are in general complex numbers) having orthogonal properties $O_{k'n}O_{m'n} = \delta_{km}$, hence this matrix belongs to the special orthogonal group of 3D rotations over field of complex numbers $O_{k'n} \in SO(3, C)$. The matrix $O_{n'k}$ can be always represented as a product of plane (or simple) rotations, irreducible representations of $SO(3, C)$. For such matrices, a special notation will be used, e.g., O_n^{Θ}, where the lower index indicates the rotation axis (the frame's unit vector) and upper index shows the rotation angle. Depending on the math nature of the angle Θ, we distinguish two types of simple rotations. If $\Theta = \alpha \in \mathbf{R}$, then we have a real simple rotation $O_n^{\Theta} \to R_n^{\alpha}$; if the angle is imaginary $\Theta = \eta \in i\ \mathbf{R}$, then we have a simple hyperbolic rotation $O_n^{\Theta} \to H_n^{\eta}$; for example Eq. (9)

$$R_3^{\alpha} \equiv \begin{pmatrix} \cos\alpha & \sin\alpha & 0 \\ -\sin\alpha & \cos\alpha & 0 \\ 0 & 0 & 1 \end{pmatrix}, \quad H_3^{\eta} \equiv \begin{pmatrix} \cos h\eta & -\mathrm{i}\sin h\eta & 0 \\ \mathrm{i}\sin h\eta & \cos h\eta & 0 \\ 0 & 0 & 1 \end{pmatrix}. \tag{9}$$

Superposition of any number (N) of real rotations (product of relevant matrices) gives a (nonplane) real rotation

$$\prod_{j=1}^{N} R_{n_j}^{\alpha_j} = R_{n_1}^{\alpha_1} \cdots R_{n_N}^{\alpha_N} \rightarrow R_{k'm} SO(3,R). \tag{10}$$

Product of multiple hyperbolic rotations is physically sensible if accompanied by real rotations in the framework of vector version of theory of relativity (see Section 3); so in general, the matrices of the type

$$\prod_{j=1}^{N}\prod_{s=1}^{M} R_{n_j}^{\alpha_j} H_{m_s}^{\eta_s} = R_{n_1}^{\alpha_1}..H_{m_1}^{\eta_1}..R_{n_N}^{\alpha_N}..H_{m_s}^{\eta_s} \rightarrow O_{k'm} \in SO(3,C) \tag{11}$$

are used in applications.

The second type of transformations is performed by an operator U and its inverse U^{-1} is given as

$$\mathbf{q}_{k'} = U\mathbf{q}_k U^{-1}. \tag{12}$$

It is evident that the transformation (12) keeps the form of the basic law (2). The operators U are known to form the (spinor) group $U \in SL(2,C)$ of special linear 2D transformations over field of complex numbers; this group is 2:1 isomorphic to $SO(3,C)$ and similarly to the Lorentz group. A special case of the transformation (12) is a real rotation made by means of the subgroup $SU(2) \in SL(2,C)$, and this spinor subgroup is 2:1 isomorphic to vector group $SO(3,R)$. It is necessary to note that the transformation of the type (12) with $U \in SU(2)$ is most frequently used for solution of a spacecraft orientation problem (see Section 5.2).

As well, in formulation of quaternion relativity (see Section 3), we shall need notion of a biquaternion (BQ-) number. Such a number has the form $b = x + y_k\mathbf{q}_{k'}$, where $x, y_k \in \mathbf{C}$ while 1, $\mathbf{q}_k$ are Q-units. BQ-numbers admit addition, multiplication, and conjugation $b = x - y_k q_k$. But the norm is not well defined since the product $b\bar{b} = x^2 + y_k y_k$ in general is not a real (and positive) number. A real number "norm" exists in the subset of vector biquaternions

$$b = (w_k + i\ z_k)\ \mathbf{q}_k \tag{13}$$

whose real and imaginary parts are mutually orthogonal

$$w_k z_k = 0 \rightarrow \|b\|^2 = b\bar{b} = w_k w_k - z_k z_k. \tag{14}$$

There are evidently zero dividers in Eq. (14), hence division is not well defined, but the subset (13 and 14) comprises basic formulas describing relative motion of arbitrary accelerated frames of reference.

3. Vector-quaternion version of the relativity theory

According to Eqs. (13) and (14), the interval of Einstein's relativity theory[1]

$$ds^2 = dx_0^2 - dx_k dx_k = dt^2 - dr^2 \tag{15}$$

admits a BQ-square root

$$d\mathbf{s} = (ie_k dt + dx_k)\ \mathbf{q}_{k'} \tag{16}$$

where displacement of observed object dx_k is orthogonal to a unit vector e_k directing change in time $dt : e_k dx_k = 0$. Under these conditions, square of Eq. (16) yields Eq. (15) $d\mathbf{s}d\mathbf{s} = ds^2$. It is convenient to explicitly relate displacement dx_k to a plane orthogonal to time-directing vector e_k with the help of metric-projector $b_{kn} \equiv \delta_{kn} - e_k e_n\ dx_k = dx_n b_{nk},$

then the orthogonality condition is fulfilled automatically $e_k dx_k = e_k dx_n b_{kn} = 0$.

The interval (15) is invariant under Lorentz transformations of coordinate system $dx^{\alpha'} = L_\lambda^{\alpha'} dx^\lambda$, $L_\lambda^{\alpha'} \in SO(1,3)$, while the Q-frame can be subject to $SO(3,C)$ rotations $\mathbf{q}_{k'} = O_{k'l}\mathbf{q}_l$; simultaneous application of the transformations, together with demand that the BQ-vector (16) form be conserved, leads to correlation between components of matrices $O_{k'l}$ and $L_{\alpha'\lambda}$[2] [5, 6]

$$ie_k O_{s'k} = ie_{s'} L_{0'0} + L_{m'0} b_{m's'}, \tag{17}$$

$$b_{nk} O_{s'k} = -ie_{s'} L_{0'm} - L_{m'k} b_{m's'}. \tag{18}$$

Eqs. (17) and (18) in particular mean that within the group $SO(3,C)$ a set of ordered simple rotations of the type (11) are distinguished, real and hyperbolic, each performed about one-unit vector of Q-triad. If for instance, direction No. 1 of $L_{\alpha'\lambda}$ is not involved in the transformation ($e_k = e_{k'} = \delta_{1k}$), then Eqs. (17) and (18) represent the matrix O as function of components of Lorentz matrix L

$$O_{k'm} = \begin{pmatrix} L^{0'}{}_0 & -iL^{0'}{}_2 & -iL^{0'}{}_3 \\ iL^{2'}{}_0 & L^{2'}{}_2 & L^{2'}{}_3 \\ iL^{3'}{}_0 & L_2^{3'} & L^{3'}{}_3 \end{pmatrix}. \tag{19}$$

The matrix (19) may describe a series of simple rotations, but real rotations should be always performed about vector $\mathbf{q}_1$ (initial or transformed), while hyperbolic rotations are allowed about

[1] Standard interval of special relativity is regarded for simplicity; similarly, interval of general relativity can be considered in tangent space $ds^2 = \theta_0^2 - \theta_k\theta_k$ with $\theta_{(\alpha)} = g_{(\alpha)\lambda}dy^\lambda$ being basic one-form and Greek indices in brackets enumerating tangent space tetrad, and those without brackets are related to curved manifold holonomic coordinates

$$\eta_{\alpha\beta} = \mathrm{diag}(1, -1, -1, -1)$$

[2] $ds^2 = \theta_0^2 - \theta_k\theta_k\theta_{(\alpha)} = g_{(\alpha)\lambda}dy^\lambda$: four-dimensional indices are raised and lowered by Minkowski metric $\eta_{\alpha\beta} = \mathrm{diag}(1, -1, -1, -1)$.

vectors $\mathbf{q}_2$ and $\mathbf{q}_3$. It is easily checked up that all matrices O of the type (19) constitute a subgroup $SO(1,2) \subset SO(3,C)$ of the ordered rotations of Q-triads.

Main idea of Q-version of relativity is to replace line element of Einstein's relativity (15) and its invariance under Lorentz group by adequate BQ-vector (16) invariant under rotational group represented by matrices $O \in SO(1,2)$. Then, instead of quadratic form of four-dimensional coordinates, an observer has at his disposal a movable Q-triad with time and distances measured along its unit vectors and dealt with the vector basement as with the Newtonian mechanics or general relativity in tetrad formulation. However, on this way, an essential peculiarity arises. Eq. (16) implies that the constructed space-time model has six dimensions, and it is a symmetric sum of two three-dimensional (3D) spaces $Q_6 = R_3 \oplus T_3$, where R_3 is the usual 3D space where coördinate and velocity change, whereas T_3 is also a 3D space but imaginary with respect to R_3. In this model, the observer works only with some sections of the 6D space; but since the objects of the observations are found in real 3D space, and imaginary time axis is distinguished, an illusion of four dimensions emerges.

Physical measurements in the Q-model are made with the help of three spatial rulers $\mathbf{q}_k$ and built-in geometric clock represented by "imaginary time rulers" (Pauli-type matrices) $\mathbf{p}_k \equiv i\mathbf{q}_k$, the two triads being obviously co-aligned. The tool-set $\Sigma \equiv \{\mathbf{p}_k, \ \mathbf{q}_k\}$ with an observer in the initial point represents full physical frame of reference, Eq. (16) can be rewritten as

$$d\mathbf{s} = e_k dt \mathbf{p}_k + dx_k \ \mathbf{q}_k. \tag{20}$$

Now, the principal statement of the Q-version of relativity follows: all physically sustainable frames of reference are interconnected by "rotational equations"

$$\Sigma' = O\Sigma, \quad O \in SO(1,2)\,. \tag{21}$$

The sustainability means form-invariance of BQ-vector (16) or (20) under transformations (21). Kinematic effects of special relativity are straightforwardly found in the Q-version; here, we demonstrate only one effect important for fractal pyramid technology accelerating a spacecraft (see Section 6).

Boost. Σ-observer always can align one of his spatial vectors (e.g., $\mathbf{q}_2$) with velocity of moving body, so basic BQ-vector can be written in the form

$$d\mathbf{s} = dt\mathbf{p}_1 \ + dr \ \mathbf{q}_2. \tag{22}$$

Let the frame Σ' be a result of a hyperbolic rotation of a constant frame Σ

$$\Sigma' = H_3^\eta \ \Sigma, \tag{23}$$

with the matrix H_3^η from Eq. (9b) (rotation about $\mathbf{q}_3$ by angle η). This simple rotation, physically a boost, obviously keeping BQ-vector (20) form-invariant

$$dt\mathbf{p}_1 + dr\mathbf{q}_2 = dt'\mathbf{p}_{1'} + dr'\mathbf{q}_{2'} \tag{24}$$

yields familiar coordinate transformations

$$dt' = dt\cosh\eta \ + dr\sinh\eta, \quad dr' = dt\sinh\eta \ + dr\cosh\eta \tag{25}$$

with respective effects of length and time segments contraction. If observed particle is the body of reference of the frame Σ', then $dr' = 0$, and one finds that the frame Σ' is moving with the velocity

$$V = \frac{dr}{dt} = \tanh\psi. \tag{26}$$

Specific features of the Q-vector version of relativity will be effectively used below in the fractal-pyramid math method to operate a spacecraft. Now, we turn to notions of a fractal space.

4. Fractal space underlying physical space

In this section, we show that a 3D space (e.g., physical space) may be endowed with a pregeometry [7] mathematically described by a complex-numbered surface, a 2D fractal space, each its vector having dimensionality half compared to that of the 3D space. We start with 2D space and construct out of its basic elements a basis of 3D space.

Let there exist a smooth 2D space (surface) endowed with a metric g_{AB} (and inverse: $g^{BC}g^{BC} \to g_{AB}g^{BC} = \delta_A^C$) and with a system of coordinates $x^A = \{x^1, \ x^2\}$; here $A, B, C = 1, 2$, δ_A^C is a 2D Kronecker symbol, summation in repeated indices is also implied. The line element of the surface is

$$ds^2 = g_{AB}dx^A dx^B; \tag{27}$$

the surface may be curved, so covariant and contravariant metric components differ. In a point, we choose a couple of unit orthogonal vectors a^A, b^B (a dyad)

$$g_{AB}a^A a^B = 1, \tag{28}$$

$$g_{AB}a^A b^B = 0. \tag{29}$$

A domain of the surface in vicinity of the dyad's initial point (together with respective part of tangent plane having the metric $\delta_{MN} = \delta^{MN} = \delta_m^N$) will be called a "2D-cell."

Considering direct (tensor) products of the dyad vectors with mixed components [8], we can construct only four such products (2 × 2 matrices): two idempotent matrices

$$G_B^A \equiv a^A a_B, \quad H_B^A \equiv b^A b_B \rightarrow G_B^A G_C^B = G_C^A, \quad H_B^A H_C^B = H_C^A, \tag{30a}$$

and two nilpotent matrices

$$D^{A_B} \equiv a^A b_B, \quad F^{A_B} \equiv b^A a_B \rightarrow D^{A_B} D^{B_C} = 0, \quad F^{A_B} F^{B_C} = 0. \tag{30b}$$

Next, we built sum and difference of the idempotent matrices

$$E \equiv E_B^A \equiv G_B^A + H_B^A = a^A a_B + b^A b_B, \quad E^2 = E, \tag{31a}$$

$$\tilde{K} \equiv \tilde{K}_B^A \equiv G_B^A - H_B^A = a^A a_B - b^A b_B, \quad \tilde{K}^2 = E, \tag{31b}$$

and sum and difference of the nilpotent matrices

$$\tilde{I} \equiv \tilde{I}_B^A = D^{A_B} + F^{A_B} = a^A b_B + b^A a_B, \quad \tilde{I}^2 = E, \tag{31c}$$

$$J \equiv J_B^A = D^{A_B} - F^{A_B} = a^A b_B - b^A a_B, \quad J^2 = -E. \tag{31d}$$

If the units Eqs. (31b) and (31c) are slightly corrected so that their product is the third unit (31d), then we obtain the basis of quaternion (and biquaternion) numbers

$$1 \equiv E, \quad \mathbf{q}_1 = -i\ \tilde{I}, \mathbf{q}_2 = -i\ J, \mathbf{q}_3 = -i\ \tilde{K}. \tag{32}$$

Now, we recall the spectral theorem (of the matrix theory) stating that any invertible matrix with distinct eigenvalues can be represented as a sum of idempotent projectors with the eigenvalues as coefficients, the projectors being direct products of vectors of a biorthogonal basis. The unit $\mathbf{q}_3$ defined in Eqs. (32), (31b) is the characteristic example

$$\mathbf{q}_3\big|_A^B = ia^A a_B - ib^A b_B = iG_B^A - iH_B^A. \tag{33}$$

Right and left eigenfunctions of $\mathbf{q}_3$ are vectors a^A, b^B and covectors a_A, b_B of the dyad, respectively; the eigenvalues are $+i$ (for a) and $-i$ (for b), and G_B^A, H_B^A are the projectors.

As mentioned above, the similarity transformation of the units

$$\mathbf{q}_{k'} = U\mathbf{q}_k U^{-1}, \quad U \in SL(2, C) \tag{34}$$

preserves the form of algebras' multiplication law (2). Therefore, vector units from Eq. (32) can be obtained from a single unit, say, q_3 by a transformation (34). Then, all vector units have same eigenvalues $\pm i$, and the eigenfunctions of the derived units are linear combinations of the eigenfunctions of the initial unit [9]. This also means that the mapping (34) is a secondary one, but the primary one is $SL(2, C)$ transformation of dyad vectors, thus forming a set of spinors from the viewpoint of the 3D space described by the triad vectors $\mathbf{q}_k$.

Hereinafter, we introduce shorter 2D-index-free matrix notations for the dyad: a vector is a column, a co-vector is a row, and a parity indicator + or − marks the sign of the eigenvalue $\pm i$

$$a^A \to \psi^+, \quad a_A \to \varphi^+, \quad b^A \to \psi^-, \quad b_A \to \varphi^-; \tag{35}$$

this helps to rewrite the above expressions more compactly. The dyad orthonormality conditions (28, 29) acquire the form

$$\varphi^\pm \psi^\pm = 1, \quad \varphi^\mp \psi^\pm = \varphi^\pm \psi^\mp = 0, \tag{36}$$

the idempotent projectors are denoted as $C^+ \equiv G = \psi^+ \varphi^+, \quad C^- \equiv H = \psi^- \varphi^-,$

and the units (32) are expressed through the single dyad vectors (co-vectors) as

$$1 = \psi^+ \varphi^+ + \psi^- \varphi^-, \tag{37a}$$

$$\mathbf{q}_1 = -i\ (\psi^+ \varphi^- + \psi^- \varphi^-), \tag{37b}$$

$$\mathbf{q}_2 = \psi^+ \varphi^- - \psi^- \varphi^-, \tag{37c}$$

$$\mathbf{q}_3 = i\ (\psi^+ \varphi^+ - \psi^- \varphi^-). \tag{37d}$$

Eq. (37) obviously demonstrates that the dyad elements are in a way "square roots" from 3D vector units. So, if we put dimensionality of any 3D line to be a unity, then dimensionality of a line on the 2D space (e.g., dimensionality of a dyad vector) must be ½; hence from the viewpoint of the 3D space, the surface determined by a dyad is fractal. The next important observation concerns transformations. The transformation (34) clearly results from the $SL(2, C)$ transformations of the dyad vectors (covectors)

$$\psi'^\pm = U\psi^\pm, \quad \varphi' = \varphi'^\pm U^{-1}. \tag{38}$$

So, apart from vector-type (8) and spinor-type (12) transformations of a Q-triad (an element of 3D space), there exists a possibility to deal with more fundamental math elements, vectors, and covectors describing "pregeometric" 2D cell of a fractal surface. These simpler math objects are subject to evidently simpler mapping (38); moreover, in the following sections, we will show that the operators of the transformations, being themselves BQ-numbers, suggest simpler and less numerous equations to solve, thus reducing degree of math load and probability of mistakes.

5. Three methods to reorient a spacecraft and fractal joystick

The orientation tasks are relevant with computations over 3D flat space modeling a local domain of the physical space. Two types of the orientation problem solutions are traditional:

(i) a series of subsequent several angles rotation and (ii) a one-angle rotation about an instant axis. Mixed variants exist, but are less productive, and they are not normally considered.

If magnitudes involved in calculations are generically measured in real numbers, then both techniques (i) and (ii) are based on the vector rotation group $SO(3, R)$. Math content of the technique (i) implies a multiple set of plane rotations [of type of Eq. (9a)] by Euler (or Krylov, or others) angles about selected axes. The technique (ii) in its turn represents a nontrivial problem of determining the instant axis of a single rotation.

Quaternions are widely known to fit better than real numbers for the orientation tasks mostly due to the fact that three vector units represent models of three mutually orthogonal gyroscope axes. As well, use of the Q-algebra formalism essentially simplifies calculations, especially for the technique (ii), since both the vector rotation group $SO(3, R)$ and its spinor "equivalent" $SU(2)$ reflection group can be used whatever enigmatic were formulas describing spinor rotations. However, the quaternion algebra reveals its unique property to split axial 3D vectors into dyad sets belonging to a fractal subspace as in Eq. (37), see also the basic work [10]. The above-described fractalization procedure, mathematically nontrivial and much less known, on the one hand clarifies "mysterious" two-side $SU(2)$ quaternion vector multiplication and on the other hand endows all algebraic objects and actions with distinct geometric sense; moreover, the calculations become most primitive. Solution of a spacecraft reorientation task as transformation of a fractal dyad represents the third math method (iii) suggested here. However, all three math methods are described in detail in this section.

5.1. Quaternion SO(3,R) approach to the reorientation problem: Technique (i)

Orientation of a spacecraft in 3D space is determined by three angles between axes of some global coordinate system and unit vectors of a frame attached to the moving body taking into account its physical symmetry. The global coordinates, e.g., are represented by a spherical system, and its local initiating vectors pointing: $\mathbf{q}_1$ to the north along the Earth's meridian), $\mathbf{q}_2$ along a parallel, and $\mathbf{q}_3$ to zenith direction. The directing vectors $\mathbf{q}_k$ are considered constant. Then, the orientation of a spacecraft bearing a frame $\mathbf{q}_{k'}$, (with $\mathbf{q}_{1'}$ along the body, $\mathbf{q}_{2'}$ a transverse one, and $\mathbf{q}_{3'}$ along gravity) is determined by three angles: "yaw" ψ, the angle between $\mathbf{q}_1$ and $\mathbf{q}_{1'}$ (rotation about $\mathbf{q}_3$); "roll" φ, angle $\mathbf{q}_2 - \mathbf{q}_{2'}$ (rotation about $\mathbf{q}_1$); and "pitch" θ, angle $\mathbf{q}_3 - \mathbf{q}_{3'}$ (rotation about $\mathbf{q}_2$). Within these notations, the spacecraft's orientation in the space is described by the matrix equation

$$\mathbf{q}_{n'} = R_{n'k}\mathbf{q}_{k'}, R \in SO(3, R). \tag{39}$$

Outlined above technique (i) demands that the matrix $R_{n'k}$ be represented as a product of simple rotations, irreducible representations of $SO(3, R)$ [a special notation for such matrix is $R^{\alpha}_{n'}$, see Section 2, Eqs. (9, 10)], each performed about a frame's unit vector. Simple rotations with the above parameters of the probe's orientations are given by the matrices

$$R_3^\psi \equiv \begin{pmatrix} \cos\psi & \sin\psi & 0 \\ -\sin\psi & \cos\psi & 0 \\ 0 & 0 & 1 \end{pmatrix}, R_2^\varphi \equiv \begin{pmatrix} \cos\varphi & 0 & -\sin\varphi \\ 0 & 1 & 0 \\ \sin\varphi & 0 & \cos\varphi \end{pmatrix}, R_1^\theta \equiv \begin{pmatrix} 1 & 0 & 0 \\ 0 & \cos\theta & -\sin\theta \\ 0 & \sin\theta & \cos\theta \end{pmatrix}. \tag{40}$$

Direct reorientation problem, i.e., reaching object's assigned orientation, can be solved by a sequence of plane rotations mathematically described by a sequent multiplication of matrices [see Eq. (10)]. This problem has no unique solution since the group $SO(3,R)$ is not commutative; i.e., different multiplication order of the matrices (40) with the same parameters (angles) generally gives different result; e.g., the products $R = R_3^\psi R_2^\varphi R_1^\theta$ and $R' = R_1^\theta R_2^\varphi R_3^\psi$ are, in general, different $R \neq R'$. Vice versa, different orders of the matrix product with other parameters may yield the same result, e.g., products $R = R_3^\psi R_2^\varphi R_1^\theta$ and $R' = R_1^{\theta'} R_2^{\varphi'} R_3^{\psi'}$ may represent equivalent rotational result $R = R'$. The possibility to represent an arbitrary $SO(3,R)$ matrix as a product of its irreducible representations given in different order in particular entails uncertainty in solution of the inverse problem when one has to determine values of angles securing an assigned reorientation of the spacecraft. Therefore, the technique (i) does not provide single-valued results.

Even with more difficult, we meet trying to use matrices from the group $SO(3,R)$in the technique (ii). As is known from the theory of matrices (see e.g., [11]) in this case, we have to solve the characteristic equation $RX = X$ searching for the matrix operator R an eigenvector X with unit eigenvalue, the vector X pointing direction of the instant rotation axis. This tough algebraic task then followed by sophisticated calculations aimed to find the instant rotation angle. The use of hypercomplex numbers essentially helped to avoid these math troubles, and about half of a century ago, quaternion algebra became a common tool serving for engineering goals of navigation and orientation. Indeed, the similarity transformation UqU^{-1} of a quaternion q performed with the help of auxiliary quaternion $U \equiv a + b\mathbf{q}$ geometrically leads to conical rotation of the vector part of q about an axis whose direction is determined by the unit Q-vector $\mathbf{q}$ (e.g., [2]); the value of the instant rotation angle is computed as 2 $\arctan(b/a)$. Below, we suggest a detailed analysis of this type of description of rotations.

5.2. Reorientation by a single rotation of the quaternion frame: Technique (ii)

Consider a 2×2matrix (with complex-number components) $U \equiv \begin{pmatrix} x & z \\ w & y \end{pmatrix}$, belonging to a special linear group $U \in SL(2,C)$, $\det U = xy - wz = 1$. The multiplication law (6) is obviously form invariant under the similarity-type transformation

$$\mathbf{q}_{n'} = U\mathbf{q}_n U^{-1}. \tag{41}$$

One readily demonstrates that the matrix U is a biquaternion with the definable norm; indeed,

$$U = \begin{pmatrix} x & z \\ w & y \end{pmatrix} = \frac{x+y}{2} + \sqrt{1 - \left(\frac{x+y}{2}\right)^2}\,\mathbf{q} \equiv a + b\mathbf{q} \tag{42}$$

where

$$a = \frac{x+y}{2}, b = \sqrt{1 - \left(\frac{x+y}{2}\right)^2}, \tag{43}$$

and $\mathbf{q}$ is a Q-vector unit

$$\mathbf{q} = \frac{1}{\sqrt{1-a^2}} \begin{pmatrix} \frac{x-y}{2} & z \\ w & -\frac{x-y}{2} \end{pmatrix}, \mathbf{q}^2 = -1. \tag{44}$$

The unit vector (44) represented through the constant basis (7) has the form; $\mathbf{q} = l_k \mathbf{q}_k = (b_k/b)\ \mathbf{q}_k$ where $l_k = b_k/b$ are components of a unit vector pointing in 3D space a vector with components b_k, then the condition $\det U = xy - wz = 1$ takes the form $a^2 + b^2 = 1$, $b^2 = b_k b_k$. This general biquaternion case will be used in subsequent studies when combined rotation-plus-translational motion is regarded (see Section 6). In this section, we consider only quaternion case: $a, b_k \in \mathbf{R}$, so the matrix U is unimodular if

$$a \equiv \cos\alpha,\ b = \sqrt{b_k b_k} \equiv \sin\alpha \tag{45}$$

therefore,

$$U = \cos\alpha + (\sin\alpha)\ l_n \mathbf{q}_n,\ U^{-1} = \cos\alpha - (\sin\alpha)\ l_n \mathbf{q}_n \tag{46}$$

with l_k representing cosines of angles between Q-vectors $\mathbf{q}_k$ and the direction determined by $\mathbf{q}$. With the help of Eqs. (46) and (2), we reproduce the transformation (42) in the developed form

$$\begin{aligned} \mathbf{q}_{k'} &\equiv U\mathbf{q}_k U^{-1} = (\cos\alpha + \sin\alpha\, l_n \mathbf{q}_n)\mathbf{q}_k(\cos\alpha - \sin\alpha\, l_m \mathbf{q}_m) = \\ &= 2\sin^2\alpha l_k l_n \mathbf{q}_n + \cos 2\alpha \mathbf{q}_k + \sin 2\alpha l_n \varepsilon_{nkm} \mathbf{q}_m = \\ &= [l_k l_n + \cos 2\alpha(\delta_{kn} - l_k l_n) + \sin 2\alpha l_m \varepsilon_{mkn}]\mathbf{q}_n. \end{aligned} \tag{47}$$

Eq. (47) in fact interlinks the $SO(3, R)$ rotation matrix components and the parameters of $SU(2)$ transformations of a Q-frame [compare with (39)]. As well, Eq. (47) helps to make the following geometric analysis.

Multiplied by l_k (with summation in index k), Eq. (47) yields the equality $l_k \mathbf{q}_{k'} = l_n \mathbf{q}_n$, meaning that vectors of the transformed frame $\mathbf{q}_{k'}$ have the same projections onto vector l_k as the initial frame $\mathbf{q}_k$; i.e., the transformation may be represented as a conical rotation about l_k, $\Phi \equiv 2\alpha$, which is angle of the rotation in the orthogonal plane with the metric $p_{kn} = \delta_{kn} - l_k l_n$ [see the second term in Eq. (47)]. Let two unit vectors e_k, n_k form this plane $p_{kn} = e_k e_n + n_k n_n$, then $l_m \varepsilon_{mkn} = e_k n_n - e_n n_k$, and the $SO(3,R)$-matrix comprised in Eq. (47) acquires the form

$$R_{k'n} = l_k l_n + \cos\Phi\ (e_k e_n + n_k n_n) + \sin\Phi\ (e_k n_n - e_k n_n). \tag{48}$$

Introducing now two artificial unit vectors with complex number components $s_k \equiv (e_k + in_k)/\sqrt{2}$ and $s_k^* \equiv (e_k - in_k)/\sqrt{2}$, we get the final (canonical) expression

$$R_{kn} = l_k l_n + e^{i\Phi}\ s_k s_n^* + e^{-i\Phi}\ s_k^* s_n. \tag{49}$$

Eq. (49) is just an explicit formulation of the spectral theorem applied on a 3D orthogonal matrix. Since its determinant differs from zero, this matrix is nonsingular, all its eigenvalues $\lambda_{(i)}$ are different, so it is simple; therefore, it can be expanded into a series of projectors $C_{(i)}$ with $\lambda_{(i)}$ as coefficients

$$R = \sum_{i=1}^{3} \lambda_{(i)} C_{(i)}. \tag{50}$$

Here, $\lambda_{(1)} = 1$, $\lambda_{(2)} = e^{i\Phi}$, $\lambda_{(3)} = e^{-i\Phi}$, $C_{(1)kn} = l_k l_n$, $C_{(2)kn} = s_k s_n^*$, $C_{(3)kn} = s_k s_n^*$; the projectors are idempotents $C_{(i)}^N = C_{(i)}$, N being a natural number, $TrC_{(i)} = 1$, $\det C_{(i)} = 0$. It is important to note that the decomposition of a matrix R into the series (49, 50) necessarily leads to appearance of the complex-numbered 2D basis $s_k\ s_k^*$; we will indicate similar features in the fractal technique (iii) below.

The value of the single rotation angle follows from computation of the trace of the matrix (49)

$$\Phi = 2\alpha = \arccos\left(\frac{O_{k'k} - 1}{2}\right); \tag{51}$$

antisymmetric part of the matrix yields the components of unit vector directing the rotation axis

$$l_j = i s_k s_m^* \varepsilon_{knj} = \frac{O_{k'm} \varepsilon_{kmj}}{\sqrt{(3 - O_{n'n})(1 + O_{n'n})}}. \tag{52}$$

Eqs. (51) and (52) represent parameters of the single rotations, the angle Φ. and components l_j of the vector pointing the rotation axis, as functions of an arbitrary $SO(3,C)$ rotation angles, e.g., yaw, roll, and pitch $\{\psi, \varphi, \theta\}$, and parameters of an equivalent single rotation, the value of the angle Φ and components (in the initial frame) l_j of the vector pointing the rotation axis.

5.3. Reorientation as transformation of a fractal surface, technique (iii)

In Section 4, we demonstrated that each vector of any Q-triad $\mathbf{q}_k$ is a linear combination of vector-covector direct products of its proper biorthogonal basis $(\psi^{\pm},\ \varphi^{\pm})$ belonging to a domain of complex-number valued 2D fractal space [see Eqs. (37)]. Then, rotation (reorientation) of the frame $\mathbf{q}_k$ by the technique (ii) on the base of the transformation (42) induces specific type of the "interior" rotation on the fractal surface level [see Eq. (38)]

$$\psi'^{\pm} = U\psi^{\pm}, \varphi'^{\pm} = \varphi^{\pm}U^{-1}. \tag{53}$$

Further on, we use for the dyad the eigenvectors $\psi^{\pm}$ [and eigencovectors as Hermitian conjugation of the vectors $\varphi^{\pm} = \left(\psi^{\pm}\right)^{T}$] of $\mathbf{q}_3$ of any Q-triad, where respective eigenvalues being $\pm i$. In the simplest case of $\mathbf{q}_3$ from Eq. (7), the constant dyad is

$$\psi^{+} = \begin{pmatrix} 0 \\ 1 \end{pmatrix}, \varphi^{+} = (0 \quad 1), \psi^{-} = \begin{pmatrix} 1 \\ 0 \end{pmatrix}, \varphi^{-} = (1 \quad 0). \tag{54}$$

Normalization and orthogonality conditions are identically satisfied. The matrix U, as a quaternion (46), is expressible in terms of the fractal basis

$$\begin{aligned} U &= \cos\alpha + l_n \mathbf{q}_n \sin\alpha = \\ &= \cos\alpha + [-l_1 i(\psi^{+}\varphi^{-} + \psi^{-}\varphi^{+}) + l_2(\psi^{+}\varphi^{-} - \psi^{-}\varphi^{+}) + l_3 i(\psi^{+}\varphi^{+} - \psi^{-}\varphi^{-})]\sin\alpha, \end{aligned} \tag{55a}$$

$$U^{-1} = \cos\alpha - [-l_1 i(\psi^{+}\varphi^{-} + \psi^{-}\varphi^{+}) + l_2(\psi^{+}\varphi^{-} - \psi^{-}\varphi^{+}) + l_3 i(\psi^{+}\varphi^{+} - \psi^{-}\varphi^{-})]\sin\alpha. \tag{55b}$$

Therefore, Eq. (53) takes the form

$$\psi'^{+} = (\cos\alpha + i l_3 \sin\alpha)\psi^{+} - \sin\alpha(i l_1 + l_2)\psi^{-}, \tag{56a}$$

$$\psi'^{-} = \sin\alpha(-i l_1 + l_2)\psi^{+} + (\cos\alpha - i l_3 \sin\alpha)\psi^{-}, \tag{56b}$$

$$\varphi'^{+} = (\cos\alpha + i l_3 \sin\alpha)\varphi^{+} - \sin\alpha(i l_1 + l_2)\varphi^{-}, \tag{56c}$$

$$\varphi'^{-} = \sin\alpha(i l_1 + l_2)\varphi^{+} + (\cos\alpha + i l_3 \sin\alpha)\varphi^{-}. \tag{56d}$$

Eq. (56) shows that the nonlinear problem formulated within the technique (ii), on the fractal surface level, is reduced to a linear task of the 2D basis rotation.

To get technological formulas convenient for fast numerical computation, we denote the final values of the new 2D basis as

$$A \equiv \cos\alpha + i l_3 \sin\alpha, \quad B \equiv \sin\alpha(i l_1 + l_2). \tag{57}$$

Then, we notice that only one new dyad vector is to be computed,

$$\psi'^{+} \equiv A\psi^{+} - B\psi^{-}. \tag{58a}$$

The second vector ψ'^{-} and the co-vectors are simply expressed through the factors (57) and their complex conjugation

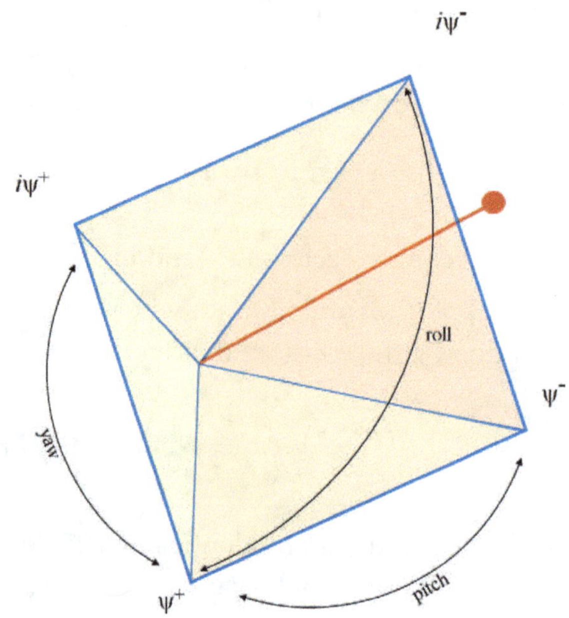

Figure 1. Fractal "joystick tool".

$$\psi'^- = B^*\psi^+ + A^*\psi^-, \varphi'^+ \equiv A^*\varphi^+ - B^*\varphi^-, \varphi'^- = B\varphi^+ + A\varphi^-. \tag{58b}$$

This helps to represent the 3D reorientation processes "subgeometrically", on the 2D fractal level, as a displacement of a "joystick" tool (see [12] and **Figure 1**).

2D complex-numbered space can be imaged as a pyramid (with no base) consisting of one real, one imaginary, and two mixed real-imaginary joined surfaces. The joystick has one of its end matched with the pyramid's top by a hinge; a certain shift of the stick gives components of a new dyad vectors and co-vectors. From these fractal elements, a new Q-frame providing the assigned reorientation of the spacecraft is straightforwardly built.

All reorientation parameters providing operations in the fractal space are in fact the components of the matrix $U \in SU(2)$; therefore, the unit vector directing the axis of instant rotation is given by Eq. (52); the fractal rotation angle is

$$\alpha = \arccos\left(\frac{\sqrt{1 + O_{k'k}}}{2}\right). \tag{59}$$

Eqs. (59), (52), (56), and (37) suggest a very simple algorithm for computation of all parameters of a single rotation and resulting matrices of a reoriented Q-triad describing new orientation of a spacecraft.

The technological scheme of the reorientation procedure can be briefly outlined as the following steps:

- A spacecraft reorientation is assigned by a series of simple rotations [Eq. (40)].
- Components of the rotation axis vector are computed [Eq. (52)].

- The angle of fractal rotation is computed [Eq. (59)].
- The dyad and resulting Q-triad are computed [Eqs. (56), (37), much simpler than in Eq. (49)].
- If the computed and assigned frames match, then the rotation parameters are sent to the operational systems realizing the reorientation.

The study suggested in Section 5 gives detailed analysis of math mechanisms linking two different approaches to solution of an object's reorientation task, a consequent 3D rotations described by matrices and a single rotation about an instant axis described by matrices. We like to emphasize importance (and original form) of Eqs. (48) and (49) explicitly demonstrating the projector-eigenvalue decomposition of any $SO(3,R)$ matrix, so immediately giving technological values of the single rotation. Another novel math feature of the problem is its connection with subgeometric properties of a fractal complex number surface.

However, thorough analysis of the Q-math reveals its additional, and important, option quite helpful in operational tasks. Namely, extension of the groups $SO(3,R)$ and $SU(2)$ to the rotations with complex parameters, $SO(3,C)$ and $SL(2,C)$, respectively, with the vector-quaternion version of relativity theory taken into account, may open a possibility not only reorient but as well simultaneously endow a spacecraft with velocity assigned in value and direction. Apparently, this math tool matching rotations and accelerations, if possible in 3D space, should exist as fractal mechanism. Designing of such original (and exotic) operational instrument is a challenging task; it is in detail analyzed in the next section.

6. Hyperbolic rotations and a fractal pyramid

In this section, we essentially extend the methods briefly described above. The crucial point of the extension is introduction of an imaginary parameter of rotation, thus involving hyperbolic functions. We assume that this action will result in possibility to control not only orientation, but as well dynamics of the spacecraft. We will prove the assumption within extended formulation of the technique (iii).

But at first, to make the picture more clear, we show it in framework of 3D serial rotations [technique (i)], and for simplicity, we implement just one supplement plane hyperbolic rotation about one axis

$$O_3^{i\eta} = \begin{pmatrix} \cos(i\eta) & -\sin(i\eta) & 0 \\ \sin(i\eta) & \cos(i\eta) & 0 \\ 0 & 0 & 1 \end{pmatrix} = \begin{pmatrix} \cosh\eta & -i\sinh\eta & 0 \\ i\sinh\eta & \cosh\eta & 0 \\ 0 & 0 & 1 \end{pmatrix} \equiv H_3^{\eta}, \tag{60}$$

so that hyperbolic functions are introduced. Then, complete rotational operator is

$$O = H_3^{\eta} R \tag{61}$$

We rewrite the operator (61) in the spinor-type form where the tilde denotes some initial basis

$$U = \left(\cosh\frac{\eta}{2} - i\sinh\frac{\eta}{2}\mathbf{q}_3\right)\left(\cos\frac{\Phi}{2} + \sin\frac{\Phi}{2} l_{\tilde{k}}\mathbf{q}_{\tilde{k}}\right), \tag{62}$$

and the components of the instant rotation axis vector given by Eq. (52). It is important to note that in the computation procedure, we have to deal with vectors belonging to the same frame. Therefore, we express $\mathbf{q}_{3'} = R_{3n}\mathbf{q}_n$ and make multiplication in Eq. (62) to obtain

$$U = \cosh\frac{\eta}{2}\cos\frac{\Phi}{2} - i\sinh\frac{\eta}{2}R_{3\tilde{n}}l_{\tilde{n}} + \left[\cosh\frac{\eta}{2}\sin\frac{\Phi}{2}l_{\tilde{n}} - i\sinh\frac{\eta}{2}\left(\cos\frac{\Phi}{2}R_{3\tilde{n}} + \sin\frac{\Phi}{2}R_{3\tilde{j}}l_{\tilde{m}}\varepsilon_{jmn}\right)\right]\mathbf{q}_n. \tag{63}$$

This expression is again a quaternion and we denote it as

$$U = \cos\Theta + (\sin\Theta)\mathbf{q},\, U^{-1} = \cos\Theta - (\sin\Theta)\mathbf{q} \tag{64}$$

where

$$\cos\Theta \equiv \cos\frac{\eta}{2}\cos\frac{\Phi}{2} - i\sinh\eta\sin\frac{\Phi}{2}R_{3\tilde{n}}l_{\tilde{n}}, \tag{65}$$

$$\sin\Theta\mathbf{q} \equiv \left[\cosh\frac{\eta}{2}\sin\frac{\Phi}{2}l_{\tilde{n}} - i\sinh\frac{\eta}{2}\left(\cos\frac{\Phi}{2}R_{3\tilde{n}} + \sin\frac{\Phi}{2}R_{3\tilde{n}}l\tilde{m}\varepsilon_{jmn},\right)\right]\mathbf{q}_{n'} \tag{66}$$

parameter Θ being a complex number. One straightforwardly verifies fulfilling the identity

$$\begin{aligned}\cos^2\Theta + (\sin^2\Theta)\mathbf{q}^2 = &\left(\cosh\frac{\eta}{2}\cos\frac{\Phi}{2} - i\sinh\eta\sin\frac{\Phi}{2}R_{3\tilde{n}}l\tilde{n}\right)\left(\cosh\frac{\eta}{2}\cos\frac{\Phi}{2} - i\sinh\eta\sin\frac{\Phi}{2}R_{3\tilde{p}}l\tilde{p}\right) + \\ &+\left[\cosh\frac{\eta}{2}\sin\frac{\Phi}{2}l\tilde{n} - i\sinh\frac{\eta}{2}\left(\cos\frac{\Phi}{2}R_{3\tilde{n}} + \sin\frac{\Phi}{2}R_{3\tilde{j}}l\tilde{m}\varepsilon_{jmn}\right)\right]\times \\ &\times\left[\cosh\frac{\eta}{2}\sin\frac{\Phi}{2}l\tilde{n} - i\sinh\frac{\eta}{2}\left(\cosh\frac{\Phi}{2}R_{3\tilde{n}} + \sin\frac{\Phi}{2}R_{3\tilde{l}}l\tilde{p}\varepsilon_{lps}\right)\right]\mathbf{q}\tilde{n}\mathbf{q}\tilde{s} = 1.\end{aligned} \tag{67}$$

Expression for the vector-directing axis of the single rotation is found from Eqs. (65) and (66)

$$l_n = \frac{\cosh\frac{\eta}{2},\, \sin\frac{\Phi}{2}l_{\tilde{n}}, -i,\, \sinh\frac{\eta}{2},\left(\cos\frac{\Phi}{2}R_{3\tilde{n}} + \sin\frac{\Phi}{2}R_{3\tilde{j}}l\tilde{m}\varepsilon_{jmn}\right.}{\sqrt{1 - (\cosh\frac{\eta}{2}\cos\frac{\Phi}{2} - i\sinh\eta\sin\frac{\Phi}{2}R_{3\tilde{n}}l_{\tilde{n}})(\cosh\frac{\eta}{2}\cos\frac{\Phi}{2} - i\sinh\eta\sin\frac{\Phi}{2}R_{3\tilde{p}}l_{\tilde{p}})}. \tag{68}$$

Eq. (61) represents an operator performing the serial rotation, and Eqs. (65), (68) give parameters of a single rotation. Physical content of this rotation is easily revealed when the mapping is made in the fractal surface format, and then returned into 3D space. Despite seeming complexity of the given expressions, the final calculation is shown to be very simple.

So, following the ideology of geometrization of the algebraic actions, we plunge into the fractal medium, and we consider the technique (iii). We rewrite fractal mapping with the operator (62) in the form

$$\psi^{\pm} = U\psi'^{\pm} = \left(\cosh\frac{\eta}{2} - i\sinh\frac{\eta}{2}\mathbf{q}_3\right)\psi'^{\pm} \tag{69}$$

where the intermediate dyad is a result of the real rotation (similar with the covectors)

$$\psi'^{\pm} = \left(\cos\frac{\Phi}{2} + \sin\frac{\Phi}{2}l_k\mathbf{q}_k\right)\tilde{\psi}^{\pm}. \tag{70}$$

We also stress that all dyad elements used in the computations are always the eigenvectors (eigencovectors) of the quaternion unit $\mathbf{q}_3$

$$\mathbf{q}_3\psi^+ = +i\psi^+, \quad \mathbf{q}_3\psi^- = -i\psi^+, \quad \varphi^+\mathbf{q}_3 = +i\varphi^+, \quad \varphi^-\mathbf{q}_3 = -i\varphi^-; \tag{71}$$

hence, Eq. (69) produces a new fractal basis simply multiplying the intermediate dyad by an exponent

$$\psi^+ = \left(\cosh\frac{\eta}{2} + \sin\frac{\eta}{2}\right)\psi'^+ = e^{\eta/2}\psi'^+, \ \psi^- = e^{-\eta/2}\psi'^-, \ \varphi^+ = e^{-\eta/2}\varphi'^+, \ \varphi^- = e^{\eta/2}\varphi'^-. \tag{72}$$

By other words, one dyad vector and one co-vector (here $\psi^+$and φ^-) become longer, and the others (ψ^- and φ^+) become shorter, all of them though preserving unit length, i.e., rescaled.

This primitive mapping has clear physical sense concerning kinematic of a spacecraft. To reveal it, we, using Eq. (75), build an "imaginary constituent" of the 3D frame vector $\mathbf{q}_{1'}$ as in Eq. (37b).

$$\mathbf{q}_{1'} = -i\left(\psi'^+\varphi'^- + \psi'^-\varphi'^+\right) = -i(e^{\eta}\psi^+\varphi^- + e^{-\eta}\psi^-\varphi^+). \tag{73}$$

However from Eqs. (37b, c), we find

$$\psi^+\varphi^- = \frac{1}{2}(i\mathbf{q}_1 + \mathbf{q}_2), \ \psi^-\varphi^+ = \frac{1}{2}(i\mathbf{q}_1 - \mathbf{q}_2); \tag{74}$$

substitution of the Eq. (74) into Eq. (73) yields

$$i\mathbf{q}_{1'} = \cosh\eta\left(i\mathbf{q}_1 + \tanh\eta\ \mathbf{q}_2\right). \tag{75a}$$

Eq. (75a) rewritten in terms of the Pauli-type matrices [as in Eqs. (20), (22)] $\mathbf{p} \equiv i\mathbf{q}$ has the form

$$\mathbf{p}_{1'} = \cosh \eta (\mathbf{p}_1 + \tanh\eta \ \mathbf{q}_2). \tag{75b}$$

Using results of Section 3, we associate the hyperbolic functions with the time ratio

$$\cosh \eta = \frac{dt}{dt'} \tag{76}$$

(linking time dt of an immobile frame and proper time dt' of moving spacecraft) and with the relative velocity ratio (c is speed of light).

$$\tanh\eta = V/c. \tag{77}$$

Then, Eq. (75b) takes the form of "vector interval" of quaternion version of relativity theory (23)

$$dt' \ \mathbf{p}_{1'} = dt \left(\mathbf{p}_1 + \frac{V}{c} \ \mathbf{q}_2 \right); \tag{78}$$

when squared, it gives the spacecraft's special relativistic space-time interval linked with the frame at rest by the Lorentz (hyperbolic) transformation

$$dt'^2 = dt^2 \ \left(1 - V^2/c^2\right) \tag{79}$$

describing kinematics of a frame moving along $\mathbf{q}_2$ with velocity V, while the vectors $\mathbf{p}_1$(or $\mathbf{p}_{1'}$) play the role of direction of time in the immobile (or moving) spacecraft. It is always possible to choose the direction $\mathbf{q}_2$ as pointing the "yaw" of a spacecraft. In particular, the velocity can be small sufficiently to reduce the calculations into classical format

$$V/c = \tanh\eta \approx \eta \tag{80}$$

besides, the velocity modulus may be variable in time; hence, the spacecraft is accelerated.

So, introducing imaginary rotation angles, we obtain a possibility to control an arbitrary space reorientation of a spacecraft with variation of its velocity in the direction that can be as well changing with time (In this sample, the vector $\mathbf{q}_2$ is in fact permanently rotating.)

This math tool has two important properties. First, a spacecraft endowed by the tool with a velocity is initially described as a relativistic system; one comes to the classical mechanics considering the hyperbolic parameter small. Second, the tool accelerates the spacecraft always in the direction of the frame vector appointed to indicate "yaw"; if this vector rotates, changing the yaw, the acceleration arrow changes with it; i.e., the spacecraft is accelerated along a curve line. These properties can be useful in real motion control.

On the 2D fractal level, the spacecraft's more complex 3D motion comprising reorientation and acceleration is accompanied by respective rotation and deformation of the mentioned above fractal pyramid. Here, this subgeometric image of the math instrument necessarily enriches a simpler model of the joystick, and moreover, to make the picture symmetric, we show positive and negative directions of the pyramid (see **Figure 2**).

Computations providing the spacecraft's reorientation and acceleration are performed on the fractal level by Eq. (58) with the functions *A*, *B* generalized as

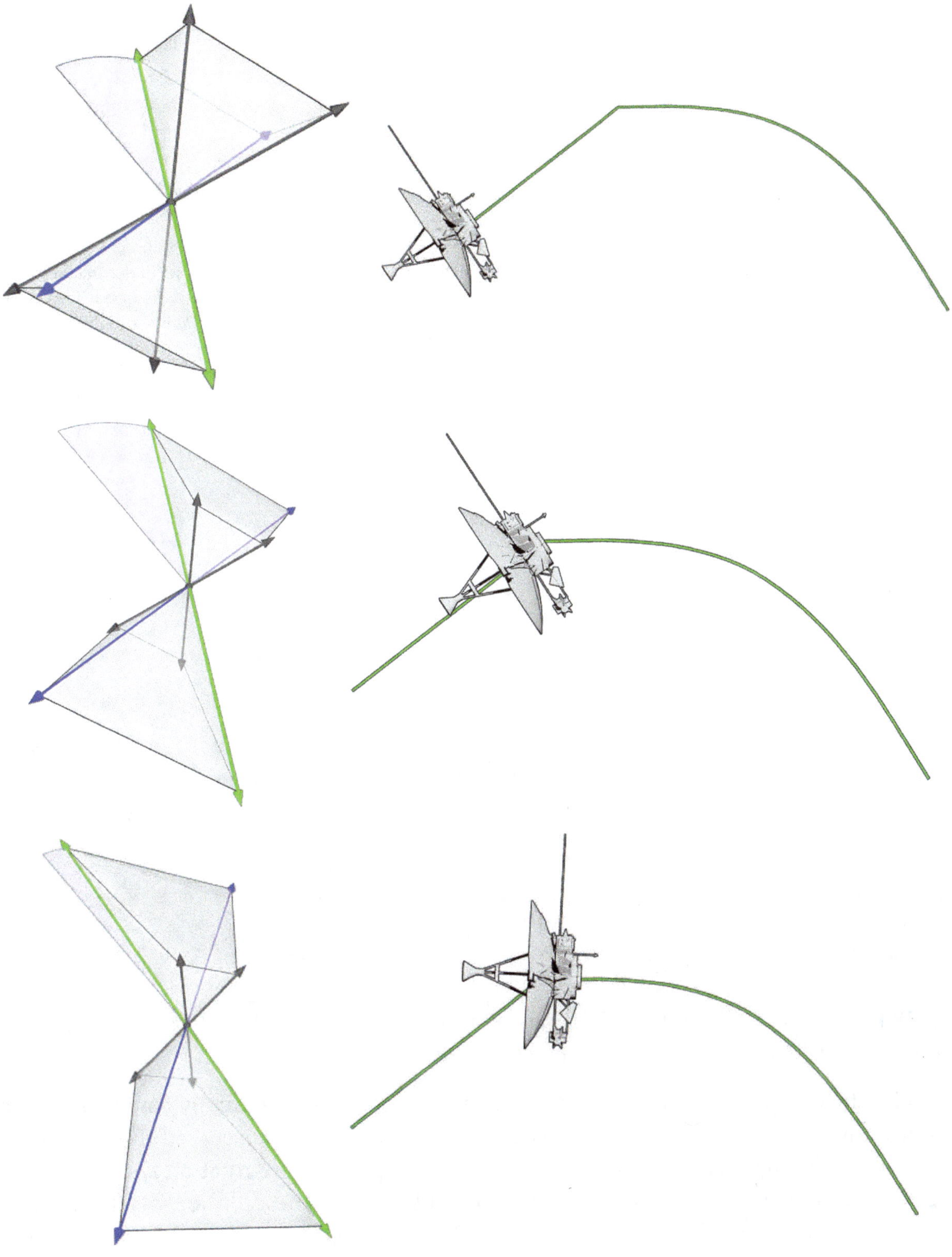

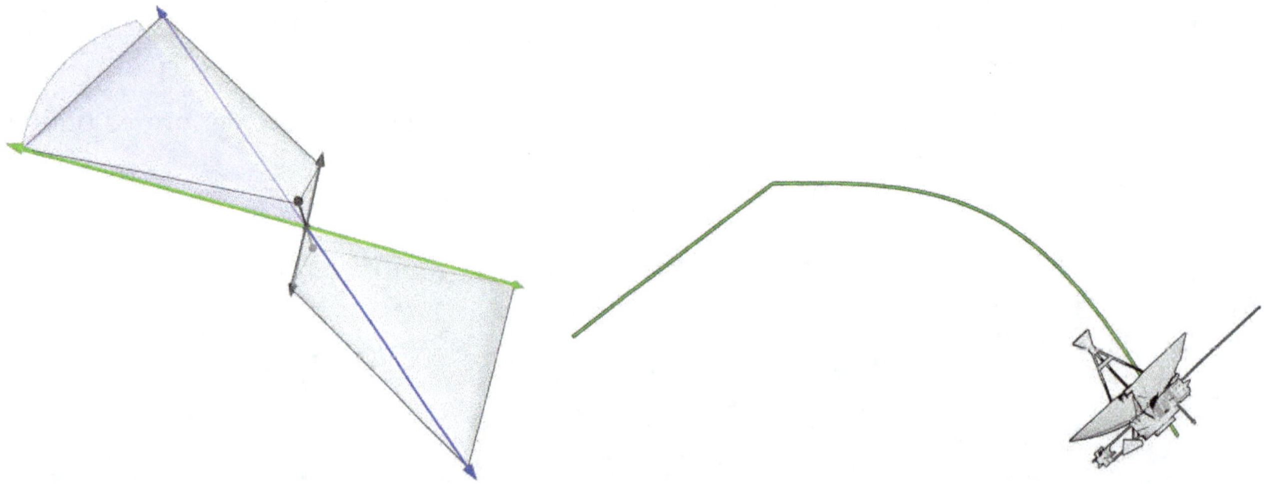

Figure 2. Case (a): The spacecraft performs a 3D rotation, the pyramid is tilted by respective halfangle. Rotations and displacements of a spacecraft (Pioneer-10) accompanied by respective 2D rotations and deformations of the fractal pyramid. Case (b): The reoriented spacecraft rectilinearly moves with some velocity, and the tilted pyramid is distorted: Two its edges become shorter, and the other two edges become longer. Case (c): The spacecraft ("frees-framed") is reoriented by another angle, and the distorted pyramid as tilted by respective halfangle. Case (d): The spacecraft moves along a curve trajectory with changing velocity (accelerated), and the pyramid is subject to permanent respective tilt and distortion.

$$A \equiv \left(\cos \frac{\Phi}{2} + i l_3 \sin \frac{\Phi}{2} \right) e^{\eta/2}, \; B \equiv \sin \frac{\Phi}{2} \; (i \; l_1 + l_2) e^{-\eta/2}, \tag{81a}$$

with hyperbolic conjugation $((\oplus : \; e^{\pm\eta/2} \rightarrow e^{\mp\eta/2}))$, similar to the complex conjugation, introduced, e.g.,

$$A^{\oplus} \equiv \left(\cos \frac{\Phi}{2} + i l_3 \sin \frac{\Phi}{2} \right) e^{-\eta/2}, \; B^{*\oplus} \equiv \sin \frac{\Phi}{2} \; (-i \; l_1 + l_2) e^{\eta/2}, \tag{81b}$$

where vector l_k directs axis of the single space rotation by angle Φ. Then (as in Section 5), only one equation is to be solved, e.g., that determining the dyad vector

$$\psi'^{+} = A \; \psi^{+} - \; B \; \psi^{-}, \tag{82a}$$

and rest of the dyad elements is found by primitive math actions

$$\psi'^{-} A^{*}\psi^{+} + B^{*}\psi^{-}, \; \varphi'^{+} = A{*}^{\oplus} \; \varphi^{+} - B{*}^{\oplus} \; \varphi^{-}, \; \varphi'^{-} = A^{\oplus}\varphi^{+} + B^{\oplus}\varphi^{-}. \tag{82b}$$

Eqs. (82), (37) immediately give expressions of all spacecraft's frame vectors, thus solving the reorientation and acceleration problem in explicit form.

One straightforwardly finds that use of the fractal technique (iii) essentially simplifies computation procedures. In paper [13], we compare math difficulty of the discussed three techniques in solution of the simple problem of the spacecraft's one-plane space rotation and acceleration. It is demonstrated there that the techniques (i) and (ii) demand solution of at least seven equations, among them are matrix equations, while the fractal technique (iii) suggests solution of only four relatively simple algebraic equations.

7. Technological scheme and concluding remarks

A sketch of technological scheme aimed to realize mixed rotation-acceleration maneuver of a spacecraft can be suggested as the following consequence of actions fit for any mentioned above approach.

- The initial and final parameters of reorientation and acceleration are assigned and memorized.
- Parameters as functions of time must be determined and input.
- Time intervals are divided into standard steps (quantized), the standard input.
- Process of computation of quantum steps starts resulting in obtaining of a series of related parameter values describing the orientation and velocity of the spacecraft's frame.
- The data of each step are transmitted to the systems changing the spacecraft orientation and velocity until the assigned values are achieved.

And we emphasize two most important results of this study.

First, we succeeded to show that an extrarotation by an imaginary angle entails endowing a spacecraft with a (relativistic) velocity, hence in addition to reorientation, to accelerate it. This math observation seems to be a novel one since no similar information is met in related literature.

Second, we show that the most mathematically economical way to compute operational parameters needed for realization of the maneuver is to utilize the "fractal pyramid" technique (definitely a new tool) comprising minimal number of math actions, where major of them are simple algorithms, other approaches having no such advantages.

Author details

Alexander P. Yefremov

Address all correspondence to: a.yefremov@rudn.ru

Institute of Gravitation and Cosmology, RUDN University, Moscow, Russia

References

[1] Halijak Ch.A. Quaternions Applied to Missle Systems, US Army Missle Research and Development Command, AD A 052232, Redstone Arsenal, Alabama 35809, Apr, 18, 1978 (Approved for public release; distribution unlimited)

[2] Branets VN, Shmyglevskiy IP. Application of Quaternions to Rigid Body Rotation Problems, (translated from the Russian in 1974). Washington D.C.: Scientific Translation Service, National Aeronautics and Space Administrationp. 352, (1973. Report No.NASA-TT-F-15414)

[3] Hamilton WR The Mathematical Papers of William Rowan Hamilton. Vol. 3. Cambridge: Cambridge University Press; 1967

[4] Yefremov AP. Quaternions and biquaternions: Algebra, geometry and physical theories. Hypercomplex Numbers in Geometry and Physics. 2004;**1**:104-119, arXiv: math-ph/050 1055

[5] Yefremov AP. Bi-quaternion square roots, rotational relativity, and dual space-time intervals. Gravitation & Cosmology. 2007;**133**(51):178-184

[6] Yefremov AP. Quaternion model of relativity: Solutions for non-inertial motions and new effects. Advanced Science Letters. 2008;**1**:179-186

[7] Wheeler JA. Pregeometry: Motivations and prospects. In: Marlov AR, editor. Quantum Theory and Gravitation. New York: Academic Press; 1080. pp. 1-11

[8] Yefremov AP. Splitting of 3D quaternion dimensions into 2D-cells and a "world screen technology". Advanced Science Letters. 2012;**5**(1):288-293

[9] Yefremov AP. Fundamental properties of quaternion Spinors. Gravitation and Cosmology. 2010;**16**(2):137-139

[10] Yefremov AP. The conic-gearing image of a complex number and a spinor-born surface geometry. Gravitation and Cosmology. 2011;**17**(1)

[11] Lancaster P, Tismenetsky M. The Theory of Matrices with Applications. 2nd ed. San Diego, London: Academic Press; 1985. p. 154

[12] Yefremov AP. Fractal Surface as the Simplest Tool to Control Orientation of a Spacecraft. Acta Astronautica. 2016;**129**:174

[13] Yefremov AP. New fractal math tool providing simultaneous reorientation and acceleration of spacecraft. Acta Astronautica. 2017;**139**:481-485

10

Consensus-Based Attitude Maneuver of Multi-Spacecraft with Exclusion Constraints

Innocent Okoloko

Abstract

Some space missions involve cooperative multi-vehicle teams, for such purposes as interferometry and optimal sensor coverage, for example, NASA Terrestrial Planet Finder Mission. Cooperative navigation introduces extra constraints of exclusion zones between the spacecraft to protect them from damaging each other. This is in addition to external exclusion constraints introduced by damaging or blinding celestial objects. This work presents a quaternion-based attitude consensus protocol, using the communication topology of the team of spacecraft. The resulting distributed Laplacians of their communication graph are applied by semidefinite programming (SDP), to synthesize a series of time-varying optimal stochastic matrices. The matrices are used to generate various cooperative attitude maneuvers from the initial attitudes of the spacecraft. Exclusion constraints are satisfied by quaternion-based quadratically constrained attitude control (Q-CAC), where both static and dynamic exclusion zones are identified every time step, expressed as time-varying linear matrix inequalities (LMI) and solved by semidefinite programming.

Keywords: attitude maneuvre, consensus, exclusion, optimization, LMI

1. Introduction

Some current space missions already demanded the deployment of teams of spacecraft which cooperate synergistically for such purposes as interferometry and sensor coverage [1, 2]; and many future missions will. Activities such as interferometry and sensor coverage require cooperative attitude control (AC)—the process of making a team of spacecraft, for example, satellites to point toward a specific direction of interest. This makes attitude control an essential part of space missions [3]. Apart from spacecraft, AC is also important in the navigation of aircraft and robots; therefore, it has been studied extensively in the literature, for example [4–11].

Basically, AC is a challenging problem, which becomes more challenging when multiple spacecraft are involved, in highly dynamic environments, and subject to external constraints such as blinding celestial objects such as the sun or some bright stars, which can damage onboard sensitive instruments. In addition, because of the close packing of spacecraft in a team, each of which has protruding appendages (e.g. thrusters and antennae), they must be careful with each other when changing attitude, in order to avoid collision with each other. When there is such a team of networked spacecraft which can communicate, then consensus theory based on graph Laplacians can be applied to achieve cooperation among them [12, 13].

The most common method of representing spacecraft attitude dynamics is by *unit quaternions*, mainly because quaternions do not encounter the singularities associated with other representations such as Euler angles and the *Modified Rodriques Parameters* (MRP). However, the non-linearity of quaternion dynamics makes it difficult to apply Laplacian-like dynamics directly to quaternions.

We shall now consider some previous work on constrained attitude control (CAC). A brief survey of the main method attitude representation is in [4]. Ref. [5] considers quadratically constrained attitude control (Q-CAC), where the exclusion problems are formulated as a quadratic optimization problem and solved using linear matrix inequalities (LMIs) and semidefinite programming (SDP). It was solved for a single-spacecraft single obstacle in [5] and for two spacecraft in [6]. In [7] an attempt was made to extend [5, 6] to more than two spacecraft and obstacles. In [7–10], was extended to multiple spacecraft multiple obstacles in *different coordinate frames* (as the case of real spacecraft will be). An attempt was made in [11] to reduce the control torques required for effective attitude stabilization from three to two. This is applicable to underactuated spacecraft. [12] applies a consensus-based approach to distributed attitude alignment of a team of communicating spacecraft flying in formation, while [14] applies a Laplacian-based protocol to *leader-follower* attitude control of a team of spacecraft using the modified Rodriquez parameters.

Among the plethora of AC algorithms, only our works [7–10] apply consensus theory directly to quaternions, and only [5–10] tackle the problem of avoidance constraints. In addition, among the works [5–10] only [8–10] were developed for spacecraft in *different coordinate frames*, which has direct practical implementation. The contributions of this chapter are therefore aspects of our previous works [7–10], which include the following: (i) the development of a quaternion consensus protocol, (ii) incorporating dynamic avoidance constraints into the consensus framework using Q-CAC, (iii) mathematical convergence analysis for the quaternion-based consensus framework and (iv) solving the problem for the realistic scenario of multiple spacecraft in different coordinate frames, thus making it more suitable for practical implementation.

Note: the words *obstacle, avoidance, exclusion* and *exclusion vector* may be used interchangeably in this chapter. **Table 1** lists frequently used notation in this chapter.

2. Problem statement

The problem of multi-spacecraft attitude control with avoidance constraints can be stated as follows. Given the initial positions $x^i(t_0) \in \mathbb{R}^3$ $i = 1 \cdots n$, initial attitudes represented by quaternions

$q^i(t_0)$, of a set of communicating spacecraft SC_i, generate a sequence of attitude consensus trajectories that drive the team to a consensus attitude $q(t_f)$ while satisfying avoidance and *norm* constraints.

The problem stated above consists of two parts: *consensus* and *avoidance*. For the consensus problem, it is desired to drive the attitudes of all SC_i to a collective consensus attitude or to various *formation attitudes*. Consensus attitude means that each SC_i should eventually point to the same direction, which is the average of the initial quaternions. Formation attitudes means SC_i should finally point to various patterns, for example, each spacecraft can point at 5° away from each other about the z-axis. This we developed by introducing *relative offset quaternions* in the consensus framework. The second problem, *avoidance* constraints, is also important, because SC_i usually have appendages, for example, some SC_i have *thrusters* that emit hot plumes (plume impingement), and some have instruments that can be damaged by blinding celestial objects or by the appendage of another team member.

However, the ordinary consensus protocol violates the non-linearity of quaternion kinematics and the quaternion norm preserving requirement and therefore cannot be applied directly with quaternion dynamics. Also, the protocol ordinarily does not solve the problem of collision avoidance in *adversarial* situations. Thus, this chapter consists of aspects of our previous works [7–10], where we developed a consensus theory of quaternions, augmented with Q-CAC-based collision avoidance mechanisms. We employed an optimization approach and cast the problems as a *semidefinite program* (SDP), augmented with some convex quadratic constraints (avoidance), written as *linear matrix inequalities* (LMI). The quaternion consensus protocol computes consensus attitude trajectories each time step, and the Q-CAC avoidance procedure decides which of the computed trajectories are safe to follow or not. Unsafe trajectories are discarded, and a new set of quaternion vectors that avoid collision is generated. The cycle repeats until consensus is achieved.

To understand the avoidance (exclusion) problem, let us illustrate with a simpler single-SC_i single-obstacle scenario as shown in **Figure 1**. In the figure, the SC_i must avoid (exclude) the Sun while rotating a photosensitive instrument from q_0 to q_f.

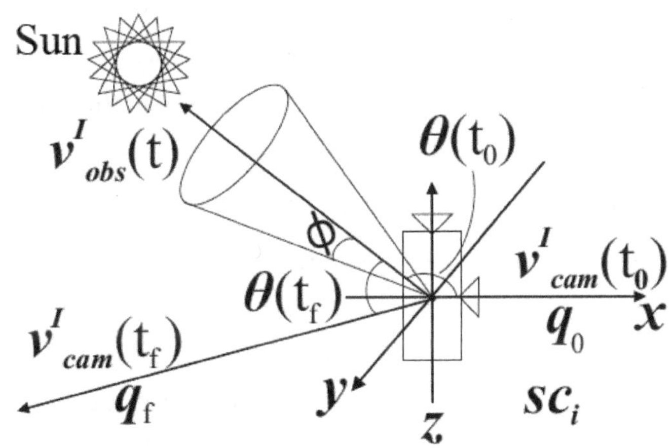

Figure 1. Constrained attitude control problem for a single-*sc* single-exclusion scenario. SC_i must avoid (exclude) the Sun while rotating a photosensitive instrument from q_0 to q_f.

Let $v^I_{cam_i}(t)$ denote the unit camera vector in $\mathcal{F}^I_{SC_i}$ corresponding to the SC_i's attitude q^i (as defined in **Table 1**), and let $v^I_{obs_i}(t)$ be the attitude quaternion of the obstacle to be avoided (in this case the Sun). Exclusion requires the time evolution of camera vector $v^I_{cam_i}$ from $v^I_{cam_i}(t_0)$ to $v^I_{cam_i}(t_f)$ to avoid $v^I_{obs_i}(t)$ all times with a minimum angular separation of $\varnothing$. The requirement is.

$$\theta(t) \geq \varnothing \tag{1}$$

or

$$\begin{aligned} & v^I_{cam_i}(t)^T v^I_{obs_i}(t) \leq \cos\varnothing, \\ & \forall t \in [t_0, t_f] \end{aligned} \tag{2}$$

The constraint is a *non-convex quadratic* constraint; it was *convexified* in [4], which made it possible to be represented as a LMI using the quaternion attitude constraint formulation developed in [3] for a single-spacecraft single-obstacle scenario. In [4], v^I_{obs} was static, while $v^I_{cam_i}(t)$ was evolving; both vectors were in the same coordinate frame. Although solving it in the same coordinate frame somewhat simplified the solution, it was not suitable for practical implementation because, in reality, the obstacle and spacecraft operate in different coordinate frames. Next, we present the basic mathematical preliminaries.

Notation	Meaning
SCi, SC_i	Spacecraft i
q^i	Attitude quaternion vector of SCi, SC_i, $q^i = [q_1\ q_2\ q_3 \mid q_4]^T$
q^{-i} or q^{i*}	Conjugate of q^i
$\overline{q}^i$	Vector part of q^i, $\overline{q}^i = [q_1\ q_2\ q_3]^T$
$\overline{q}^{i\times}$	Antisymmetric of q^i
$\boldsymbol{q}$	Stacked vector of more than one quaternion vectors
q^{off}	Stacked vector of more than one offset quaternion vectors
Ω, Π	Quaternion dynamics plant matrix
$\mathbf{P}$	Quaternion dynamics Laplacian-like plant matrix
ω	Angular velocity
τ	Control torque
J	Inertia matrix
$\mathbf{L}$	Laplacian matrix
$\mathbf{P}$	Laplacian-like stochastic matrix
$\mathbf{I}_n$	Then $n \times n$ identity matrix
S^m	The set of $m \times m$ positive definite matrices

Notation	Meaning
$\tilde{A}$	Cone avoidance constraint matrix
$\mathcal{R}^i$	Rotation matrix corresponding to q^i
$\mathcal{F}^I_{SC_i}$	Fixed coordinate (Inertial) frame with origin at SC_i's center
$\mathcal{F}^B_{SC_i}$	Rotational coordinate (Body) frame with origin at SC_i's center
$v^B_{obs_i}$	Vector of obstacle in $\mathcal{F}^B_{SC_i}$
$v^I_{obs_i}$	Vector of obstacle in $\mathcal{F}^I_{SC_i}$
$v^I_{obs_i\text{-}j}$	Vector of the jth obstacle in $\mathcal{F}^I_{SC_i}$
$v^B_{cam_i}$	Vector of the SC_i's camera in $\mathcal{F}^B_{SC_i}$
$v^I_{cam_i}$	Vector of the SC_i's camera in $\mathcal{F}^I_{SC_i}$
$\otimes$	Kronecker multiplication operator
$\odot$	Quaternion multiplication operator
$\ominus$	Quaternion difference operator
t_0	Initial time
t_f	Final time
x^i	Position vector of SCi, SC_i
x	Stacked vector of n position vectors
$(x^{ij})^{off}$	Offset vector between i and j
x^{off}	Stacked vector of n offset vectors
$\mathcal{C}$	The consensus space for $\mathbf{q}$, $\mathcal{C} = \{\mathbf{q} \vert q^1 = q^2 =, \cdots, = q^n\}$

Table 1. Frequently used notations in this chapter.

3. Mathematical background

In this section, we shall briefly consider the two basic mathematical theories relevant to this chapter. More comprehensive study and analysis are in [10].

3.1. Quaternion-based rotational dynamics

Because quaternions are free from the problems of singularities inherent in Euler angles and most other ways of representing rotations, it is convenient to use unit quaternions to represent the attitude of a rigid body rotating in three-dimensional space (such as spacecraft or satellite) [15]. The quaternion is a four-element vector:

$$q = [q_1\ q_2\ q_3|q_4]^T. \tag{3}$$

Here, $[q_1\ q_2\ q_3]^T$ is the vector part, representing the axis of rotation in the Cartesian (x, y, z) coordinates, and q_4 is a scalar part, representing the angle of rotation of the quaternion in degrees. The difference between two quaternions q^1 and q^2 can be represented in multiplication terms as.

$$\begin{aligned} q^d &= q^1 \odot q^{-2} = q^1 \odot [-q_1^2 - q_2^2 - q_3^2 - q_4^2]^T \\ &= Q^2 q^1, \end{aligned} \tag{4}$$

where q^{-2} is the *conjugate* of q^2. We used $\odot$ here as a quaternion multiplication operator. And Q^2 is defined as

$$Q^i = \begin{bmatrix} q_4^i & q_3^i & -q_2^i & -q_1^i \\ -q_3^i & q_4^i & q_1^i & -q_2^i \\ q_2^i & -q_1^i & q_4^i & -q_3^i \\ q_1^i & q_2^i & q_3^i & q_4^i \end{bmatrix} \tag{5}$$

Eq. (4) means that q^d is the rotation quaternion that originally transformed q^1 to q^2 or, alternatively, q^d is a rotation quaternion that can transform q^1 to q^2.

The rotational dynamics for the ith quaternion is.

$$q^i = \frac{1}{2}\Omega^i q^i = \frac{1}{2}\Pi^i \omega^i \tag{6}$$

where

$$\Omega^i = \begin{bmatrix} 0 & \omega_3^i & -\omega_2^i & \omega_1^i \\ -\omega_3^i & 0 & \omega_1^i & \omega_2^i \\ \omega_2^i & -\omega_1^i & 0 & \omega_3^i \\ -\omega_1^i & -\omega_2^i & -\omega_3^i & 0 \end{bmatrix} \tag{7}$$

$$\Pi^i = \begin{bmatrix} -q_4^i & q_3^i & -q_2^i \\ -q_3^i & -q_4^i & q_1^i \\ q_2^i & -q_1^i & -q_4^i \\ q_1^i & q_2^i & q_3^i \end{bmatrix} \tag{8}$$

are the *plant* matrices of quaternion dynamics.

Euler's first-order discretization of Eq. (6) yields

$$q^i(k+1) = I_4 q^i(k) + \frac{\Delta t}{2}\Omega^i(k) q^i(k) = q^i(k) + \frac{\Delta t}{2}\Pi^i(k)\omega^i(k). \tag{9}$$

The dynamics of the rotational (angular) velocity ω^i of q^i is

$$\begin{bmatrix} \dot{\omega}_1^i \\ \dot{\omega}_2^i \\ \dot{\omega}_3^i \end{bmatrix} = \begin{bmatrix} \left((J_2^i - J_3^i)\omega_2^i\omega_3^i + \tau_1^i\right)/J_1^i \\ \left((J_3^i - J_1^i)\omega_3^i\omega_1^i + \tau_2^i\right)/J_2^i \\ \left((J_1^i - J_2^i)\omega_1^i\omega_2^i + \tau_3^i\right)/J_3^i \end{bmatrix}$$

$$= \underbrace{\begin{bmatrix} 0 & \frac{J_2^i}{J_1^i}\omega_3^i & -\frac{J_3^i}{J_1^i}\omega_2^i \\ \frac{J_3^i}{J_2^i}\omega_3^i & 0 & -\frac{J_1^i}{J_2^i}\omega_1^i \\ \frac{J_1^i}{J_3^i}\omega_2^i & -\frac{J_2^i}{J_3^i}\omega_1^i & 0 \end{bmatrix}}_{Y^i} \begin{bmatrix} \omega_1^i \\ \omega_2^i \\ \omega_3^i \end{bmatrix} + \underbrace{\begin{bmatrix} 1/J_1^i & 0 & 0 \\ 0 & 1/J_2^i & 0 \\ 0 & 0 & 1/J_3^i \end{bmatrix}}_{(J^i)^{-1}} \begin{bmatrix} \tau_1^i \\ \tau_2^i \\ \tau_3^i \end{bmatrix}. \tag{10}$$

Euler's first-order discretization of Eq. (10) is

$$\omega^i(k+1) = \left(\mathbf{I}_3 + \Delta t \Upsilon^i(k)\right)\omega^i(k) + \Delta t \left(J^i\right)^{-1}\tau^i(k), \tag{11}$$

where ω_j^i is the rotational velocity, J_j^i is the moment of inertia, and τ_j^i is the control torque, of the ith rigid body along the three principal axes $j = 1, 2, 3$. Combining Eqs. (9) and (11) in stacked vector form yields.

$$\underbrace{\begin{bmatrix} -\Delta t\left(J^i\right)^{-1} & \mathbf{I}_3 & \mathbf{0}_{3\times 4} \\ \mathbf{0}_{4\times 3} & -\frac{\Delta t}{2}\Pi^i(k+1) & \mathbf{I}_4 \end{bmatrix}}_{F^i(k)} \underbrace{\begin{bmatrix} \tau^i(k) \\ \omega^i(k+1) \\ q^i(k+2) \end{bmatrix}}_{\mathfrak{T}^i(k+1)} = \underbrace{\begin{bmatrix} \left(\mathbf{I}_3 + \Delta t \Upsilon^i(k)\right)\omega^i(k) \\ q^i(k+1) \end{bmatrix}}_{y^i(k)} \tag{12}$$

The typical task of controller synthesis is to determine the torque τ^i that stabilizes the system.

3.2. Basic consensus theory

The problem of *consensus* theory is to create distributed protocols based on *communication graphs* which can drive the states of a team of communicating agents to a common state or an agreed state. Where the agents i $(i = 1, \cdots, n)$ are represented by vertices of the communication graph; the edges of the graph are the communication links between them. Let the state of agent (vehicle) i be x^i, and $\mathbf{x}$ is the stacked vector of all the states of the vehicles. For systems modeled by first-order dynamics, the following first-order consensus protocol (or similar protocols) has been proposed, for example [16, 17]:

$$\dot{\mathbf{x}}(t) = -\mathbf{L}\left(\mathbf{x}(t) - \mathbf{x}^{off}\right). \tag{13}$$

We know that consensus has been achieved when $\|x^i - x^j\| \rightarrow (x^{ij})^{off}$ as $t \rightarrow \infty$, $\forall i \neq j$. A more comprehensive analysis of the mathematical basis of graph theoretic consensus theory can be found in [10].

Now we state the limitations of consensus theory that motivates our work. First, the basic consensus protocol Eq. (13) does not admit quaternions directly because quaternion dynamics are highly nonlinear. It violates quaternion unit norm requirements, and therefore we cannot practically apply Eq. (6) with consensus directly. To extend Eq. (13) to attitude quaternions, we proposed the following consensus protocol for quaternions [7–10]:

$$\dot{\mathbf{q}}(t) = -\mathbf{P}(t)\left(\mathbf{q}(t)\Theta\mathbf{q}^{-off}\right). \tag{14}$$

Here, $\mathbf{P}(t)$ is a Laplacian-like stochastic matrix whose values are partially unknown, but a Laplacian-like structure is imposed on it by optimization, and $\mathbf{q}(t) = [q^1(t), q^2(t) \cdots q^n(t)]^T$. We present more analysis of $\mathbf{P}(t)$ in the "Solutions" section.

4. Solutions

We present a four-step solution to the problem statement in Section 2 [7–10], listed as follows: (1) development of a consensus protocol for quaternions, (2) development of collision avoidance behavior for quaternion consensus, (3) determining obstacle vectors in different coordinate frames and (4) integration of quaternion consensus with Q-CAC avoidance.

4.1. Development of a consensus protocol for quaternions

To handle the difficulty of non-linearity in quaternion kinematics, we develop a consensus protocol especially for quaternions. We adopt an optimization approach and cast the problem as a semidefinite program, which is subject to convex quadratic constraints, stated as linear matrix inequalities (LMI). Based on the current communication graph of any SC_i, a series of Laplacian-like matrices $\mathbf{P}^i(t)$ are synthesized each time step to drive $q^i(t)$ to consensus while satisfying quaternion kinematics:

$$\dot{q}^i(t) - \mathbf{P}^i(t)\left[q_1^T(t)q_2^T(t) \cdots q_y^T(t)\right], \tag{15}$$

where $q_1^T(t)q_2^T(t) \cdots q_y^T(t)$ are the quaternions of the y other neighboring SC which SC_i can communicate with at time t. Euler's first-order discretization of Eq. (15) is

$$q_{k+1}^i = q_k^i - \Delta t \underbrace{\left[y\Lambda_1^i(\mathrm{t}) - \Lambda_2^i(\mathrm{t}) \cdots - \Lambda_y^i(\mathrm{t})\right]}_{\mathbf{P}^i(\mathrm{t})} \left[q_1^T(t) q_2^T(t) \cdots q_y^T(t)\right], \tag{16}$$

where $\Lambda^i(t) > 0$ is an unknown positive definite optimization matrix variable, whose components are chosen by the optimization process. For analysis purposes, we shall now reconsider the collective quaternion consensus dynamics Eq. (14). The components of $\mathbf{P}(t)$ are

$$\mathbf{P}(t) = \underbrace{\begin{bmatrix} \Lambda^1(t) & \cdots & 0 \\ \vdots & \ddots & \vdots \\ 0 & \cdots & \Lambda^n(t) \end{bmatrix}}_{\Lambda(\mathrm{t})} \underbrace{\begin{bmatrix} l_{11}\mathbf{I}_4 & \cdots & l_{1n}\mathbf{I}_4 \\ \vdots & \ddots & \vdots \\ l_{n1}\mathbf{I}_4 & \cdots & l_{nn}\mathbf{I}_4 \end{bmatrix}}_{\Gamma = \mathbf{L} \otimes \mathbf{I}_4} \tag{17}$$

where Γ is composed of components of the Laplacian $\mathbf{L} = [l_{ij}]$ $(i, j = 1, \cdots, n)$, which gives $\mathbf{P}(t)$ its Laplacian-like behavior, and $\Lambda^i(t) > 0$ is as previously defined.

We now present the proof of stability of $\mathbf{P}(t)$, that is, that Eq. (14) does indeed achieve consensus. Different versions of all the theorems, lemmas and proofs in this section had been presented in [7–10]. Let us begin by recalling the following standard result on a *matrix pencil* [18].

Theorem 1: For a *symmetric-definite* pencil $\mathbf{A} - \lambda\mathbf{B}$, there exists a nonsingular $Z = [z_1, \cdots, z_n]$ such that

$$Z^T\mathbf{A}Z = \mathrm{diag}(a_1, \cdots, a_n) = D_A, \tag{18}$$

$$Z^T\mathbf{B}Z = \mathrm{diag}(b_1, \cdots, b_n) = D_B. \tag{19}$$

Moreover, $\mathbf{A}z_i = \lambda_i\mathbf{B}z_i$ for $i = 1, \cdots, n$, where $\lambda_i = a_i/b_i$.

Lemma 1: For any time t, the eigenvalues of $\mathbf{P}(t)$ are $\gamma_i\eta_i(t)$. Here, γ_i are the eigenvalues of Γ and $\eta_i(t)$ the eigenvalues of $\Lambda(t)$. It can therefore be observed that $\mathbf{P}(t)$ has only four zero eigenvalues; the rest of its eigenvalues are strictly positive.

Proof: To find the eigenvalues of $\mathbf{P}(t)$, consider a scalar λ such that for some nonzero vector z:

$$\Gamma z = \lambda\Lambda^{-1}(t)z. \tag{20}$$

Eq. (20) defines a *symmetric-definite generalized eigenvalue problem* (SDGEP), where $\Gamma - \lambda\Lambda^{-1}(t)$ defines a matrix pencil. Theorem 1 therefore immediately implies that the eigenvalues of $\mathbf{P}(t)$ are $\gamma_i\eta_i(t)$. It is also easy to observe (or show numerically) that due to the property of the Laplacian matrix $\mathbf{L}$, $\mathbf{P}(t)$ has positive eigenvalues except for four zero eigenvalues. This proves the claim.

Theorem 2: The time-varying system Eq. (14) achieves consensus.

Proof: For simplicity, we shall assume no offsets are defined, that is, $\mathbf{q}^{off} = \mathbf{0}$ (or $(q^{off})^i = [0\ 0\ 0\ 1]^T\ \forall i$). By consensus theory, when $\mathbf{q}$ has entered the *consensus space* $\mathcal{C} = \{\mathbf{q} \mid q^1 = q^2 =, \cdots, = q^n\}$, then $\dot{\mathbf{q}} = \mathbf{0}$ (i.e. no vehicles are moving anymore). $\mathcal{C}$ is the *nullspace* of $\mathbf{P}(t)$, that is, the set of all $\mathbf{q}$ such that $\mathbf{P}(t)\mathbf{q} = 0$. Therefore, $\mathbf{q}$ stays in $\mathcal{C}$ once it enters there.

Suppose that $\mathbf{q}$ has not entered $\mathcal{C}$ (i.e. $\dot{\mathbf{q}} \neq \mathbf{0}$), then consider a Lyapunov candidate function $V = \mathbf{q}^T \Gamma \mathbf{q}$; $V > 0$ unless $\mathbf{q} \in \mathcal{C}$. Then:

$$\begin{aligned} \dot{V} &= \mathbf{q}^T \Gamma \dot{\mathbf{q}} + \dot{\mathbf{q}}^T \Gamma \mathbf{q}, \\ &= -\mathbf{q}^T \Gamma \mathbf{P}(t) \mathbf{q} - \mathbf{q}^T \mathbf{P}(t) \Gamma \mathbf{q}, \\ &= -\mathbf{q}^T \Gamma \Lambda(t) \Gamma \mathbf{q} - \mathbf{q}^T \Gamma \Lambda \Gamma \mathbf{q}, \\ &= -2\mathbf{q}^T \Gamma \Lambda(t) \Gamma \mathbf{q}, \\ &= -2s^T \Lambda(t) s, \end{aligned} \tag{21}$$

where $s = \Gamma q \neq 0$ for $\mathbf{q} \notin \mathcal{C}$, which implies that $\mathbf{q}$ approaches a point in $\mathcal{C}$ as $t \to \infty$. This proves the claim. Eq. (21) is true as long as $\mathbf{L}$ is nonempty, that is, some vehicles can sense, see or communicate with each other all the time.

4.2. Development of collision avoidance behavior for quaternion consensus

Eq. (15) or (16) will indeed generate a consensus $q^i(t)$ for any SC_i, but the system still needs to determine whether the trajectory is safe or not. This brings us to the issue of avoidance. Any rigid appendage attached to the body of SC_i, for example, a camera, whose direction vector is $v^I_{cam_i}$ in *inertial frame*, can be transformed to the spacecraft fixed *body frame* by the rotation:

$$v^B_{cam_i}(t) = \mathcal{R}_i^{-1}(t) v^I_{cam_i}(t). \tag{22}$$

where

$$\mathcal{R}_i(t) = \left(\left(2q^i_4(t) \right)^2 - 1 \right) \mathbf{I}_3 + 2\overline{q}^i(t) \overline{q}^i(t)^T - 2q^i_4(t) \overline{q}^i(t)^\times \tag{23}$$

is the rotation matrix corresponding to the $q^i(t)$ at time t; $\overline{q}^i(t)^\times$ is the *antisymmetric matrix* [19]. For a simpler analysis, let us consider a single SC_i with a single camera, $v^I_{cam_i}$, and m (possibly, time-varying) obstacles, $v^I_{obs_i \cdot j} (j = 1, \cdots, m)$, defined in $\mathcal{F}^I_{SC_i}$. We want $v^I_{cam_i}$ to avoid all $v^I_{obs_i \cdot j}$ when SC_i is re-orientating. Then following Eq. (3), the resulting attitude constraint of Eq. (2) can be written as

$$q^i(t)^T \tilde{A}^i_j(t) q^i(t) \leq 0. \tag{24}$$

Its LMI equivalent [5] is

$$\begin{bmatrix} \mu & q^i(t)^T \\ q^i(t) & \left(\mu \mathbf{I}_4 + \tilde{A}_j^i(t)\right)^{-1} \end{bmatrix} \geq 0. \tag{25}$$

where

$$\tilde{A}_j^i(t) = \begin{bmatrix} A_j(t) & b_j(t) \\ b_j(t)^T & d_j(t) \end{bmatrix} \in \mathbb{R}^{4\times 4}, \tag{26}$$

and

$$A_j(t) = v^B_{cam_i}(t) v^I_{obs_i.j}(t)^T + v^I_{obs_i.j}(t) v^B_{cam_i}(t)^T - \left(v^B_{cam_i}(t)^T v^I_{obs_i.j}(t) + \cos\theta\right)\mathbf{I}_3, \tag{27}$$

$$b_j(t) = -v^B_{cam_i}(t) \times v^I_{obs_i.j}(t), \tag{28}$$

$$d_j(t) = v^B_{cam_i}(t)^T v^I_{obs_i.j}(t), \tag{29}$$

for $j = 1, \cdots, m$.

Eq. (24) defines the set of attitude quaternions $q^i(t)$ to satisfy the constraint $v^I_{cam_i}(t)^T v^I_{obs_i.j}(t) \geq \varnothing \forall t \in [t_0, t_f]$, so it is used to find a collision-free $v^I_{cam_i}(t)$. In Eq. (25), μ is chosen to ensure that $\mu \mathbf{I}_4 + \tilde{A}_j^i(t)$ is positive definite.

However, the solution presented above assumes that $v^B_{cam_i}(t)^T$ and $v^I_{obs_i.j}(t)$ are in the same coordinate frame and that $v^I_{obs_i.j}(t)$ is static, so t is constant. In reality, this is not so. To address such a practical issue, we present a mechanism to calculate $v^I_{obs_i.j}$ (defined in $\mathcal{F}^I_{SC_i}$) corresponding to $v^I_{obs_j}$ (defined in $\mathcal{F}^I_{SC_j}$) ($v^I_{obs_i.j}$ means the obstacle vector originated from the rotating frame of SC_j but defined in $\mathcal{F}^I_{SC_i}$). This is essentially a mechanism to determine the intersection point of $v^I_{obs_j}(t)$ with the sphere of radius r, centerd on SC_i. If indeed such an intersection exists, it defines $v^I_{obs_i.j}$ which can be used to define an attitude constraint represented as Eq. (24) to be avoided by SC_i.

The scenario is illustrated in **Figure 2**, whereSC_1 and SC_2 are shown in their different coordinate frames relative to Earth. A thruster attached to SC_1 body frame is at $v^I_{obs_1}$, while the circles around SC_1 and SC_2 are spheres representing the coordinate frames from which their attitude evolves. If both spacecraft are close enough, then vector $v^I_{obs_1}$ may intersect a point on the sphere of SC_2, whereby the intersection defines $v^I_{obs_2.1}$ in the frame of SC_2. The requirement is that as SC_2 changes its attitude from q_0 to q_f, $v^I_{cam_2}$ must avoid the cone created around $v^I_{obs_2.1} \forall t \in [t_0, t_f]$.

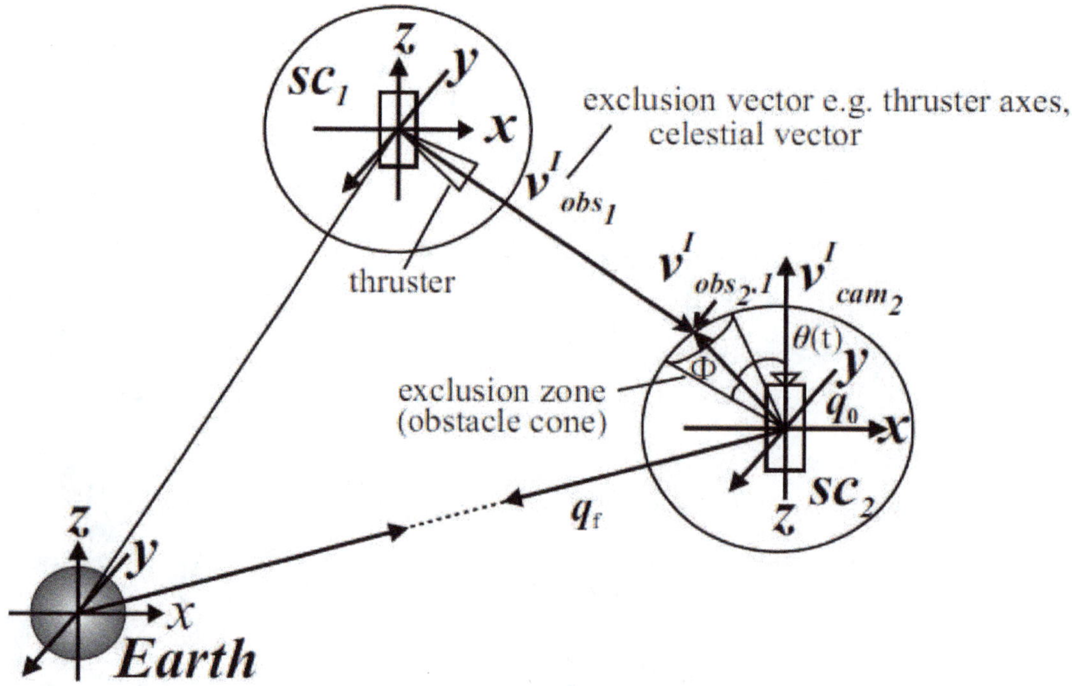

Figure 2. Q-CAC problem in different frames. SC_2 must maneuver from q_0 to q_f, while $v^I_{cam_2}$ must avoid $v^I_{obs_2.1}$ by at least $\varnothing \forall t \in [t_0, t_f]$.

4.3. Determination of obstacle vectors in different coordinate frames

Pursuing the issue of practicality further, given SC_i in $\mathcal{F}^I_{SC_i}$ and SC_j in $\mathcal{F}^I_{SC_j}$ with emanating vectors, an intersection between vectors emanating from $\mathcal{F}^I_{SC_j}$ with the sphere centered on $\mathcal{F}^I_{SC_i}$ can be determined, either by using onboard sensors or by application of computational geometry. Given a line segment $[p_1, p_2]$, originating at p_1 and terminating at p_2, a point $p = [p_x\ p_y\ p_z]^T$ on $[p_1, p_2]$ can be tested for intersection with a sphere centered at an external point p_3 with radius r [20]. Therefore, for any $v^I_{obs_j}(t)$ in $\mathcal{F}^I_{SC_j}$, if an intersection point $p(t)$ exists at time t with the sphere centered on $\mathcal{F}^I_{SC_i}$ with radius r, then $v^I_{obs_i.j}(t) = p(t)$; otherwise, one can set $v^I_{obs_i.j}(t) = -v^I_{cam_i}(t)$ to show that no constraint violation has occurred. The value of r will thus depend on the current application but must be proportional to the urgency of avoiding obstacle vectors originating from other spacecraft. The above formulation effectively completes the decentralization of the avoidance problem which has already been partly decentralized by Eq. (16). Eq. (16) will be written in a semidefinite optimization program, which gives us the privilege to apply further constraints. Therefore, the *norm constraints* required by quaternion kinematics can be enforced as follows:

$$q^{i^T}_k \left(q^i_{k+1} - q^i_k\right) = 0 \tag{30}$$

Essentially, Eq. (30) is the discrete time version of $q^i(t)^T \dot{q}^i(t) = 0$ or $\mathbf{q}(t)^T \dot{\mathbf{q}}(t) = 0$. This guarantees that $q^i(t)^T q^i(t) = 1$ or $\mathbf{q}(t)^T \mathbf{q}(t) = n$ for nSC, iff $\|q^i(0)\| = 1\ \forall\, i$.

4.4. Integration of quaternion consensus with Q-CAC avoidance

The integration of the quaternion consensus protocol with the Q-CAC collision avoidance in different coordinate frames is a two-stage process. First, the quaternion consensus protocol generates a set of consensus quaternion trajectories using Eq. (15) or (16). Then Eq. (25) tests whether the generated sequence is safe or not. If the next safe quaternion trajectory q^i_{safe} has been determined, the control torque τ^i and angular velocity ω^i to rotate the SC_i optimally to q^i_{safe} can be determined by using the normal quaternion dynamics Eq. (12). Otherwise, Eq. (25) adjusts the q^i_{unsafe} to generate a q^i_{safe}, which will be close to but not be exactly q^i_{safe}. The cycle repeats until consensus is achieved.

Using semidefinite programming, the solutions presented previously are cast as an optimization problem, augmented with a set of LMI constraints and solved for collision-free consensus quaternion trajectories. We consider the algorithm in discrete time. Given the initial attitude $q^i(0)$ of SC_i, $(i=1,\cdots,n)$, find a sequence of consensus quaternion trajectories that satisfies the following constraints:

$$\begin{aligned}
&q^i_{k+1} = q^i_k - \Delta t\mathbf{P}^i(\mathrm{t})\mathrm{q}^i_k,\\
&\mathrm{q}^i_k T(\mathrm{q}^i_{k+1} - q^i_k) = 0,\\
&\begin{bmatrix} \mu & q_i(t)^{\mathrm{T}} \\ q_i(t) & \left(\mu\mathbf{I}_4 + \tilde{A}^i_j(t)\right)^{-1} \end{bmatrix} \geq 0.\\
&\underbrace{\begin{bmatrix} -\Delta t(J^i)^{-1} & \mathbf{I}_3 & \mathbf{0}_{3\times4} \\ \mathbf{0}_{4\times3} & -\dfrac{\Delta t}{2}\Pi^i_{k+1} & \mathbf{I}_4 \end{bmatrix}}_{F^i_k} \underbrace{\begin{bmatrix} \tau^i_k \\ \omega^i_{k+1} \\ q^i_{k+2} \end{bmatrix}}_{\mathfrak{T}^i_{k+1}} = \underbrace{\begin{bmatrix} (\mathbf{I}_3 + \Delta t Y^i_k)\omega^i_k \\ q^i_{k+1} \end{bmatrix}}_{y^i_k}
\end{aligned} \tag{31}$$

5. Simulation results

We shall present only three results for attitude multi-path planning in different coordinate frames due to limitation of space. These results will partly be found in [7–10]. For the SDP programming and simulation, we used the available optimization software tools SeDuMi [21] and YALMIP [22] running inside Matlab®.

5.1. Q-CAC avoidance in different coordinate frames without consensus

In this experiment $SC_1$and SC_2 are changing their orientation to point an instrument to Earth. They are close to each other, and their thrusters can cause plume impingements to damage each other. Their initial quaternions are $q^1_0 = q^2_0 = [0\,0\,0\,1]^T$. The desired final quaternions are.

$$\begin{aligned}
q^1_f &= [0.2269\ 0.0421\ 0.9567\ 0.1776]^T\\
q^2_f &= [0\ 0\ 0.9903\ 0.1387]^T.
\end{aligned} \tag{32}$$

Three thrusters of SC_1 in $\mathcal{F}^B_{SC_1}$ are

$$v^B_{obs_1.1} = [-0.2132 - 0.0181\ 0.9768]^T$$
$$v^B_{obs_1.2} = [0.314\ 0.283 - 0.906]^T \qquad (33)$$
$$v^B_{obs_1.3} = [-0.112 - 0.133 - 0.985]^T.$$

A single thruster of SC_2 in $\mathcal{F}^B_{SC_2}$ is

$$v^B_{obs_2} = [0.02981\ 0.0819\ 0.9962]^T. \qquad (34)$$

It is desired that $v^I_{obs_2}$ avoid $v^I_{obs_1.1}$ by 50° and avoid $v^I_{obs_1.2}$ and $v^I_{obs_1.3}$ by 30°, while both are maneuvering to their desired final attitudes. The trajectories obtained are shown in **Figure 3 (a)** and **(b)**. This experiment demonstrates that when both constraints are in conflict, the avoidance constraint is superior to the desired final quaternion constraint. As seen from (a), SC_2 cannot reconfigure exactly to the desired q^2_f due to the satisfaction of the avoidance constraints. This can be resolved by changing either the position of SC_2 or SC_1.

5.2. Consensus with Q-CAC avoidance in different coordinate frames

In this experiment SC_i $(i=1, 2, 3)$ will maneuver to a consensus attitude. Each carries a sensitive instrument $v^I_{cam_i}$, pointing in the direction SC_i's initial attitude quaternion. In addition, each SC_i has only one thruster pointing to the opposite (rear) of SC_i's initial attitude. It is desired that the time evolution of the attitude trajectory of the sensitive instrument avoids the thruster plumes emanating from each of the two other SC by 30°. From the generated initial quaternions, there is possibility of intersection of the thrusters of SC_1 and SC_3, with SC_2, and the thruster of SC_2 may impinge on SC_1 or SC_3 at any time k.

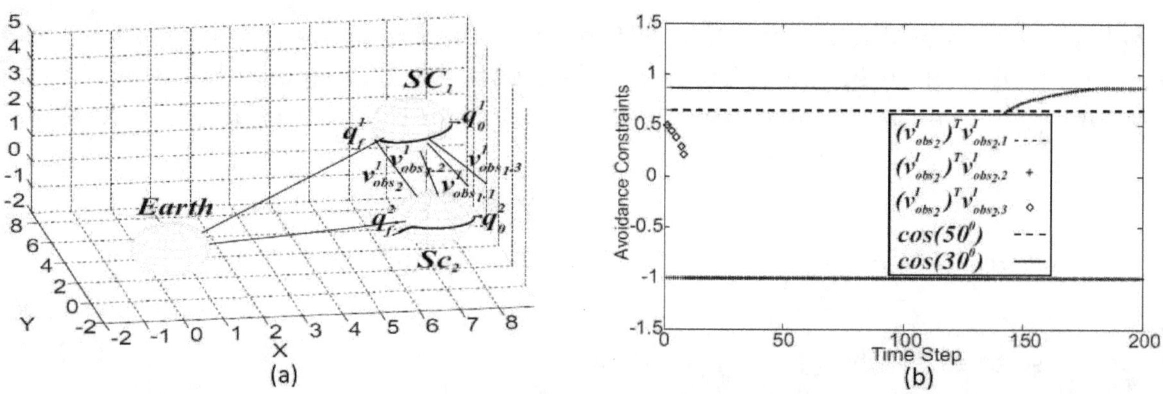

(a) (b)

Figure 3. (a) shows the avoidance between thrusters of $SC_1$and SC_2 during reorientation to Earth: SC_2 cannot reconfigure to the desired q^2_f due to the avoidance constraints. Note that $v^I_{obs_2.1}$, $v^I_{obs_2.2}$ and $v^I_{obs_2.3}$ are the points of intersections of $v^I_{obs_1.1}$, $v^I_{obs_1.2}$ and $v^I_{obs_1.3}$ with SC_2. (b) Satisfaction of avoidance constraints: the sudden jumps to and from -1 indicate times when any of $v^I_{obs_1.1}$, $v^I_{obs_1.2}$ and $v^I_{obs_1.3}$ lost intersection with the sphere of SC_2 and therefore was replaced with $-v^I_{obs_1.i}$, $i = 1, \cdots, 3$.

The initial positions are

$$\begin{aligned}\mathcal{F}^I_{SC_1} &= [-2\ 0\ 2]^T\\ \mathcal{F}^I_{SC_2} &= [0.5\ 0\ 2]^T\\ \mathcal{F}^I_{SC_3} &= [3\ 0\ 2]^T.\end{aligned} \tag{35}$$

A set of initial quaternions were randomly generated, with the following data:

$$\begin{aligned}q^1_0 &= [-0.5101\ 0.6112 - 0.3187 - 0.5145]^T\\ q^2_0 &= [-0.9369\ 0.2704 - 0.1836 - 0.124]^T\\ q^3_0 &= [0.1448 - 0.1151\ 0.1203\ 0.9753]^T.\end{aligned} \tag{36}$$

Figure 4 (a) shows the solution trajectories while (b) shows the avoidance graph; no constraints are not violated; (c) shows the consensus graph. The final consensus quaternion is $q_f = [-0.8167\ 0.4807 - 0.2396\ 0.2112]^T$, which is the normalized average of the initial attitude quaternions. This proves that consensus is indeed achieved by Eq. (16).

5.3. Consensus-based attitude formation acquisition with avoidance

This experiment is to test the capability of the quaternion consensus algorithm in attitude formation acquisition. SC_i $(i=1,2,3)$ will maneuver to a consensus formation attitude, with relative offset quaternions defined to enable the sensitive instruments to point at 30°offsets from each other about the z-axis. The previous set of initial data for q^i_0 and $\mathcal{F}^I_{SC_i}$ were used. Like the previous experiment, it is desired that the sensitive instruments avoid the thruster plumes emanating from each of the two other SC by an angle of 30°.

The relative offsets are defined as

$$\begin{aligned}q^{off}_1 &= [0\ 0\ 0\ 1]^T\\ q^{off}_2 &= [0\ 0\ 0.2588\ 0.9659]^T\\ q^{off}_3 &= [0\ 0\ 0.5\ 0.866]^T.\end{aligned} \tag{37}$$

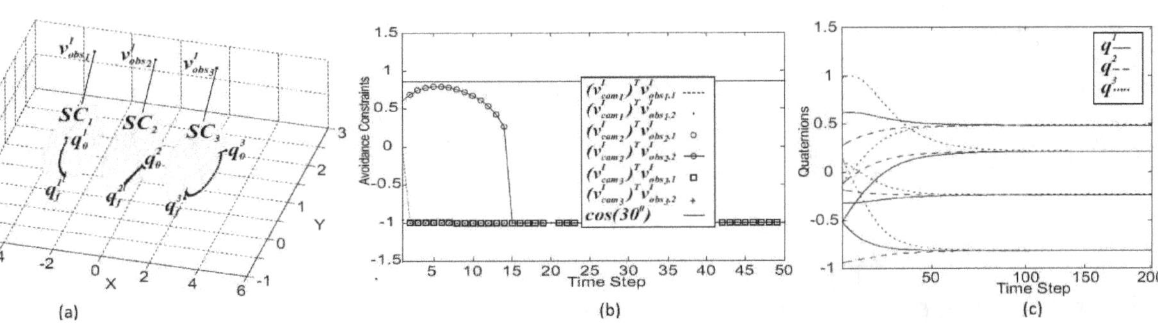

Figure 4. (a) Reorientation to consensus attitude with intervehicle thruster plume avoidance, (b) avoidance constraints graph and (c) attitude consensus graph.

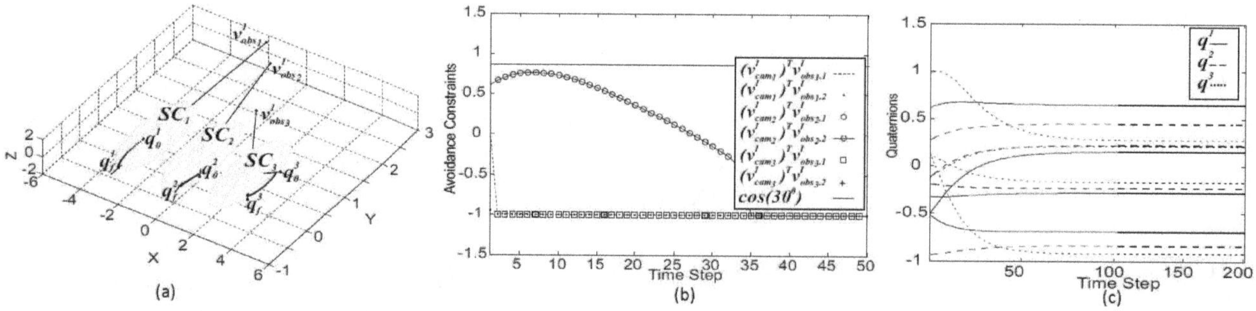

Figure 5. (a) Reorientation to consensus formation attitude with intervehicle thruster plume avoidance, (b) avoidance constraints graph and (c) attitude consensus graph.

Figure 5 (a) shows the trajectories, while (b) shows the avoidance graph; no constraints are violated. Finally, (c) shows the consensus graph. The final consensus quaternions are.

$$\begin{aligned} q_f^1 &= [-0.6926\ 0.6468 - 0.2798\ 0.1541]^T \\ q_f^2 &= [-0.8364\ 0.4455 - 0.2303\ 0.2212]^T \\ q_f^3 &= [-0.9232\ 0.2138 - 0.1652\ 0.2733]^T. \end{aligned} \tag{38}$$

The differences of these quaternions are 30° apart about the same axis. Clearly, the algorithm is capable of attitude formation acquisition with avoidance.

6. Conclusion

In this chapter, a method of consensus with quaternion-based attitude maneuver with avoidance, of multiple networked communicating spacecraft, was presented. The presentation is composed of aspects of solutions we previously developed, by combining consensus theory and Q-CAC optimization theory. The solutions enable a team of spacecraft to point to the same direction or to various formation patterns, while they avoid an arbitrary number of attitude obstacles or exclusion zones in any coordinate frames. The proof of stability of the Laplacian-like dynamics was also presented. Simulation results also demonstrated the effectiveness of the algorithm. We hope to implement the algorithms using rotorcraft and specialized hardware.

Author details

Innocent Okoloko

Address all correspondence to: okoloko@ieee.org

Department of Electrical Engineering, Universidad de Ingenieria y Tecnologia, Lima, Peru

References

[1] Blackwood G, Lay O, Deininger B, Gudim M, Ahmed A, Duren R, Noeckerb C, Barden B. The StarLight mission: A formation-flying stellar interferometer. In: SPIE 4852, Interferometry in Space; 22 August; Waikoloa, Hawaii. SPIE Digital Library; 2002. DOI: 10.1117/12.460942

[2] Beichman CA. NASA's terrestrial planet finder. In: Darwin and Astronomy: The Infrared Space Interferometer; 17–19 November; Stockholm, Sweden. The Netherlands: Noordwijk; 2000. p. 29-30. DOI: ISSN/ISBN:03796566

[3] Ahmed A, Alexander J, Boussalis D, Breckenridge W, Macala G, Mesbahi M, Martin MS, Singh G, Wong E. Cassini Control Analysis Book. Pasadena, CA: Jet Propulsion Laboratory, CALTECH technical report; 1998 NA p. DOI: NA

[4] Wen JT, Kreutz-Delgado K. The attitude control problem. IEEE Transactions on Automatic Control. 1991;**36**(10):1148-1162. DOI: 10.1109/9.90228

[5] Kim Y, Mesbahi M. Quadratically constrained attitude control via semidefinite programming. IEEE Transactions on Automatic Control. 2004;**49**:731-735. DOI: 10.1109/TAC.2004.825959

[6] Kim Y, Mesbahi M, Singh G, Hadaegh FY. On the convex parameterization of constrained spacecraft reorientation. IEEE Transactions on Aerospace and Electronic Systems. 2010;**46**(3):1097-1109. DOI: 10.1109/TAES.2010.5545176

[7] Okoloko I, Kim Y. Distributed constrained attitude and position control using graph laplacians. In: ASME Dynamic Systems and Control Conference; 13–15 September; Cambridge, Massachusetts. ASME; 2010. p. 377-383. DOI: 10.1115/DSCC2010-4036

[8] Okoloko I, Kim Y. Attitude synchronization of multiple spacecraft with cone avoidance constraints. In: IEEE Aerospace Conference; 3–10 March; Big Sky, Montana. IEEEXplore; 2012. p. 1-10. DOI: 10.1109/AERO.2012.6187119

[9] Okoloko I, Kim Y. Attitude synchronization of multiple spacecraft with cone avoidance constraints. Systems & Control Letters. 2014;**69**:73-79. DOI: https://doi.org/10.1016/j.sysconle.2014.04.008

[10] Okoloko I. Multi-Path Planning and Multi-Body Constrained Attitude Control [dissertation]. Stellenbosch, South Africa: PhD Thesis: Stellenbosch University; 2012. p. 185. Available from: http://hdl.handle.net/10019.1/71905

[11] Bullo F, Murray R M, Sarti A. Control on the sphere and reduced attitude stabilization. In: IFAC symposium on nonlinear control systems; 25-28 June; Tahoe City, CA. Elsevier; 1995. p. 495-501. DOI: https://doi.org/10.1016/S1474-6670(17)46878-9

[12] Ren W. Distributed attitude alignment in spacecraft formation flying. International Journal of Adaptive Control and Signal Processing. 2006;**21**(2–3):95-113. DOI: 10.1002/acs.916

[13] Fax AJ. Optimal and Cooperative Control of Vehicle Formations [thesis]. Pasadena, CA: PhD Thesis: CALTECH; 2002. p. 135. Available from: thesis.library.caltech.edu/4230/1/Fax_ja_2002.pdf

[14] Dimarogonas DV, Tsiotras P, Kyriakopoulos KJ. Leader-follower cooperative attitude control of multiple rigid bodies. Systems and Control Letters, DOI. 2009;**58**(6):429-435 https://doi.org/10.1016/j.sysconle.2009.02.002

[15] Kuipers JB. Quaternions and Rotation Sequences: A Primer with Applications to Orbits, Aerospace and Virtual Reality. 1st ed. Princeton, NJ: Princeton University Press; 2002. p. 371. ISBN-13:978–0691102986

[16] Peng L, Zhao Y, Tian B, Zhang J, Bing-Hong W, Hai-Tao Z, Zhou T. Consensus of self-driven agents with avoidance of collisions. Physical Review. 2009;**79**(E). DOI: 10.1103/PhysRevE.79.026113

[17] Olfati-Saber R. Flocking for multi-agent dynamic systems: Algorithms and theory. IEEE Transactions on Automatic Control. 2006;**51**(3):401-420. DOI: 10.1109/TAC.2005.864190

[18] Golub GH, Van Loan CF. Matrix Computations. 3rd ed. Baltimore, MD: Johns Hopkins University Press; 1996. p. 699. ISBN-13: 978–0801854149

[19] Hughes PC. Spacecraft Attitude Dynamics. 2nd ed. Mineola, NY: Dover Publications Inc; 2004. p. 592. ISBN13:9780486439259

[20] Eberly DH. 3D Game Engine Design: A Practical Approach to Real-Time Computer Graphics. 2nd ed. London, UK: Taylor & Francis; 2012. p. 1015. DOI: ISBN: 978-0-12-229063-3

[21] Sturm JF. Using SeDuMi 1.02, a Matlab toolbox for optimization over symmetric cones. Optimization methods and software. 1998;**11**(12):625-653. DOI: http://dx.doi.org/10.1080/10556789908805766

[22] Lofberg J. Yalmip: A toolbox for modelling and optimization in Matlab. In: IEEE CACSD Conference; 2–4 Sept.; Taipei, Taiwan. IEEEXplore; 2004. p. 284-289. DOI: 10.1109/CACSD.2004.1393890

11

Space Thermal and Vacuum Environment Simulation

Roy Stevenson Soler Chisabas, Geilson Loureiro and
Carlos de Oliveira Lino

Abstract

The space simulation chambers are systems used to recreate as closely as possible the thermal environmental conditions that spacecraft experience in space, as well as also serve to space components qualification and material research used in spacecraft. These systems analyze spacecraft behavior, evaluating its thermal balance, and functionalities to ensure mission success and survivability. The objective of this chapter is to give a broad overview on space simulation chambers, describe which are the environmental parameters of space that can be simulated in this type of ground test facilities, types of the space environment simulators, class of phenomena generated inside, and the technological evolution of these systems from its conception. This chapter describes the basic systems and devices that compose the space simulation chambers.

Keywords: space environment simulation, space simulation chamber, thermal vacuum chamber

1. Introduction

The spacecrafts are developed for various applications such as space science, navigation, communications, technology testing and verification, earth observation, weather observation, military applications, human space flight, planetary exploration, and others [1]. According to the type of mission destined to fulfill by the spacecraft, it is possible to classify them into several types such as Flyby spacecraft, Orbiter spacecraft, Atmospheric spacecraft, Lander spacecraft, Rover spacecraft, Penetrator spacecraft, Observatory spacecraft, and Communications spacecraft [2].

To start the operation phase, the spacecraft need to meet all conventional space project life cycle development phases such as Concept Studies, Concept and Technology Development, Preliminary Design & Technology Completion, Final Design & Fabrication, System Assembly Integration and Test (AIT), Launch Campaign, Operations & Sustainment and Closeout [3]. In the integration and test phases, the spacecraft is assembled, integrated, and tested. **Figure 1** shows the usual activities that comprise assembly, integration, and test of spacecraft.

For execution, spacecraft tests program is required for the uses in different types of facilities such as Vibration test facility, Acoustics test facility, Mass properties test facility, space simulation chamber, EMC test facility, Magnetics test facility, and others [4]. These facilities are designed to research, develop, test, and verify the performance of the spacecrafts. In these facilities, it is possible to simulate the environmental conditions experienced by the spacecrafts during the launch phase and their exposure to the space environment.

1.1. Thermal tests

Several types of thermal tests are required for development, performance validation, and to ensure the survivability of the spacecraft in operation. These tests can be performed in components, subsystems, and systems levels. The thermal testing usually includes a Thermal Cycle Test (TCT), Thermal Vacuum Test (TVT), Thermal Balance Test (TBT), and Vacuum Bake-out test [5].

Thermal Cycle Test (TCT): This is generally executed in ambient pressure through the use of environmental chambers. This test is usually executed to subsystem or system level. The test article will be exposed to a series of cycles of hot and cold temperatures. The thermal cycling generates an environmental stress in the test article that allows to identify material and workmanship defects.

Thermal Vacuum Test (TVT): This type of test submit the specimen to a series of cycles of hot and cold temperatures in a high vacuum environment. Space simulation chambers are used to perform this type of test. This test is executed to subsystem or system level. During the development of TVT, functional tests for the performance verification of the subsystem or system are performed.

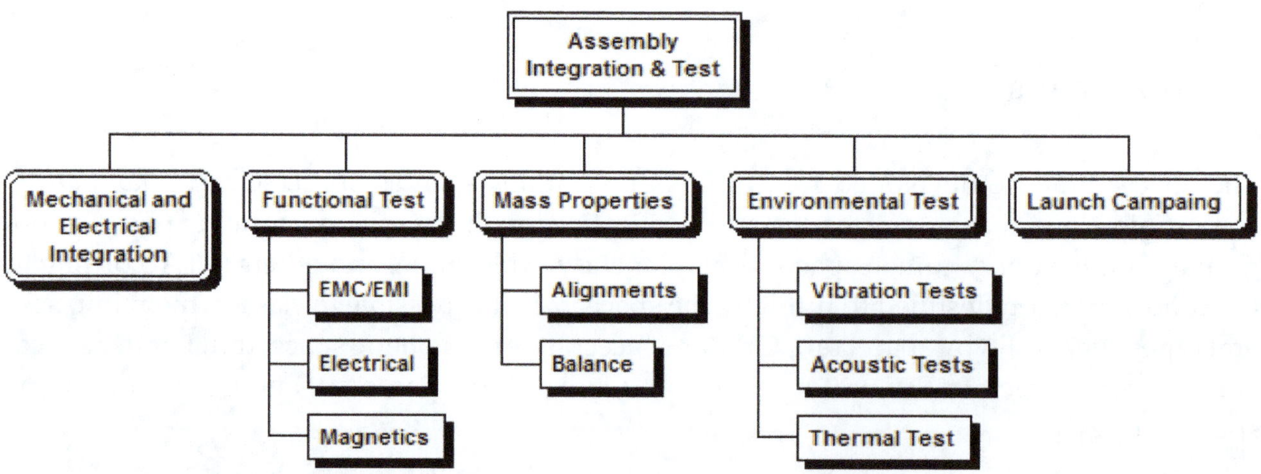

Figure 1. Activities in the assembly integration and test process.

Figure 2. (Left) Chamber A – NASA Johnson Space Center's and (Right) large space simulator (LSS) ESTEC test Centre, ESA.

Thermal Balance Test (TBT): Space simulation chambers are used to perform this type of test. The purpose of this test is: Demonstrate the performance of the thermal control system to maintain the temperatures within the operational limits. This test is necessary to verify the performance of the spacecraft thermal design when it is exposed the space thermal environment conditions. This test is also used to measure the thermal deformations in the system.

The TBT and TVT are used to demonstrate the capability of the subsystem or system to tolerate the consequences of the continuous thermal cycling during operation in the space thermal environment. The selection of number of thermal cycles in TBT and TVT depends on the type of test level.

Vacuum Bake-out Test: The spacecraft is exposed to high temperature in a high vacuum environment during a determined time to stimulate their outgassing. This test is executed to subsystem or system levels. Space simulation chambers are used to perform this type of test.

The thermal tests are performed at various temperature levels established in each test levels. The test levels are: Development Tests, Qualification Tests, Acceptance Tests, and Protoflight Qualification Tests. Considering the type test level, general rules and standard are available to determine the temperature and pressure levels of thermal tests in subsystems and spacecrafts. Some rules and standards are: GSFC-STD-7000, MIL-STD-1540D, MIL-HDBK-340A, ECSS-E-ST-10-03C, TR-2004(8583)-1, NASA LSP-REQ-317.01, among others.

During spacecraft environmental testing, which is part of the Assembly, Integration and Test process (AIT), Space Simulation Chambers play a key role to spacecraft systemic models qualification (e.g. Structural Model, Thermal Model, Engineering Model, Qualification Model, Flight Model, and Protoflight Model). **Figure 2** demonstrates two types of space simulation chambers.

2. Space simulation chambers

The space simulation chambers are systems used to recreate as close as possible the environment conditions that spacecrafts experiences into space, as well as serves to space components qualification and material research used in spacecrafts. These systems allow the spacecrafts thermal behavior to be analyzed [6]. There are two types of space environment simulators, the ones with solar simulator and the ones without. Systems without solar simulator are known as Thermal Vacuum Chambers [7]. These systems also recreate the space environment conditions, including solar radiation, by using different devices in the test setup. Space simulation chambers are designed to serve as a test medium for various types of spacecrafts and their subsystems.

3. Space environment

The ambient which experience the spacecraft consist in the combination of the space environment in function of the orbit where the mission will be developed (Low Earth Orbit, Medium Earth Orbit, Polar Orbit, Geosynchronous Orbit, and Interplanetary Orbit) and the environment generated by same spacecraft in operation [8]. The space environment main characteristics experienced by spacecraft orbiting the Earth are: high vacuum, cold space environment, and different sources of radiation. The space environmental phenomena are showed in **Figure 3**.

A spacecraft in space experiences an intense radiation when it is exposed to the sun. When the spacecraft is into the umbra (without sunlight), it experiences an environment of extreme coldness. These conditions allow to calculate the spacecraft temperature during operation, which is determined by a balance between spacecraft internal heat, radiant energy absorbed by spacecraft, and radiant energy emitted to space by spacecraft surfaces [5, 9].

3.1. Pressure

The pressure experienced by spacecraft varies from 1×10^{-3} mbar near Earth atmosphere to 1×10^{-12} mbar in deep space. In a pressure of more than 1×10^{-6} mbar, the molecular mean free path is very wide, which reduces heat transfer to solar radiation.

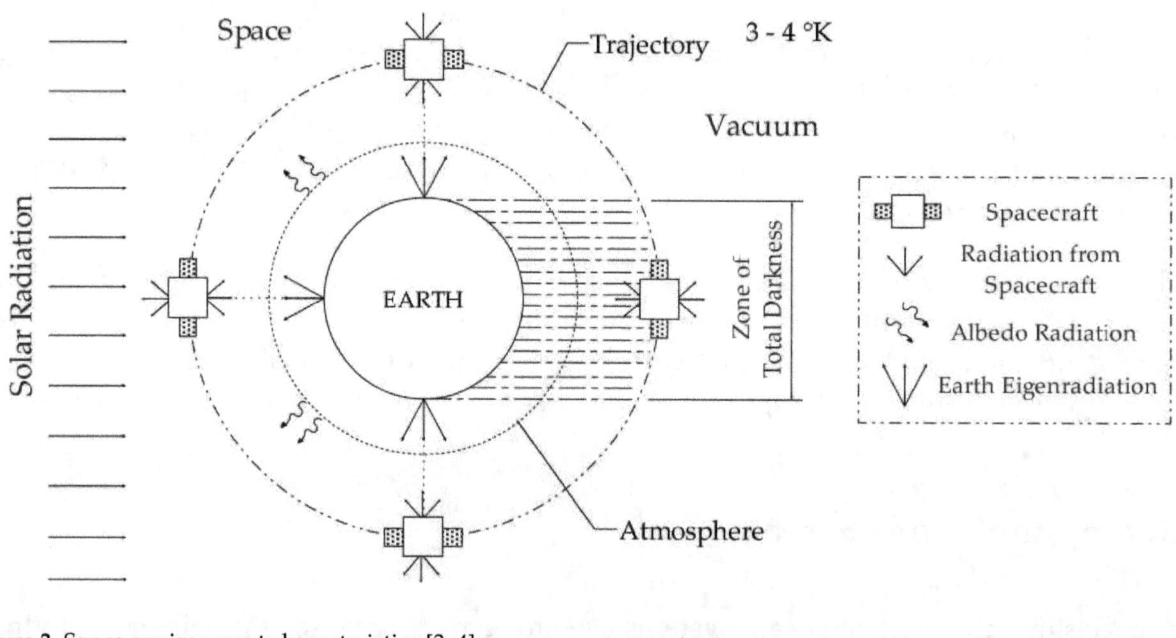

Figure 3. Space environment characteristics [3, 4].

3.2. The solar radiation

The solar radiation is a high intensity energetic phenomenon, which represents an approximate 1400 W/m^2 heat flux in the spacecraft. The heat flux change depending on the distance from the spacecraft to the sun. The absorption of such energy would generate a very high temperature inside the spacecraft, however, just a fraction of heat is absorbed due to space environment characteristics and spacecraft surfaces physical properties [5, 10].

3.3. Cold temperature (space heat sink)

Deep space is similar to an infinite dissipation black body, where a passive body experiences a balance temperature between −270.15°C (3 K) and −260.15°C (4 K) [10]. This concept implies that the heat emitted by a spacecraft will not return to it [9].

3.4. Albedo and eigenradiation of the earth

Albedo is the fraction of incident solar radiation reflected by the Earth or the moon, which reaches spacecraft depending on its position and distance. The eigenradiation is the Earth's thermal radiation, which allows the balance between absorbed solar radiation and the Earth's generated heat [1]. Albedo is approximately 0.48 kW/m^2, and the Earth's radiation is approximately 0.23 kW/m^2. The values that can take both forms of radiation depend on the relative position of the spacecraft to the Earth and Sun [10].

4. Space environment simulation

Space simulation chambers simulate the space thermal environment with close proximity, because to generate a temperature of −269.15°C (4 K), without any reflectivity as in space, would be economically unviable. Therefore, after analyzing chambers data since its invention and also Stefan Boltzmann law analysis, it was historically opted to generate temperatures from −195.85 to −173.15°C (77.3–100 K), which only represent a small error percentage to assess spacecraft in low temperatures, without significantly affecting thermal balance study [7–9, 10]. Due to this reason, it was established the trend of using heat transfer surfaces which generate the minimal temperature of −173.15°C (100 K).

For thermal balance study and analysis, it is essential to ensure the thermal loads that the spacecraft will receive from several sources of radiation in space. This radiation sources are transformed in high temperatures experienced by spacecraft according to its position in space and materials characteristics. The thermal loads can be simulated through solar simulators or using heat transfer surfaces. The solar simulator is a compounded system with an artificial light source adjusted through optical mechanisms and filters that provide intensity and spectral composition similar to sunlight for the spacecraft test. Solar simulators can generate thermal loads similar to the Sun using high intensity infrared lamps, but with an excessive cost due to high power consumption, preventing their use in some simulation systems. Therefore it is used to replace them by heat transfer surfaces that can generate temperatures greater than 126.85°C (400 K) [7]. Albedo and eigenradiation are not simulated in thermal vacuum chambers since their values are diffuse and depend on the spacecraft position relative from the Earth and Sun, among other characteristics [10].

Given these restrictions and limitations, the thermal vacuum chambers simulate with closeness the vacuum and cold space environment. Beyond this, through the use of other devices to the system (electrical heaters, infrared heaters, or Cal-Rods), it is possible to simulate the thermal loads that will be experienced by spacecraft when exposed to solar radiation during its operation. It should be noted that the spacecraft is mathematically modeled using software, which use the exact values of all phenomena experienced in space.

In the space thermal environment simulation, it is not necessary to duplicate an ultrahigh vacuum level as that owns by outer space, but it is necessary to duplicate the effects that this environment generates in the materials, components, subsystems, and spacecraft systems. Due to the above mentioned, for the space environment simulation test, it is necessary to achieve a level pressure less than 1×10^{-6} mbar, because to this level is possible to properly evaluate the specimen and eliminate some undesirable effects such as the gas thermal conduction and arc and glow discharges [11].

5. Systems of the space simulation chambers

Through the study and analysis of environment conditions that shall be created by the space simulation chambers, it is possible to establish their basic composition. The conditions to be simulated are transformed into functions assigned to systems or a set of systems that will permit their generation. The basic systems that compose the space simulation chambers are shown in **Figure 4**.

The structure of the chamber, also known as vacuum chamber or vacuum vessel, allows the conservation of vacuum and thermal radiation phenomena, which is very important characteristics to simulate the space thermal environment; this also houses the test specimen.

The vacuum system function is to produce a desirable vacuum level in a reasonable time and maintain such vacuum level during all test time.

The thermal system function is to reproduce as close as possible the heat sink of space (cold environment). The decontamination system function is to achieve a significant reduction of

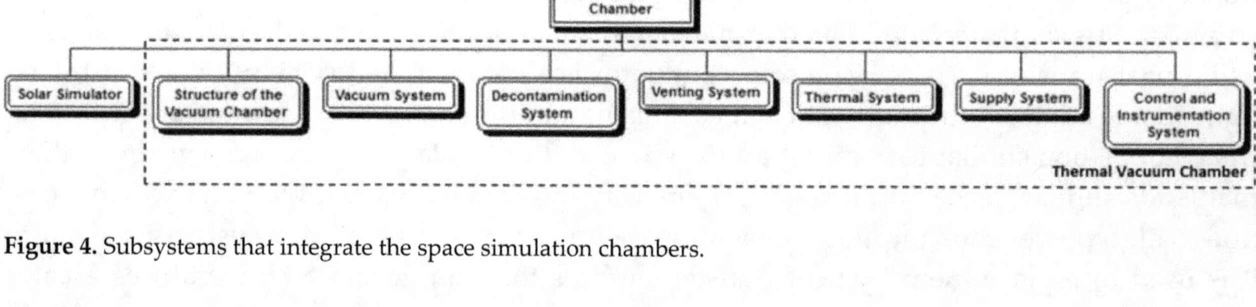

Figure 4. Subsystems that integrate the space simulation chambers.

the contamination due to outgassing generated by compounds and materials inside the vacuum chamber during an environmental test. The venting system permits the pressure inside the vacuum chamber to return to atmospheric pressure.

The supply system provides and manages the necessary resources (water, electricity, compressed air, specific substances, etc.) to operate the devices integrated in the systems of the space simulation chambers. The control and instrumentation system provide the mechanisms and interfaces to control and monitor the different mechanical, electronic, and electromechanical devices that compose the systems of the space simulation chambers.

The following sections will performed a description of each system that integrates the space simulation chambers, also some basic criteria and requirements to its function and interaction with the spacecraft or test specimen.

6. Structure of the vacuum chamber

The chamber structure allows the conservation of vacuum and thermal radiation phenomena, which are very important characteristics to simulate the space environment. There are several structural shapes for thermal vacuum chambers, but not all of them have a good structural rigidity which prevent their collapse by pressure changes (internal/external difference) and other stresses. **Figure 5** shows the different chamber shapes and their rigidity level.

A very common way to increase the structural rigidity of these shapes is through the use of stiffening rings. Stiffening rings, which are welded into the body extension, can reinforce structures that lack stiffness or have a considerable size. A cylindrical structure with dome ends is a typical design for a space simulation chamber. One criterion for defining the size of a chamber is the minimum operation pressure (vacuum level it would support). Another criterion is the thermal system size inside chamber and the maximum dimension of test specimen [11]. Taking into account the vacuum and thermal cycling generation processes, the materials for space simulation chambers manufacturing should meet certain requirements.

The material selection requirements for space simulation chamber manufacturing are stated below: the system materials shall preserve its mechanical properties under radiation, extreme

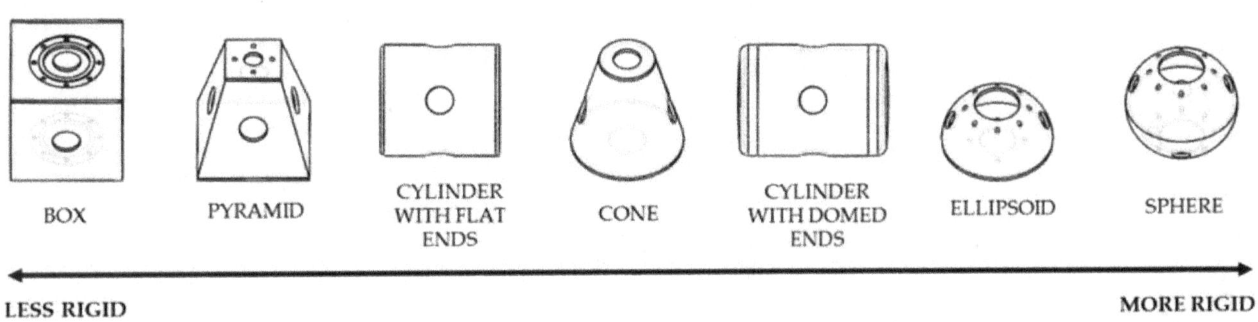

Figure 5. Chamber shapes and rigidity of the shapes [12].

temperature changes (≤ −173.15°C and ≥126.85°C), and high vacuum and ultra vacuum (1×10^{-7} to 1×10^{-12} mbar). The system materials vapor pressure shall be minimal when it is exposed to high temperatures (≥126.85°C) during operation. The system structural materials shall be impermeable to gases, with a surface to prevent impurities and substances retention. The structure of the vacuum chamber shall be designed to maintain a high structural rigidity. The system materials shall not react in vacuum and with other adjacent materials. The adjacent materials thermal expansion shall match the system without generating undesirable distortions and mechanical interactions. The materials of the system shall not excessively emanate gases under high-energy particles interaction. The system materials shall have a low outgassing potential (less than 10^{-6} mbar ls^{-1} cm^{-2}.) under vacuum. The system materials shall have proper degassing properties for manipulation. The system materials shall be suited to minimize or cancel the presence of sources of steam and undesirable gases (see **Figure 6**). The system shall be designed to be installed in cleanrooms and clean zones.

The basic criteria for the materials selection for space simulation chambers fabrication is the compliance of the previous defined requirements. The majority of space simulation chambers are fabricated from 300 series stainless steel [11]. For the chamber structure, type 304 stainless steel is used most frequently in vacuum systems [11, 12]. The 304 stainless steel is an appropriate material for vacuum chambers (also known as 18/8 stainless steel by its composition of 18% chrome and 8% nickel) given its properties such as low thermal conductivity, ductility, corrosion resistance, stiffness, weldability, and no magnetic reaction. Its surface shall be polished by several techniques (electro-polished, grained, bead blasted, machined/ground all over, and others) to homogenize, reducing the effective surface area, and adsorption capacity [14].

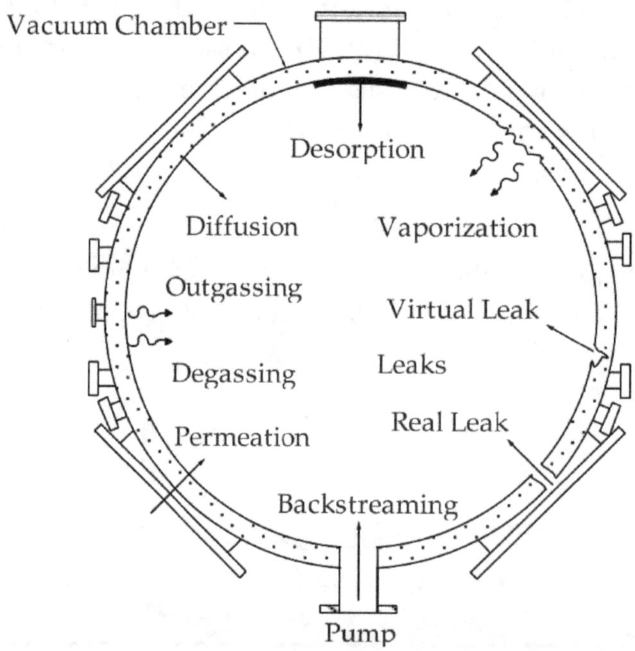

Figure 6. Potential source of gases and vapors in a vacuum chamber [13].

6.1. Penetrations

The chambers structure surfaces have mechanical interfaces called Flanges that allow the connection between several devices or units such as pumps, valves, sensors, filters, residual gas analyzers, electrical feedthroughs, mechanical feedthroughs, and others. These flanges can be rotatable or non-rotatable type. Flanges are designed from international organizations codes (ISO, ANSI, and DIN) that determine their dimensions, performance, materials, application, and usages [14]. **Figure 7** shows some types of flanges usually used in space simulation chambers. To realize specific measurements or to monitor specific equipment processes and observe internal vacuum chamber phenomena, the chamber structures provide viewports flanges or observation windows. The viewports are specifically designed for vacuum, and to resist mechanical and thermal stresses generated by simulation system operation. The viewports discs are usually manufactured with a special glass, quartz, sapphire, or borosilicate. Depending on the viewport material and vacuum level generated inside the chamber, special materials are adopted for sealing. The viewport shall not contact any other surface than the sealing materials and the simulated environment.

6.2. Note

The sealing materials and techniques, types of welding for the structure union, and the general feedthroughs characteristics that can be used in space simulation chambers are not described in this chapter, given the textual extension that would be generated. However, author believe these are important topics and relevant for space environment simulators design. These topics will be addressed to future publications.

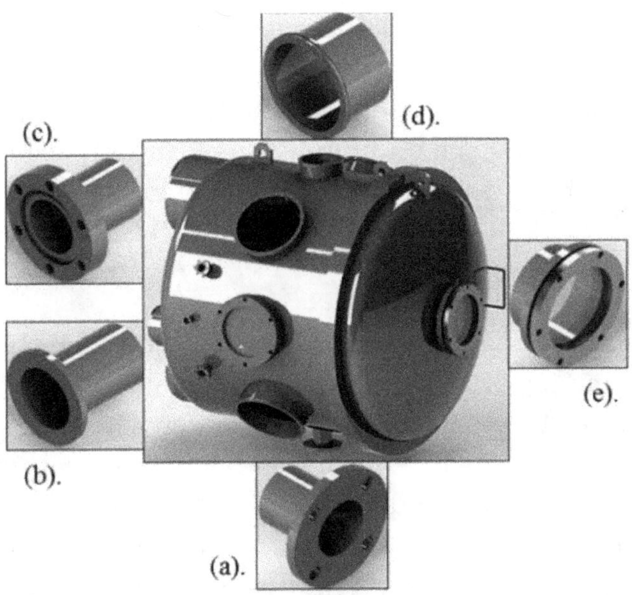

Figure 7. Common flange systems. (a) ASA flange, (b) KF flange, (c) CF flange, (d) ISO flange, (e) Viewport.

6.3. Rules and codes

There are no standards or specific rules that describe criteria to build space simulation chambers, however, pressure vessels international design standards are generally used for reference making the appropriate adjustments considering a vacuum chamber operation. The following standards define pressure vessels material selection, design, manufacturing, inspection, test, and certification:

- American Society of Mechanical Engineers (ASME) Boiler and Pressure Vessel (BPV) Code, Section VIII: Rules for Construction of Pressure Vessels, Divisions I & II;
- PD 5500 Specification for Unfired Fusion welded pressure vessels, Sections 2 & 3 published by British Standard Institute (BSI);
- Code De Construction des Apparelis a Pression (CODAP) French code for construction of unfired pressure vessels issued by SNCT (Syndicat National de la Chaudronnerie, de la Tolerie et de la tuyauterie industrielle);
- The European Standard EN 13445 Unfired pressure vessels issued by CEN (European Committee for Standardization);
- The European Standard EN 13458-2 Cryogenic Vessels – Statics Vacuum, Insulated Vessels – Part 2: Design, Fabrication, Inspection, and Testing issued by CEN (European Committee for Standardization).

7. Vacuum system

The function of this system is to reduce the pressure inside the chamber by means of a controlled evacuation of particles in gaseous and suspension state, which generally comprises the air found inside the chamber. The typical constituents of dry air are: nitrogen (N_2), oxygen (O_2), argon (Ar), carbon dioxide (CO_2), neon (Ne), helium (He), methane (CH_4), krypton (Kr), and hydrogen (H_2) among others. The pumping systems has capacity of evacuate to ambient these types of gases to generate vacuum inside the chamber.

7.1. Pumping systems

The gases removal to achieve a specific level of vacuum within the chamber is executed step-by-step using different pumping systems, which can operate individually or in specific cases in an interconnected way. The interconnected pumping systems (roughing pump and/or backing pump with high vacuum pump) are used to achieve different levels of vacuum. Usually two types of units are used: one to decrease pressure inside the chamber from 1 to 10^{-3} mbar (rough vacuum/medium vacuum), and another to decrease the pressure from 10^{-3} 3 to 10^{-8} mbar or less (high vacuum). This pumping process is necessary due to mechanical performance limits existing in the pumping units. **Figure 8** shows the range of vacuum and vacuum pumps.

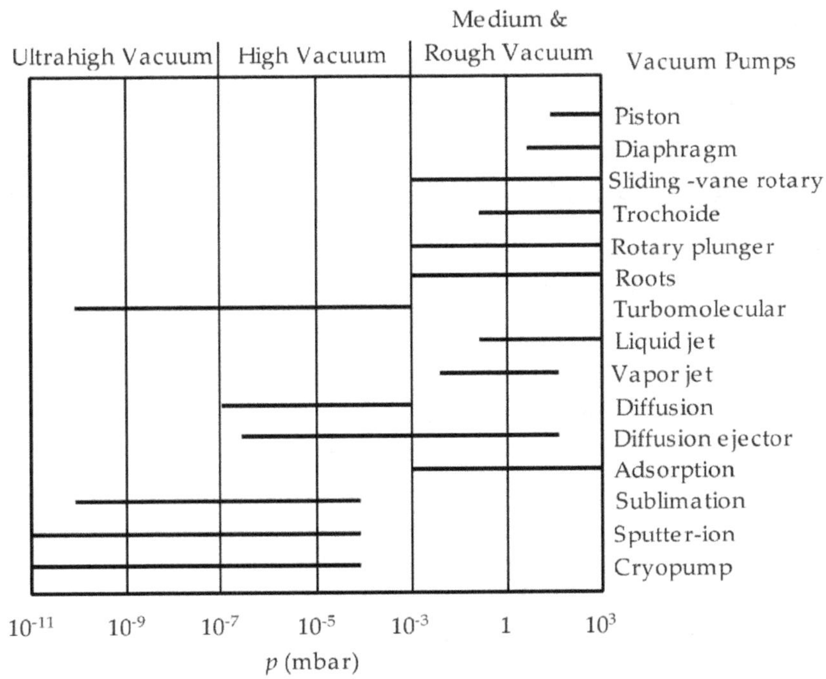

Figure 8. Ranges of vacuum and common working ranges of vacuum pumps [15].

The maximum level of vacuum that can be generated inside the space simulation chamber depends of the efficiency of the pumping units, the level of conductance in lines and appropriate control of cleaning, which avoids the presence of undesirable gases. It is worth to mention that not all pumping systems are suitable or entirely adequate to use in thermal vacuum systems for space simulation, given that some of these by the nature of their operation use lubricant components for cooling or for vacuum generation. This condition represents a risk due to probability of migration of polluting vapors into the chamber. When this happens, using filter elements as traps or a cold trap, it can reduce the effective capacity of the pumping unit [11]. Since the early 1980s, pumping systems have evolved mechanical and electronically, improving their performance, facilitating their operation, increasing their safety levels, reliability, cleaning, removing or replacing (in some cases) the use of consumables for operation and cooling of moving parts. The most used pumping systems for gases removal in space simulation chambers are: positive displacement pumps, cryogenic pumps, adsorption pumps, diffusion pumps, ion pumps and turbo-molecular pumps [11].

7.2. Traps and cold traps

The vacuum pumps that use oil for refrigeration or operation increase the oil vapor pressure control importance since the oil is exposed to the gases that are pumped from the vacuum chamber. If the oil vapor pressure is too high, it vaporizes when exposed to vacuum and may migrate to chamber, contaminating its interior [16]. The vacuum system is composed by foreline traps, refrigerated baffles and cold traps. Such devices are installed between the pumps and the vacuum chamber. They are used to remove contaminant particles and prevent backstreaming of oil vapors generated by pumps operation [16]. Foreline traps are devices composed by several types of filtering elements such as fibrous stainless steel sieve, fiberglass, or synthetic zeolite. These elements capture solid and gaseous substances that are potential system contaminants.

The refrigerated baffles and cold traps devices have cryogenic substances flowing to generate very low temperatures in their structure. These mechanisms restrain gaseous molecules that impact on their surface and transform them into crystalized solid particles (inverse sublimation or deposition) [16]. A significant temperature reduction at any part of the vacuum system reduces the vapor pressure, allowing a clean control of the process and faster achieving different vacuum levels inside the chamber. This principle is applied through the use of cold plates, decontamination plates, cryogenic pumps as well as the previously mentioned devices.

7.3. Pressure measurement

To measure the pressure inside the chamber, a variety of sensors for each vacuum level are used. Usually the sensors are divided into total pressure and partial pressure gauges. The total pressure gauges are classified from the operation of its internal mechanisms, which determine the pressure in a specific space, using hydrostatic pressure phenomena, thermal conductivity, or electrical ionization [14]. The partial pressure gauges determine the pressure of a gas mixture identifying their composition in a vacuum environment. In **Figure 9**, gauges for different ranges of vacuum are identified. Thermal conductivity and capacitive gauges (types of total pressure gauges) are generally used to determine the pressure in regions of low and medium vacuum. Usually it is used as hot and cold cathode ionization gauges and penning gauge to determining the pressure in regions of high and ultrahigh vacuum [14]. The adequate calibration of the vacuum gauges is fundamental to determine correctly the pressure level inside the vacuum chamber.

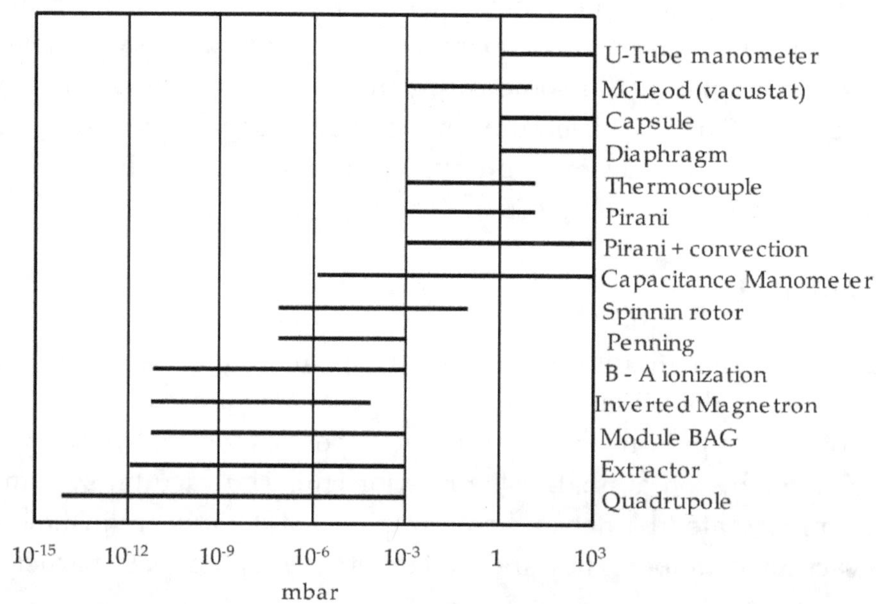

Figure 9. Pressure ranges for different gauges [14].

8. Decontamination system

This system generate a significant temperature reduction in a specific part inside the vacuum chamber, reducing the vapor pressure and capturing the molecules in suspension, allowing a clean control of the process, obtaining different vacuum levels inside the chamber more faster. This process considerably reduces the work of the pumping units. Usually, a cryopanel or a cold plate installed inside the vacuum chamber is flooded with a cryogenic substance and captures gaseous molecules that collide on their surface and turns them into crystalized solid particles (Deposition). Other types of device to collect condensable vapors are cold Fingers.

8.1. Contamination

Some techniques for the quantitative and qualitative analysis of the contaminants outgassed during space environment simulation test are: residual gas analysis, infrared spectroscopy, gas chromatography, mass spectrometry, quartz-crystal microbalance gravimetric, ultraviolet reflectance, and others [17]. The methods for detection sources of the contamination and measuring contamination during the development of the space environment simulation test are the direct and indirect methods [18]. A direct method for detection of the contamination is using the witness plates or reflective windows that are placed inside the space simulation chamber or in spacecraft. After the thermal vacuum test execution, the plates are analyzed through spectrometers to detect particulate pollutants. Residual gas analyzers and mass spectrometers are used to determine the gaseous composition inside the space simulation chamber and their pressure during operation. An indirect method for detection of the contamination is by means of the cleaning the witness plates after the thermal vacuum test execution. The film resulting from the cleaning process of the plates is analyzed through spectrometers [18]. The correct temperature and humidity level shall maintain the ideal environment for the space simulation chamber functioning.

9. Venting system

When testing procedures are finished and is necessary that the ambient pressure inside the chamber is returned, the vacuum system have a circuit of connected components, which allow access of filtered dry gas to increase the number of particles, reducing the mean free path. This procedure is used to open the chamber ensuring safety for system operators through the increase of chamber's internal pressure normalizing the environmental conditions. An inert gas is used, such as dry nitrogen gas (GN2), to vent the vacuum chamber to avoid pollution of objects that are housed therein. If ventilation is performed with other substances such as air or other compound gases, such substances can react with the surrounding temperature, causing water vapor molecules, and undesirable phenomenon of condensation which can be allocated in various areas within the chamber, contaminating exposed areas and the test specimen [5]. Usually, special filters are installed on the gas inlet line preventing access of impurities or microparticles.

10. Thermal system

The thermal system represents the mechanism whereby is possible simulate in a cycling manner the solar radiation effects and total darkness experienced by spacecrafts [7]. The cold heat sink of space can also be simulated. All phenomena described above are simulated with an approximation. To produce such conditions, the system uses a set of surfaces installed inside the chamber called "Thermal shrouds", and a number of external pipes connected in a hydraulic control circuit. The thermal shrouds, also known as cryoshrouds, are surfaces that are installed between the test object and the inner walls of the chamber in all directions. Thermal shrouds are mechanisms that provide a similar environment to the cold heat sink of space. This is obtained (between other factors) with the surface shroud high absorption coefficient for radiation in the parts that interact with test specimen. The shrouds have flowing circuits that are usually flooded by cryogenic substances, which can vary in temperature before reaching its interior, through an external thermal control process. The thermal platens are thermally controlled surfaces which have a similar operation to the shrouds, and generally the specimens are installed. The difference between shroud and platens are the thermal transfer characteristics. In **Figure 10**, the distribution of heat transfer surfaces in a space simulation chamber are identified.

Thermal shrouds and platens are generally manufactured in aluminum alloys 6000, 5000, and 1100 series. This series of aluminum alloys have low outgassing rates. The aluminum alloys are most frequently used in the manufacture of cryoshrouds surfaces, since it has a high thermal conductivity and high strength. However, nowadays is possible find on the market thermal shrouds made of 300 series stainless steel, which provides good characteristics for use in space environment simulation. Thermal shrouds need manufacturing treatments and special surface finish [19, 20].

10.1. Types of thermal shrouds and substances

There are various types of cryoshrouds profiles. **Figure 11** shows some of them.

Each of these profiles types can be adopted as thermal shrouds for space simulation systems. These profiles offers specific features of heat transfer, which depend mainly on: the type of material selected for the manufacture of its ducts and panels; the effective area occupying

Figure 10. Thermal-vacuum chamber parts.

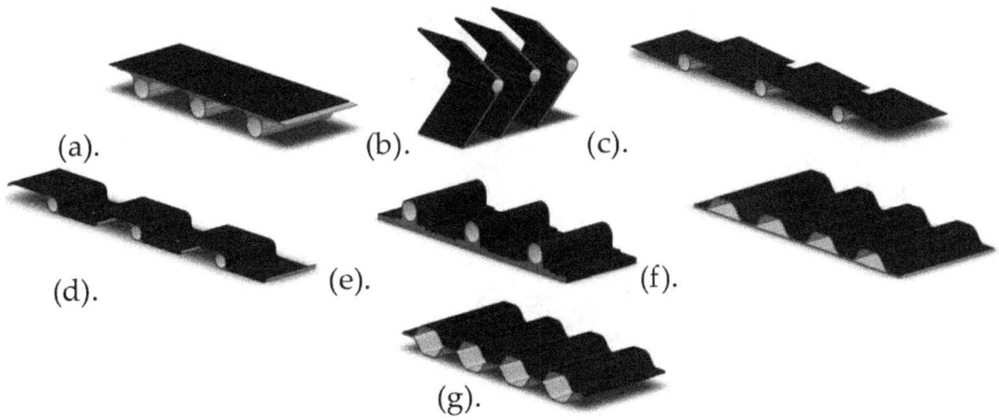

Figure 11. Different types of thermal shrouds. (a) Flat wing, (b) Chevron, (c) Bat wing, (d) Relieved bat wing, (e) D-tube on sheet, (f) Single embossed, (g) Double embossed.

each of their layers; the volume of the internal circuit; the type of fluid or substance to travel within; and surface finish.

Within the systems, external hydraulic circuit as well as the flow circuit of the internal thermal shrouds, flows a series of substances which by their properties can generate approximately −173.15°C (100 K), which is the minimum operating temperature required to develop a space thermal environment simulation [9]. Some substances as carbon dioxide (CO_2), oxygen (O_2), argon (A), carbon monoxide (CO), nitrogen (N_2), neon (Ne), hydrogen (H_2), and helium (He) have qualities thermal appropriates for using in the generation of the cold environment as the outer space. Some of these substances are difficult to handle, such as hydrogen and oxygen, which are highly flammable. Helium is an expensive gas, and the acquisition of high volumes in many cases is restricted. Nitrogen is a good choice for use in space simulators in liquid and gaseous state [9, 19]. In the liquid state (LN2), it has a temperature of −195.8°C, and in its gaseous state (GN2) under thermal/pressure control can reach temperatures from −180 to 150°C. In addition, nitrogen is relatively cheap and is commonly used in cryogenic processes. It should be noted that the nitrogen temperature ranges and the thermal load transmitted by the thermal shroud to the test object basically depend on the thermal control efficiency of the cryogenic substance and thermal shroud performance in terms of heat transfer and optical proprieties [19].

10.2. Generation of temperatures and considerations for environment simulation

By controlling the temperature that the fluid travels inside the thermal shrouds, it is possible to produce high and low temperatures in function of time following a thermal test profile [20]. These profiles are designed according to the nature of the spacecrafts mission and the type of thermal test: Thermal Balance Test, Thermal Vacuum Test, Vacuum Bake-out Test, and Functional Performance Test [5, 6]. During cycles, the distribution of heat transfer surfaces (thermal shrouds and platens) generates a series of desirable and undesirable thermal phenomena inside the vacuum chamber. The surfaces transfer heat to the specimen inside the chamber. The thermal shrouds do this transfer through radiation, and platens through conduction. In **Figure 12**, the basic components forms of heat transfer and other phenomena generated are identified.

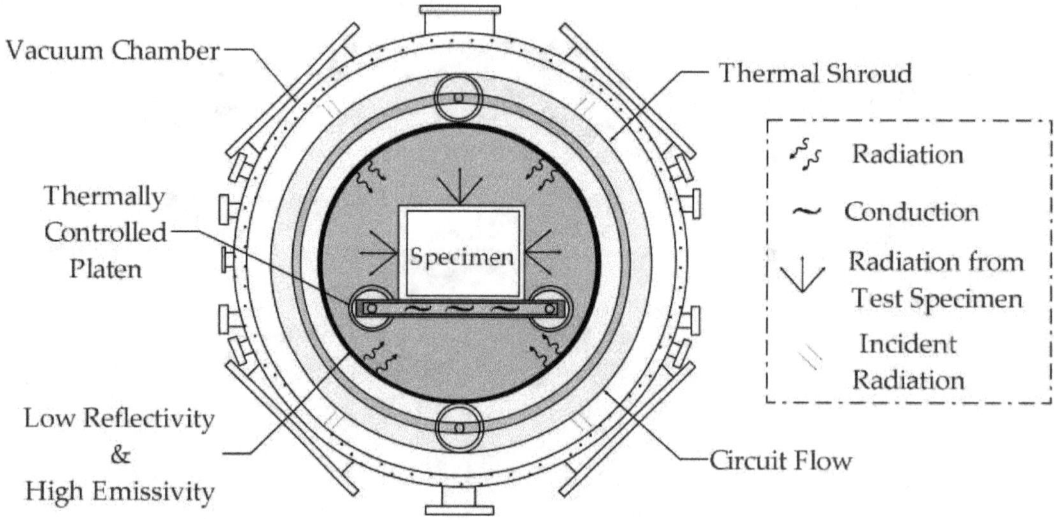

Figure 12. Basic components and heat transfer in the phenomena generated in the chamber.

It is important to note that to simulate the absence of thermal reflectivity, such as experienced by spacecraft in the space, and without influencing the test profile, the internal vacuum chamber areas, and the thermal shrouds receive a special surface finish. In addition, on account of the operation of thermal shrouds, a number of undesirable phenomena that alter the expected simulation environment are generated, and therefore is necessary to take into account a number of considerations to override such unwanted effects.

To simulate the space thermal conditions, the heat transfer surfaces must have a high radiation absorption coefficient, besides a characteristic of low reflectivity while remaining highly emissive. To accomplish this, the thermal shrouds area exposed to the spacecraft is coated with special black paint, which has low reflectivity in the visible spectrum, with a high capacity of radiation absorption [9, 19]. The coatings developed for this purpose have the capacity to withstand the vacuum and cyclic temperature changes, preserving its physical properties without distortion or outgassing in the test environment. To minimize the incident radiation that impacts the chamber by shroud functioning, both the chamber inner area and the exposed shroud surface shall have a special finish. Similarly, the internal chamber area shall be polished as part of the manufacturing process, giving a high gloss aspect to its surface, and the thermal shroud usually has a natural shine because of its manufacturing material. The amount of radiation between these surfaces is canceled by the surfaces finishes (emissivity between two walls).

10.3. Control temperature systems

To produce temperature changes within the space simulation chamber, there are a variety of forms the thermal control of the cryogenics substances that circulate through the internal thermal shrouds ducts. Some thermal control systems for space simulation chambers are: Gas-Bypass Flow System, Liquid Nitrogen Injection Systems, Gaseous Nitrogen injection Systems, Liquid and Gaseous Nitrogen Systems Combined, Thermal Conditioning Units (TCU), Helium refrigeration systems among others. Some of these systems mentioned above are classified as mechanical refrigeration systems, which use a closed-loop structure for recycling the cryogenic substances used as heat transfer vehicle in the thermal shrouds [20]. External piping system which circulates cryogenic substances, generally occupy considerable areas due to its size, and have a special thermal insulation to prevent energy loss during storage and fluid flowing into the system. In order to prevent loss of the substances cryogenic properties, a mechanical or vacuum jacketed insulation pipes is commonly used.

11. Solar simulator

In the space, environment simulation systems with solar simulators, the vacuum chamber design, vacuum system, and thermal system are designed from the solar simulation system technical and morphological characteristics [9]. Solar simulators are systems that reproduce the solar radiation phenomena experienced by spacecraft in space environment. The flux of incident solar energy on the spacecraft can be defined by intensity, spectral distribution, uniformity, and collimation. Collimation is an important effect for the spacecraft thermal balance, since it determines the spacecraft surfaces incident amount of energy, which needs to align with sunlight. Collimation also influences the behavior of spacecraft parts reflection [9]. The collimation of the solar radiation is determined by the sun incidents vector angle, which is 32 minutes of arc from earth distance. The solar simulator is composed by a set of lamps and a projector with a lens system (integrator) that points the generated light through a window to the chamber interior (See **Figure 13**). After that, the collimator mirror reflects the light to the test area in order to simulate the sun's natural collimation and light intensity [1]. The light source used by solar simulators is a set of lamps usually of Carbon-Arc, Xenon or Mercury-Xenon Compact-Arc types [9]. These simulators generate at least 1.35 KW m^{-2}. The unintended radiation, product from light sources, is reduced through refrigeration mechanisms that are usually cryogenic substances or water. The collimator mirror is composed by a number of segments of mirrors that reflect the integrator light. To maintain their proper functioning, the mirror segments inside vacuum chamber are cooled and heated with an isolated nitrogen circuit from the space simulation chamber thermal system [1]. Solar simulators can be classified in "modular type" or "simple reflector type" [9]. Both types of simulators have a set of mirrors that may be disposed as off-axis or on-axis position (see **Figure 13**).

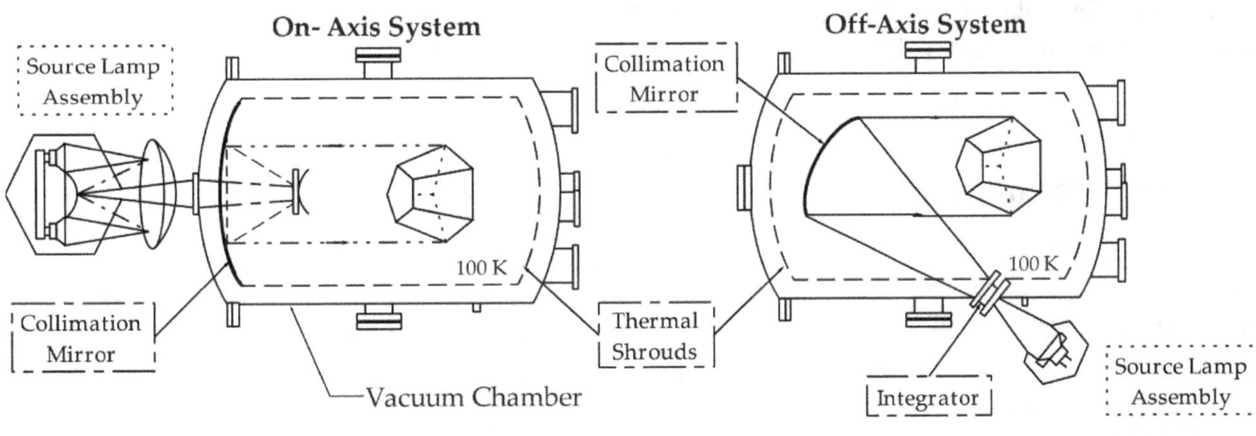

Figure 13. Two types of solar simulator configuration.

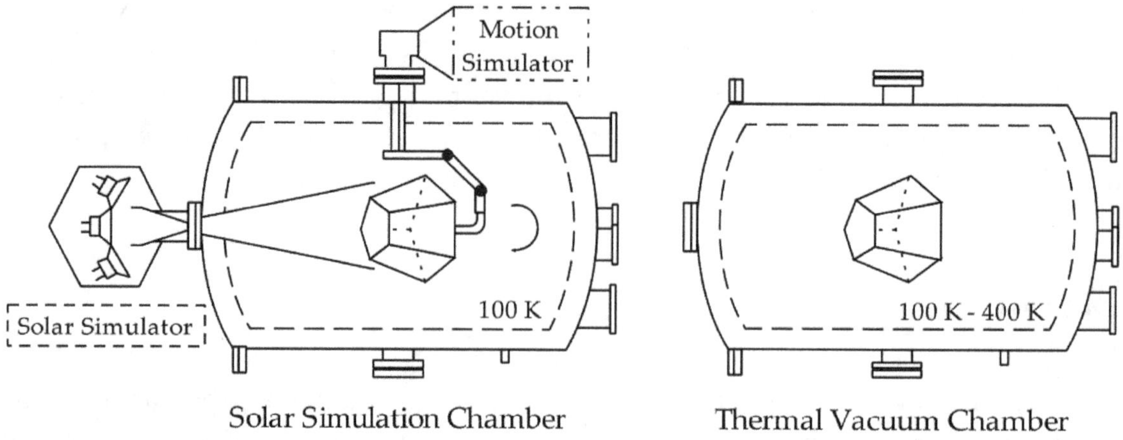

Figure 14. Space simulation chambers types.

The difference between them is that the test specimen emitted radiation does not return directly to the simulation system in an off-axis type configuration, decreasing the source of error by reflection in a test execution with solar simulator [9].

11.1. Motion simulator

A motion simulator allows to guide the spacecraft with respect to artificial solar beam [1]. The space simulation chamber with solar simulator may have a motion simulator system attached with external mechanisms or installed within the vacuum chamber (see **Figure 14**). Such system provides rotation (vertical or/and horizontal) to the tested spacecraft in order to distribute the solar radiation loads such as in the real operational environment. The motion simulator provides spacecraft orientation according to light direction or gravity vector [1]. In order to avoid undesirable thermal effects during tests, the motion simulator structure has the same temperature and optical characteristics than chamber thermal shrouds (see Section 10). The actuators and interior simulator mechanical components are maintained at atmospheric pressure through a special venting system [1].

The space simulation chambers with solar simulators have their vacuum chamber dimensions determined by solar simulator optical geometry, motion simulator size, and the volume used by cold environment simulation mechanisms [6].

12. Supply system

This system manages the necessary supplies for the operation of each of the devices in the space simulation chamber. The supply systems usually control water, compressed air, cryogenic substances, and electricity. The supply system consists of pneumatic and hydraulic lines, where are installed flow control valves, relief valves, filters, lubricators, pressure regulators, and different gauges.

13. Controls and instrumentation system

The controls and instrumentation system represents the interface through which the operator can exercise control and acquire information about the status of systems and devices that form the space simulation chamber. This system allows the operator to control the components involved in each stage of vacuum generation, and allows to control, monitoring, and intervention in the processes of temperature conservation and cycling [20].

The primitive control systems are coupled consoles in multi-bays or racks, which have mimic panels that depict the distribution of the components that are part of the vacuum and thermal systems. The mimic panels have lights that inform the status of devices, as well as pushbuttons and selector switches through which the operator can control. In addition, these control racks have several sections that houses command transmission mechanisms, indicators of operational data, and analogic displays that report on the status of the systems in operation. These control systems have interlocks, which protect the integrity of the systems, controlling unwanted decisions that can be made by operators. Racks have multiple connection lines inside them, analog gauges, microprocessors, controllers with operational architecture Proportional – Integral – derivative (PID), and I/O modules. The I/O modules allow discrete data processing signals generated by the buttons/switches on mimic panels, and transform these signals into decisions to electropneumatic and electromechanical devices that are hosted on the extension of thermal and vacuum systems. Generally, the control racks have a section of light and sound type alarms, which communicate extreme temperature, pressure, blackouts, and low flow conditions of supplies for normal operation of components [20].

Modern control systems for space simulation chambers are computing platforms powered by the use of programmable logic controller (PLC), which enables automation for some stages of operation and safe manual control of vacuum processes acquisition and thermocycling inside the chamber. The PLCs control units of main and auxiliary pumping, as well as vacuum valves, safety valves, thermal devices, and other components are part of chamber systems. Such controllers are connected to a central processor where their operating status is displayed on a Supervisory Control and Data Acquisition Program (SCADA) [20]. These systems consist of robust information platforms, and modern graphical unit interface/human-machine interface that allows active interaction between the operator and the system. The chamber operator communicates control decisions using computers, in which the systems distribution is shown.

14. Conclusions

The space simulation chambers serve as a test medium for various types of spacecraft and their subsystems. There are two types of space environment simulators, the ones with solar simulator and the ones without. Systems with solar simulator are known as solar simulation chambers. Systems without solar simulator are known as thermal vacuum chambers. The space simulation chambers with solar simulators have their vacuum chamber dimensions determined by solar simulator optical geometry, motion simulator size, and the volume used by cold environment simulation mechanisms. The basic systems that compose the space simulation chambers are: Structure of the Chamber, Vacuum System, Decontamination System, Venting System, Thermal System, Solar Simulator, Supply System, and Controls and Instrumentation System. The space simulation chambers have several vacuum pumps and thermal shrouds. The function of the pumps is to produce a desirable vacuum level in a reasonable time, conserving such level during development all test. The thermal shrouds are mechanisms that provide a similar environment to the cold heat sink of space. By controlling the temperature of the thermal shrouds, it is possible to produce high and low temperatures in function of time following a specific test profile for the spacecraft. In the space thermal environment simulation, it is not necessary duplicate exactly environment of the outer space, however, it is necessary to duplicate the effects that this environment generates in the materials, components, subsystems, and spacecraft systems. Using liquid nitrogen at 77°K as the cryogenic fluid for the thermal system operation, the space thermal radiation environment can be duplicated with an error of <1%. For the space environment simulation test is necessary achieved a level pressure less than 1×10^{-6} mbar, because to this level is possible properly evaluate the specimen and eliminate some undesirable effects. A contamination controlled environment is required in order to prevent damages to specimens and contamination to the vacuum chamber interior caused by airborne particles. The composition of the ambient inside the space simulation chamber can be identified and analyzed through equipments such as: gas chromatograph, mass spectrometers, quartz crystal microbalance (QCM), thermoelectric quartz crystal microbalance (TQCM), and witness plates.

Author details

Roy Stevenson Soler Chisabas[1,2,3]*, Geilson Loureiro[1] and Carlos de Oliveira Lino[1]

*Address all correspondence to: roy.soler@lit.inpe.br; rssolerc@gmail.com

1 National Institute for Space Research, INPE, São Paulo, Brazil

2 Integration and Testing Laboratory, LIT, São Paulo, Brazil

3 Systems Concurrent Engineering Laboratory, LSIS, São Paulo, Brazil

References

[1] Ley W, Wittmann K, Hallmann W. Handbook of Space Technology. 1st ed. John Wiley & Sons Ltd: Chichester; 2009 ISBN: 978-0-470-69739-9

[2] NASA. NASA—Spacecraft Classification [Internet]. 2017. Available from: https://www.nasa.gov/audience/forstudents/postsecondary/features/F_Spacecraft_Classification.html [Accessed: 2017-10-10]

[3] NASA. NASA Systems Engineering Handbook (NASA/SP-2016-6105). Rev 2. USA: CreateSpace Independent Publishing Platform; 2017 ISBN: 978-1977821966

[4] Vincent L. Pisacane: Fundamentals of Space Systems. 2nd ed. New York: Oxford University Press; 2005 ISBN 978-0-19.516205-9

[5] David G. Spacecraft Thermal Control Handbook – Volume I: Fundamental Technologies. 2nd ed. California: The Aerospace Press; 2002. ISBN 1-884989-11-X (v. 1)

[6] Nuss HE. Space Simulation Facilities and Recent Experience in Satellite Thermal Testing. Vacuum Pergamon Journals Ltd: Germany; 1987. DOI: 10.1016/0042-207X (87)90013-3

[7] Mercer S. Cryogenics: A technological tool for space scientist. Cryogenics. 1968. DOI: 10.1016/0011-2275(68)90044-1

[8] Hastings D, Garrett H. Spacecraft-Environment Interactions. 1st ed. Cambridge: Cambridge University Press; 1996. DOI: 10.1017/CBO9780511525032

[9] NATO - AGARD. Space Simulation Chambers and Techniques. 1st ed. Charlotte: Technical Editing and Reproduction Ltd; 1964. DOI:10.14339

[10] Haefer RA. Vacuum and Cryotechniques in space research. Elsevier. August 1972. DOI: 10.1016/0042-207X(72)93789-X

[11] Santeler DJ, Holkeboer DH, Jones DW, Pagano F. Vacuum Technology and Space Simulation. 1st ed. Washington D.C: NASA; 1966. NASA SP-105

[12] Harrison K: Engineering a Better Vacuum Chamber [Internet]. 2017. Available from: http://www.gnbvalves.com/pdf/EngineeringaBetterVacuumChambe.pdf [Accessed: 2017-10]

[13] O'Hanlon JF. A User's Guide to Vacuum Technology. 3rd ed. Unites States of America: John Wiley & Sons Inc; 2003. ISBN: 978-0-471-27052-2

[14] Chambers A, Fitch RK, Halliday BS. Basic Vacuum Technology. 2nd ed. Bristol: IOP Ltd; 1998. ISBN: 0 7503 0495 2

[15] Umrath W. Fundamentals of Vacuum Technology. Cologne: LEYBOLD; 1998. ISBN: 978-5881359867

[16] Herring D. Vac Aero International Inc. Cold Traps [Internet]. 2017. https://vacaero.com/information-resources/vac-aero-training/1225-cold-traps.html [Accessed: 2017-10-12]

[17] Goldsmith JC, Nelson ER. Molecular contamination in environmental testing at Goddard Space Flight Center. In: Fifth Space Simulation Conference Session I, Contamination; 14-16 September 1970; Gaithersburg, Maryland

[18] European Cooperation for Space Standardization, ECSS-Q-ST-70-05C Space Product Assurance – Detection of organic contamination of surfaces by infrared spectroscopy. Noordwijk, Netherlands. Requirements & Standards Division; 2009

[19] Gary S. Ash. Manufacturing of Cryoshrouds Surfaces for Space Simulation Chambers. In: 25th Space Simulation Conference. Environmental Testing, Session IV: New Capabilities and Facilities; 20-23 October 2008; Maryland: 2015

[20] Chisabas RSS. Space simulation chambers state-of-the-art. In: 67th International Astronautical Congress (IAC); 26-30 September 2016; Guadalajara, Mexico. 2016

12

Code Optimization for Strapdown Inertial Navigation System Algorithm

Ivana Todić and Vladimir Kuzmanović

Abstract

Inertial navigation systems are in common use for decades due to its advantages. Since INS outputs are usually used for inputs in different control algorithms (depending on applications), INS will induce certain errors and limitations. This chapter deals with optimization of the inertial navigation algorithm against limitations due to the accuracy and stability of signals from the sensors and constraints resulting from the integration step and processor speed used for embedded applications. Inertial navigation considered here is "strapdown" inertial navigation system (SINS) which assumes a fixed inertial measurement unit (IMU). In this chapter, fundamentals of strapdown inertial navigation will be presented as well as three different algorithms which will be analyzed in regard to numerical stability, time consumption and processor load criteria.

Keywords: strapdown inertial navigation system, quaternions, forward Euler integration, code optimization, code analyses

1. Introduction

INS is inertial navigation system, the system that determines the position based on the output of the motion sensors: accelerometers and gyroscopes. The first INS was based on accelerometers mounted on gimbal platform, to ensure measurement of acceleration in navigational frame. Nowadays "strapdown" inertial navigation system (SINS) is in common use, due to its mechanical simplicity, reduced size and price compered to platform INS. Strapdown inertial navigation system implies a fixed inertial measurement unit (IMU), whereby the analytical picture of the navigation system is obtained from the integration of the gyroscope rates.

The main problem that arises when SINS is used is the exact determination of the orientation based on the gyroscopes outputs. Every error made in this stage will affect the error of projection of the gravitational acceleration. Accelerations are integrated twice in order to determine the position, so any errors made when determining the orientation will cause the error in position determination to increase exponentially with integration time.

Errors when determining orientation are caused by the gyroscope performance and precision, as well as signal processing methods used for processing gyroscope outputs. Besides hardware limitations of the gyroscopes, algorithms used for orientation calculation also cause errors. This chapter focuses only on errors caused by applied algorithms and on optimization of these algorithms in terms of time consumption and processor load.

2. Fundamentals of inertial navigation

The basic idea of inertial navigation is based on the integration of acceleration measured by the accelerometers; see [1]. The accelerometers measure the specific force that can be represented as:

$$\mathbf{f} = \mathbf{a} - \mathbf{g} \tag{1}$$

where a is the absolute acceleration, acceleration in relation to the inertial coordinate frame, g is the gravitational acceleration.

In this chapter, the effect of the rotation of the Earth (which can simply be introduced into equations for the needs of systems operating in a longer time interval) is neglected.

In accordance with the previous assumption, the following relationship between acceleration and velocity in relation to the inertial coordinate frame is:

$$\begin{aligned} \mathbf{a} &= \left.\frac{d\mathbf{V}}{dt}\right|_I \\ \left.\frac{d\mathbf{V}}{dt}\right|_I &= \left.\frac{d\mathbf{V}}{dt}\right|_N + \boldsymbol{\omega}_N \times \mathbf{V} \end{aligned} \tag{2}$$

where $\left.\frac{d\mathbf{V}}{dt}\right|_N$ is the speed derivative relative to the navigation coordinate frame, $\boldsymbol{\omega}_N$ is the absolute angular velocity of the navigation coordinate frame.

In the inertial navigation algorithm, for the navigation coordinate frame, the ENUp coordinate frame has been adopted; see [2]. This choice is made due to the desire to have the height coordinate positive and on the other hand in order to more accurately determine the azimuth numerically.

In accordance with the ENUp coordinate frame, the following relations apply:

$$\begin{aligned} f_E &= \frac{dV_E}{dt} + \omega_N V_{up} - \omega_{up} V_N \\ f_N &= \frac{dV_N}{dt} - \omega_E V_{up} + \omega_{up} V_E \\ f_{up} &= \frac{dV_{up}}{dt} + \omega_E V_N - \omega_N V_E + g \end{aligned} \tag{3}$$

As a result of the WGS84 standard for the Earth shape (see [3]), projection of angular speeds of the ENUp coordinate frame has been adopted in the following form:

$$\omega_E = -\frac{V_N}{R_\phi + h}$$
$$\omega_N = \frac{V_E}{R_\lambda + h} \qquad (4)$$
$$\omega_{up} = \frac{V_E}{R_\lambda + h}\tan\phi$$

where h is the height above the reference ellipsoid, R_ϕ, R_λ is the radius of the curvature of the reference ellipsoid in the north-south and east-west directions, respectively.

$$R_\phi = \frac{R_e(1-e^2)}{(1-e^2\sin^2\phi)^{\frac{3}{2}}}$$
$$R_\lambda = \frac{R_e}{(1-e^2\sin^2\phi)^{\frac{1}{2}}} \qquad (5)$$

where R_e is the equatorial radius of the Earth, $e^2 = 1 - \frac{b^2}{a^2}$ is the eccentricity of the reference ellipsoid.

As the accelerometers measure acceleration in the coordinate frame related to the object, it is necessary to determine the transformation matrix from the body frame (see [4]) into the navigation frame, using information from the gyroscopes.

The navigation algorithm adopted here can be divided into two parts. The first part that works with higher frequency plays the role of determining velocity and angle increments, while the other part of the algorithm that works eight times slower provides information on the position and the speed in the navigation coordinate frame (usually required by the guidance law in the case of the missile application). Such algorithm is advantageous from the point of optimization of the calculation time in the control computer, which can be divided into eight different steps. Also, this SINS algorithm proved to be mathematically more stable in relation to others, in determining the quaternion position at the same sampling time. Namely, when integrating angular velocities in order to obtain the angular position, depending on the size of the integration step, the quaternion error increases over time, and in addition to renormalization, it also affects the overall error in position and velocity. This error does not occur with this algorithm.

2.1. Determination of angular increments and transformation matrix

The first step in determining the transformation matrix is the determination of angular inclusions, and as explained above, this process is repeated with the basic integration step which in this example is $t_s = 2$ ms:

$$\alpha_{x_b, y_b, z_b} = \int_{t_k}^{t_k + t_s} \omega_{x_b, y_b, z_b} dt \qquad (6)$$

where $\omega_{x_b, yb, zb}$ is the gyroscope signals in the body coordinate frame.

The process of calculating the position quaternion or the transformation matrix is also divided into two parts.

The first part is the calculation of the quaternion between the navigation coordinate frame and the body frame, assuming that the navigation coordinate frame can be considered inert during one step of integration.

The second part is used for the quaternion correction due to the rotation of the navigation coordinate frame.

If we compare these two transformations, we can conclude that the first transformation is the rotation of "fast" motion. One of the reasons why this algorithm proved to be numerically more stable is the separation of the integration of the "fast" rotation from the integration of the "slow" rotation.

If we compare the angular rates of those two motions, we can conclude that the "slow" rotation rates are four or more times lower than the "fast" rotation rates which leads to numerical integral errors when these two rotations are combined.

In accordance with the above, the following relations apply:

$$\begin{aligned} \mathbf{q}_{n+1}^{I} &= \mathbf{q}_n \Delta \mathbf{q}_f \\ \mathbf{q}_{n+1} &= \Delta \mathbf{q}_s \mathbf{q}_{n+1}^{I} \end{aligned} \tag{7}$$

where $\mathbf{q}^I$ is the quaternion of rotation from the body to the inertial coordinate frame, $\mathbf{q}$ is the quaternion of rotation from the body to the navigational coordinate frame, $\Delta\mathbf{q}_f$ is the quaternion of fast rotation increment, $\Delta\mathbf{q}_s$ is the quaternion of slow rotation increment.

The quaternion of fast rotation can be represented in the form of a rotary vector as follows:

$$\Delta \mathbf{q}_f = \begin{bmatrix} \Delta q_{f0} \\ \Delta q_{f1} \\ \Delta q_{f2} \\ \Delta q_{f3} \end{bmatrix} = \begin{bmatrix} \cos \frac{\Delta \Phi}{2} \\ \frac{\Delta \Phi_{xb}}{\Delta \Phi} \sin \frac{\Delta \Phi}{2} \\ \frac{\Delta \Phi_{yb}}{\Delta \Phi} \sin \frac{\Delta \Phi}{2} \\ \frac{\Delta \Phi_{zb}}{\Delta \Phi} \sin \frac{\Delta \Phi}{2} \end{bmatrix} \tag{8}$$

The following relationship holds for small angles:

$$\Delta \mathbf{\Phi} = \int_{t_n}^{t_n+t_m} \boldsymbol{\omega} dt + \frac{1}{2} \int_{t_n}^{t_n+t_m} (\mathbf{\Phi} \times \boldsymbol{\omega}) dt \tag{9}$$

where $t_m = 8t_s$ is the slow integration step.

To solve the previous equation, a four-step algorithm will be used (Conning correction [5–7]):

$$\Delta\mathbf{\Phi} = \begin{bmatrix} \Delta\Phi_{xb} \\ \Delta\Phi_{yb} \\ \Delta\Phi_{zb} \end{bmatrix} = \begin{bmatrix} \sum_{j=1}^{4}\alpha_{xb}(j) \\ \sum_{j=1}^{4}\alpha_{yb}(j) \\ \sum_{j=1}^{4}\alpha_{zb}(j) \end{bmatrix} + \frac{2}{3}\left(\mathbf{P}_1\begin{bmatrix} \alpha_{xb}(2) \\ \alpha_{yb}(2) \\ \alpha_{zb}(2) \end{bmatrix} + \mathbf{P}_3\begin{bmatrix} \alpha_{xb}(4) \\ \alpha_{yb}(4) \\ \alpha_{zb}(4) \end{bmatrix}\right)$$
$$+\frac{1}{2}(\mathbf{P}_1 + \mathbf{P}_2)\left(\begin{bmatrix} \alpha_{xb}(3) \\ \alpha_{yb}(3) \\ \alpha_{zb}(3) \end{bmatrix} + \begin{bmatrix} \alpha_{xb}(4) \\ \alpha_{yb}(4) \\ \alpha_{zb}(4) \end{bmatrix}\right) + \frac{1}{30}(\mathbf{P}_1 - \mathbf{P}_2)\left(\begin{bmatrix} \alpha_{xb}(3) \\ \alpha_{yb}(3) \\ \alpha_{zb}(3) \end{bmatrix} - \begin{bmatrix} \alpha_{xb}(4) \\ \alpha_{yb}(4) \\ \alpha_{zb}(4) \end{bmatrix}\right) \tag{10}$$

where

$$\alpha(j) = \alpha^{k}(t_s) + \alpha^{k-1}(t_s)$$

$$\mathbf{P}_j = \begin{bmatrix} 0 & -\alpha_{zb}(j) & \alpha_{yb}(j) \\ \alpha_{zb}(j) & 0 & -\alpha_{xb}(j) \\ -\alpha_{yb}(j) & \alpha_{xb}(j) & 0 \end{bmatrix}$$

If we return to the quaternion of slow rotation, the following relationship is valid:

$$\Delta\mathbf{q}_s = \begin{bmatrix} \cos\frac{\Omega t_m}{2} \\ -\frac{\Omega_x}{\Omega}\sin\frac{\Omega t_m}{2} \\ -\frac{\Omega_y}{\Omega}\sin\frac{\Omega t_m}{2} \\ -\frac{\Omega_z}{\Omega}\sin\frac{\Omega t_m}{2} \end{bmatrix} \tag{11}$$

where. $\Omega_x, \Omega_y, \Omega_z$ is the projections of the absolute angular velocity of the navigation coordinate frame on its axes.

If we neglect the rotation of the Earth, the following applies:

$$\begin{aligned} \Omega_x &= -\frac{V_y}{R_y} - \frac{V_x}{a}e^2 b_{13} b_{23} \\ \Omega_y &= \frac{V_x}{R_x} + \frac{V_y}{a}e^2 b_{13} b_{23} \\ \Omega_z &= 0 \\ \frac{1}{R_x} &= \frac{1}{a}\left(1 - e^2\frac{b_{33}{}^2}{2} + e^2 b_{13}{}^2 - \frac{h}{a}\right) \\ \frac{1}{R_y} &= \frac{1}{a}\left(1 - e^2\frac{b_{33}{}^2}{2} + e^2 b_{23}{}^2 - \frac{h}{a}\right) \end{aligned} \tag{12}$$

where b_{ij} are members of the transformation matrix from the Earth-coordinate frame (ECEF) into the navigation coordinate frame $\mathbf{B}_{ECEF}^{n}$.

The Poisson equation for the transformation matrix from the coordinate frame related to the Earth (ECEF) in the navigation coordinate frame can be written in the following form:

$$\begin{aligned} \dot{\mathbf{B}}_{n}^{ECEF} &= \mathbf{B}_{n}^{ECEF}\Delta\omega_{n-ECEF} \\ \mathbf{B}_{ECEF}^{n} &= \left(\mathbf{B}_{n}^{ECEF}\right)^{T} \\ \Delta\omega_{n-ECEF} &= \begin{bmatrix} 0 & 0 & \Omega_y \\ 0 & 0 & -\Omega_x \\ -\Omega_y & \Omega_x & 0 \end{bmatrix} \end{aligned} \quad (13)$$

The recursive solution of the Poisson equation can be represented in the following way:

$$\begin{aligned} b_{12}(N) &= b_{12}(N-1) - \Omega_y b_{32}(N-1)t_m \\ b_{22}(N) &= b_{22}(N-1) + \Omega_x b_{32}(N-1)t_m \\ b_{32}(N) &= b_{32}(N-1) + \left(\Omega_y b_{12}(N-1) - \Omega_x b_{22}(N-1)\right)t_m \\ b_{13}(N) &= b_{13}(N-1) - \Omega_y b_{33}(N-1)t_m \\ b_{23}(N) &= b_{23}(N-1) + \Omega_x b_{33}(N-1)t_m \\ b_{33}(N) &= b_{33}(N-1) + \left(\Omega_y b_{13}(N-1) - \Omega_x b_{23}(N-1)\right)t_m \\ b_{31}(N) &= b_{12}(N)b_{23}(N) - b_{22}(N)b_{13}(N) \end{aligned} \quad (14)$$

With the quaternion of fast and the quaternion of slow rotations defined above, on the basis of Eq. (7), the quaternion of total rotation can be determined and with its direct cosine matrix representing the transformation from the body to the navigation coordinate frame. This matrix will be updated with the time step of the slow integration:

$$\mathbf{C}_b^n = \begin{bmatrix} 1-2({q_2}^2+{q_3}^2) & 2(q_1q_2-q_0q_3) & 2(q_0q_2+q_1q_3) \\ 2(q_1q_2+q_0q_3) & 1-2({q_1}^2+{q_3}^2) & 2(q_2q_3-q_0q_1) \\ 2(q_1q_3-q_0q_2) & 2(q_0q_1+q_2q_3) & 1-2({q_1}^2+{q_2}^2) \end{bmatrix} \quad (15)$$

2.2. Determination of speed and position in space

Previously defined method used for determining the angle increments based on measured gyroscope signals can now be used in the same way to define the speed increments based on signals from the accelerometer. These increments are also determined by the fast integration step t_s:

$$\Delta W_{x_b, y_b, z_b} = \int_{t_k}^{t_k+t_s} a_{x_b, y_b, z_b} dt \quad (16)$$

where $a_{x_b, yb, zb}$ is the signals from the accelerometer in the body coordinate frame.

The absolute acceleration can be written in the following form:

$$\left.\frac{d\mathbf{V}}{dt}\right|_I = \left.\frac{d\mathbf{V}}{dt}\right|_b + \boldsymbol{\omega}_b \times \mathbf{V} \tag{17}$$

where $\left.\frac{d\mathbf{V}}{dt}\right|_b$ is the total speed derivatives with respect to the body coordinate frame, $\left.\frac{d\mathbf{V}}{dt}\right|_I$ is the total speed derivatives with respect to the inertial coordinate frame, $\boldsymbol{\omega}_b$ is the absolute angular velocity of the body coordinate frame.

The specific force projections acting in the body coordinate frame are obtained from the accelerometer. Accordingly, the integration will be performed in the body coordinate frame, and the previous equation can be written like

$$\left.\frac{d\mathbf{V}}{dt}\right|_b = \left.\frac{d\mathbf{V}}{dt}\right|_I - \boldsymbol{\omega}_b \times \mathbf{V} \tag{18}$$

If we apply integration with the slow integration step to the previous equation, we obtain the following:

$$\begin{aligned}
\int_{t_k}^{t_k+t_m} \frac{d\tilde{V}_{x_b}}{dt} dt &= \int_{t_k}^{t_k+t_m} \frac{d\overline{V}_{x_b}}{dt} dt + \int_{t_k}^{t_k+t_m} \left(\omega_{z_b} V_{y_b} - \omega_{y_b} V_{z_b}\right) dt \\
\int_{t_k}^{t_k+t_m} \frac{d\tilde{V}_{y_b}}{dt} dt &= \int_{t_k}^{t_k+t_m} \frac{d\overline{V}_{y_b}}{dt} dt + \int_{t_k}^{t_k+t_m} \left(\omega_{x_b} V_{z_b} - \omega_{z_b} V_{x_b}\right) dt \\
\int_{t_k}^{t_k+t_m} \frac{d\tilde{V}_{z_b}}{dt} dt &= \int_{t_k}^{t_k+t_m} \frac{d\overline{V}_{z_b}}{dt} dt + \int_{t_k}^{t_k+t_m} \left(\omega_{y_b} V_{x_b} - \omega_{x_b} V_{y_b}\right) dt
\end{aligned} \tag{19}$$

The recursive solution of the previous equations is done in eight steps (sculling correction; see [5–7]) from which the step of slow integration was adopted as $t_m = 8t_s$:

$$\begin{aligned}
W_{x_b,k} &= W_{x_b,k-1} + W_{y_b,k-1}\alpha_{z_b,k} - W_{z_b,k-1}\alpha_{y_b,k} + \Delta W_{x_b,k} \\
W_{y_b,k} &= W_{y_b,k-1} + W_{z_b,k-1}\alpha_{x_b,k} - W_{x_b,k-1}\alpha_{z_b,k} + \Delta W_{y_b,k} \\
W_{z_b,k} &= W_{z_b,k-1} + W_{x_b,k-1}\alpha_{y_b,k} - W_{y_b,k-1}\alpha_{x_b,k} + \Delta W_{z_b,k} \\
W_{z_b,k} &= W_{z_b,k-1} + W_{x_b,k}\alpha_{y_b,k} - W_{y_b,k}\alpha_{x_b,k} + \Delta W_{z_b,k} \\
W_{y_b,k} &= W_{y_b,k-1} + W_{z_b,k}\alpha_{x_b,k} - W_{x_b,k}\alpha_{z_b,k} + \Delta W_{y_b,k} \\
W_{x_b,k} &= W_{x_b,k-1} + W_{y_b,k}\alpha_{z_b,k} - W_{z_b,k}\alpha_{y_b,k} + \Delta W_{x_b,k}
\end{aligned} \tag{20}$$

The initial values in each new step of slow integration are $W_{x_b} = W_{y_b} = W_{z_b} = 0$.

After calculating the velocity increments in the body coordinate frame, it is possible to determine the increment of the velocities in the navigation coordinate frame, since the matrix of transformation between the body and the navigational coordinate frame has already been defined:

$$\begin{bmatrix} \Delta W_x \\ \Delta W_y \\ \Delta W_z \end{bmatrix} = \mathbf{C}_b^n \begin{bmatrix} W_{x_b} \\ W_{y_b} \\ W_{z_b} \end{bmatrix} \tag{21}$$

The speed of the object relative to the Earth in the navigation coordinate frame can now be represented by the following relations, with the remark that the Earth's rotation that is neglected:

$$\begin{aligned} V_x &= W_x - \int_{t0}^{t} V_z \Omega_y dt \\ V_y &= W_y + \int_{t0}^{t} V_z \Omega_x dt \\ V_z &= W_z - \int_{t0}^{t} (V_y \Omega_x - V_x \Omega_y + g) dt \end{aligned} \tag{22}$$

where Ω_x, Ω_y, Ω_z is the projections of the absolute angular velocity of the navigation coordinate frame on its axes, W_x, W_y, W_z is the sums of projections of velocity increments in the navigational frame.

The determination of the position in the navigation coordinate frame can be solved in two ways: by integration of the velocities, which is the case in determining the height, or by the relationship between the matrix defined by Poisson's equation and its definitions:

$$\mathbf{B}_{ECEF}^n = \begin{bmatrix} -\sin\varphi\cos\lambda\sin\varepsilon - \sin\lambda\cos\varepsilon & \sin\varphi\sin\lambda\sin\varepsilon + \cos\lambda\cos\varepsilon & \cos\varphi\sin\varepsilon \\ -\sin\varphi\cos\lambda\cos\varepsilon + \sin\lambda\sin\varepsilon & -\sin\varphi\sin\lambda\cos\varepsilon - \cos\lambda\sin\varepsilon & \cos\varphi\cos\varepsilon \\ \cos\varphi\cos\lambda & \cos\varphi\sin\lambda & \sin\varphi \end{bmatrix} \tag{23}$$

where φ is the latitude, λ is the longitude, ε is the azimuth.

Geographical navigation parameters can be determined from the relation of the preceding equation and Eq. (14):

$$\begin{aligned} \phi &= \arctan\frac{b_{33}}{b_0} \quad [-90, +90] \\ \lambda &= \arctan\frac{b_{32}}{b_{31}} \quad [-180, 180] \\ \varepsilon &= \arctan\frac{b_{13}}{b_{23}} \quad [0, 360] \\ b_0 &= \sqrt{{b_{13}}^2 + {b_{23}}^2} \end{aligned} \tag{24}$$

As the azimuth is now defined, projections of speed in the ENUp coordinate frame can be determined:

$$\begin{aligned} V_N &= V_y\cos\varepsilon + V_x\sin\varepsilon \\ V_E &= -V_y\sin\varepsilon + V_x\cos\varepsilon \end{aligned} \tag{25}$$

The position in the ENUp frame can be determined as

$$
\begin{aligned}
E &= \frac{180}{\pi}(\lambda - \lambda_0)\cos(\varphi_0)a \\
N &= \frac{180}{\pi}(\varphi - \varphi_0)a \\
h &= \int_{t_0}^{t} V_z dt
\end{aligned}
\tag{26}
$$

Similarly, using the matrix definition from the navigation coordinate frame and the body frame, we can get to the relations for angular positions:

$$
\begin{aligned}
\psi &= \arctan\left(\frac{C_b{}^n(1,1)}{C_b{}^n(2,1)}\right) \\
\varphi &= \arctan\left(\frac{C_b{}^n(3,2)}{C_b{}^n(3,3)}\right) \\
\theta &= \arcsin(C_b{}^n(3,1))
\end{aligned}
\tag{27}
$$

3. Strapdown INS (SINS) algorithms

Three SINS algorithms based on previously defined mathematical model will be presented here.

The basic solution of SINS is forward Euler method applied to the main equations for rotation and translation. Block diagram of this method is presented in **Figure 1**. In this algorithm there is no division to the fast and the slow rotation, and all calculation is done in each step.

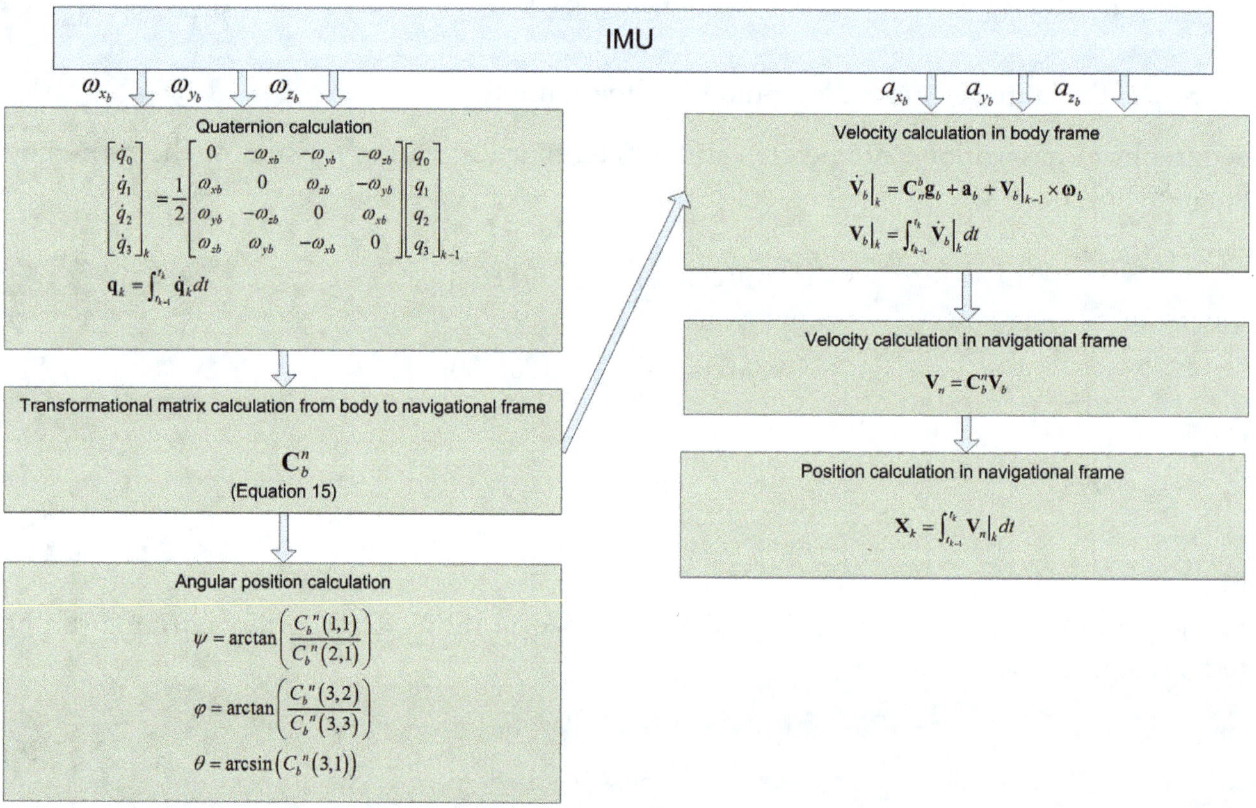

Figure 1. Forward Euler SINS algorithm block diagram.

The other solution of SINS algorithm—the regular SINS—based on mathematical model previously defined is presented in **Figure 2** as block diagram. The regular SINS algorithm calculates the velocity and angle increments eight times, and in the last step, Conning and Sculling corrections are implemented including all the other equations in **Figure 2**.

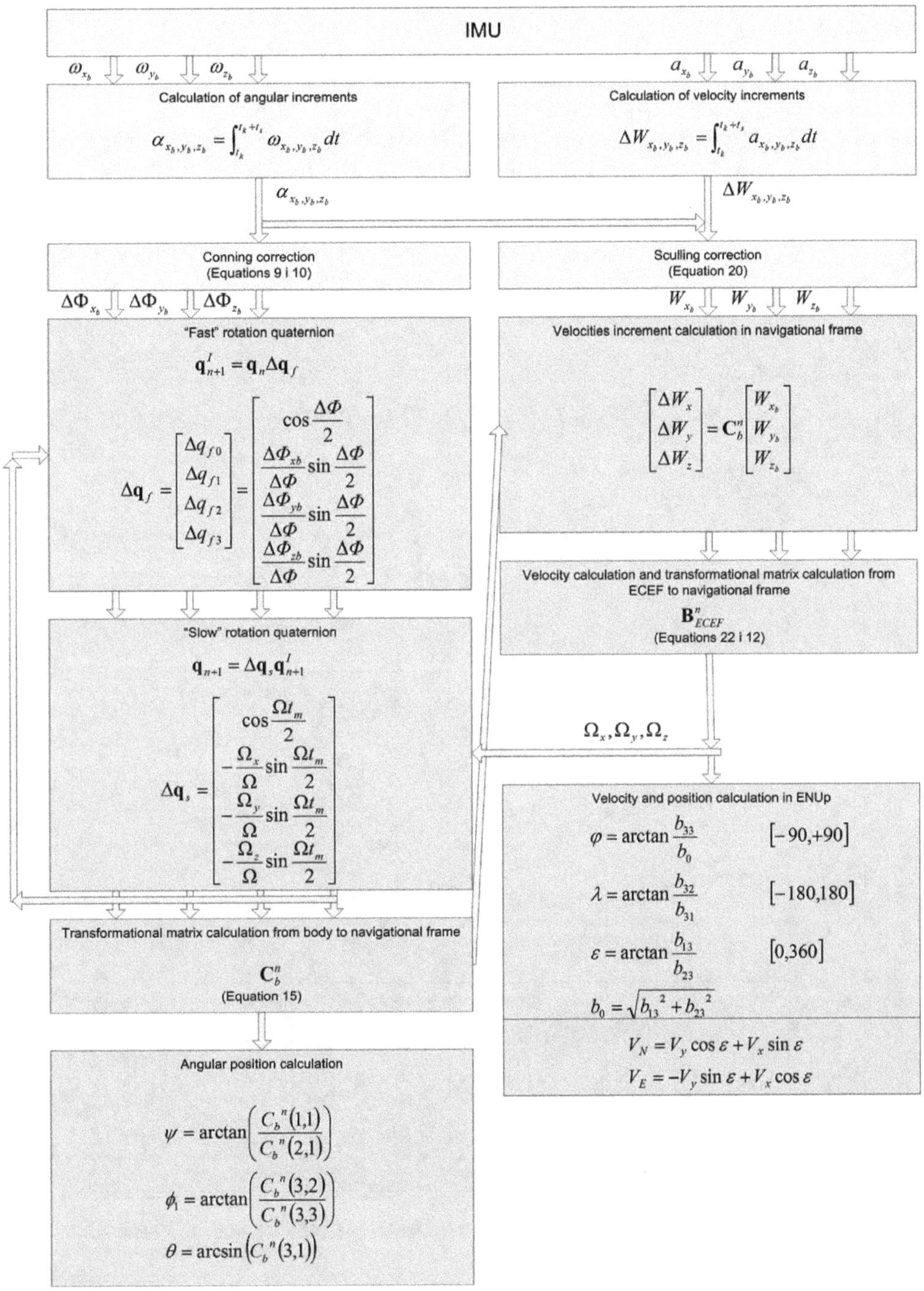

Figure 2. Regular SINS algorithm block diagram.

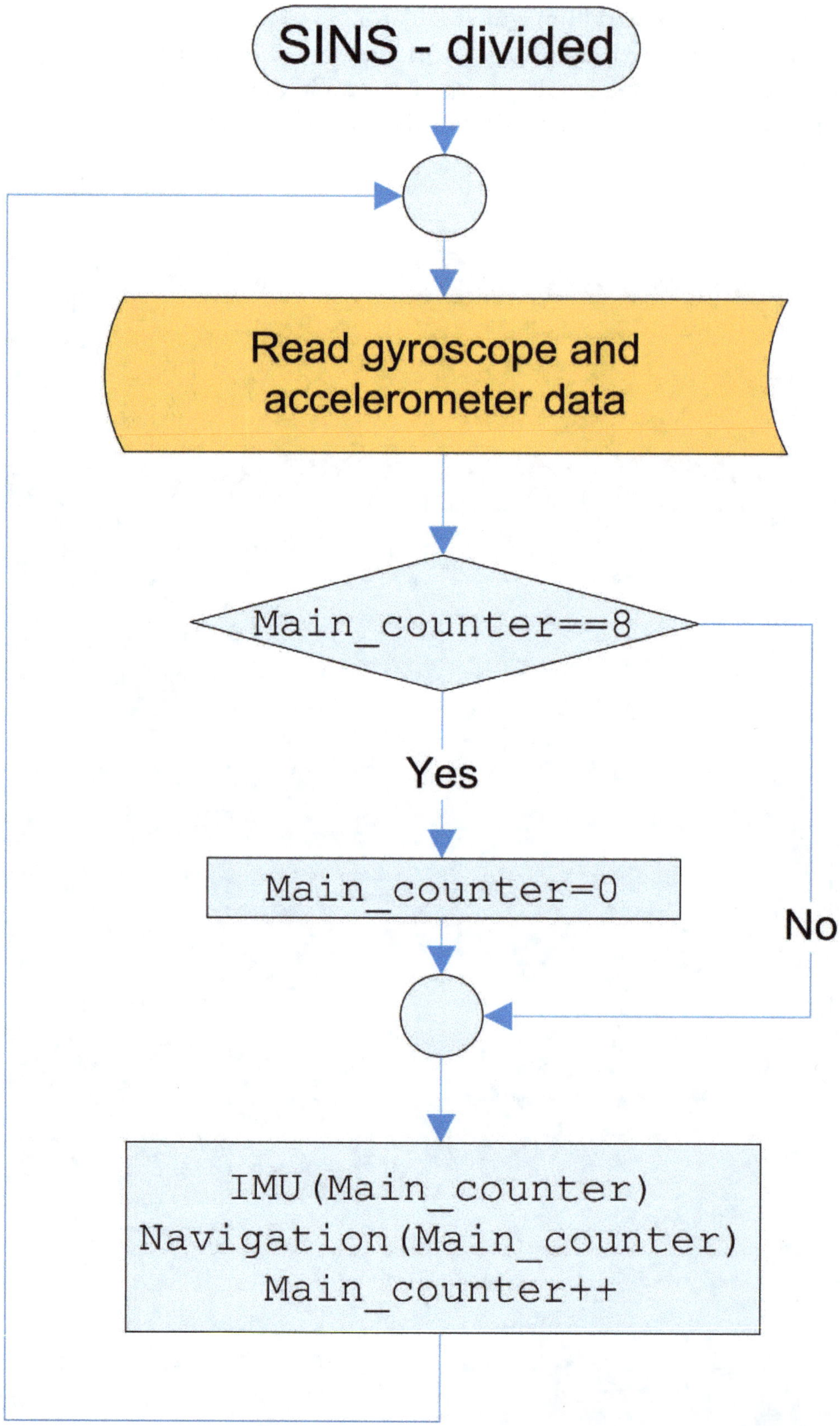

Figure 3. Divided SINS algorithm main flowchart.

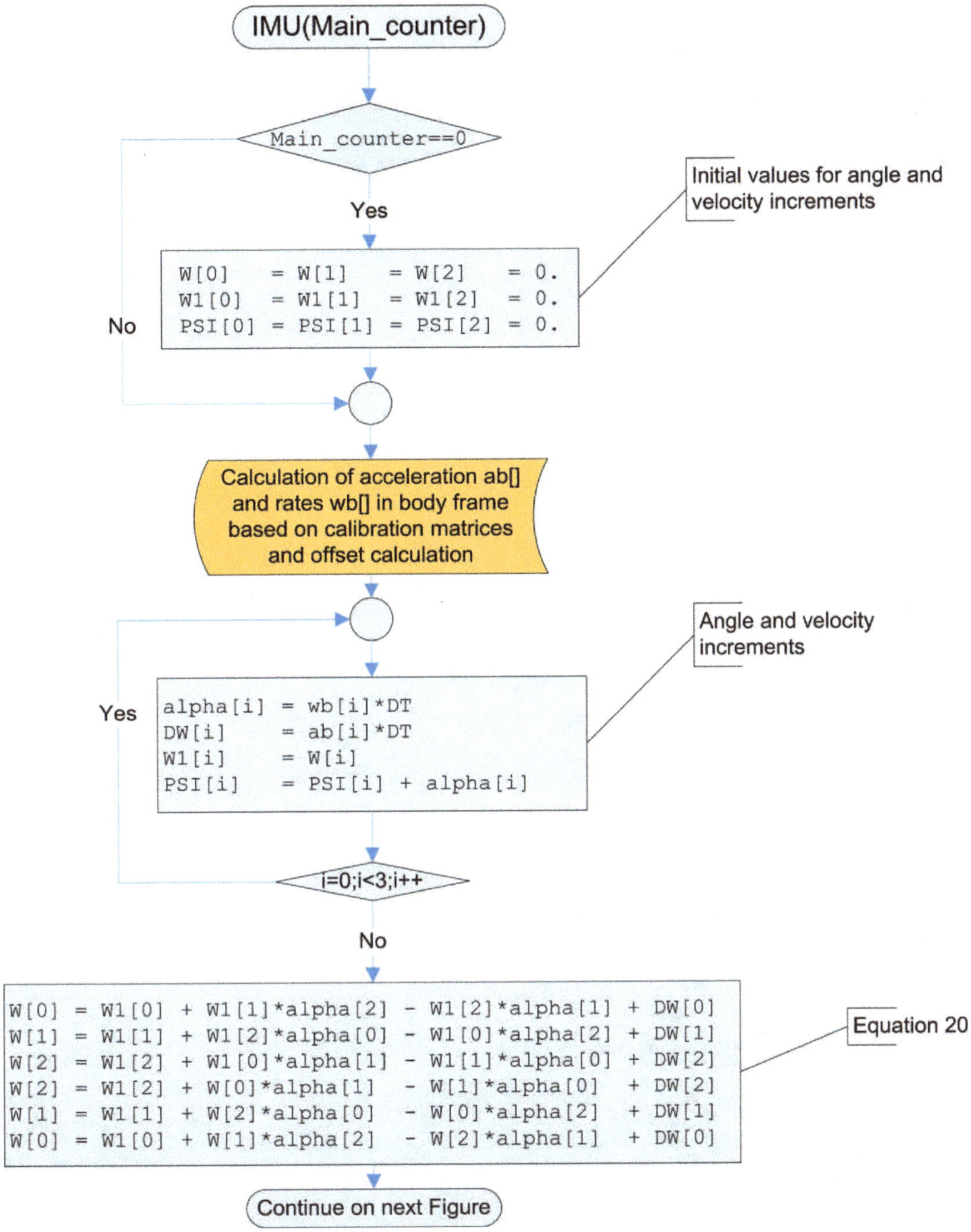

Figure 4. IMU procedure flowchart—part one.

The last solution that is considered is SINS algorithm divided in eight steps. This eight-step algorithm naturally arose as a consequence of Conning equation, and it is presented in the following flowcharts.

From **Figure 3**, it can be seen that in each step, the main algorithm will call IMU and navigation procedures. This means that in each step, some part of calculation will be completed.

From **Figure 4**, it can be seen that sculling correction will be calculated in each step (**Figures 5** and **6**).

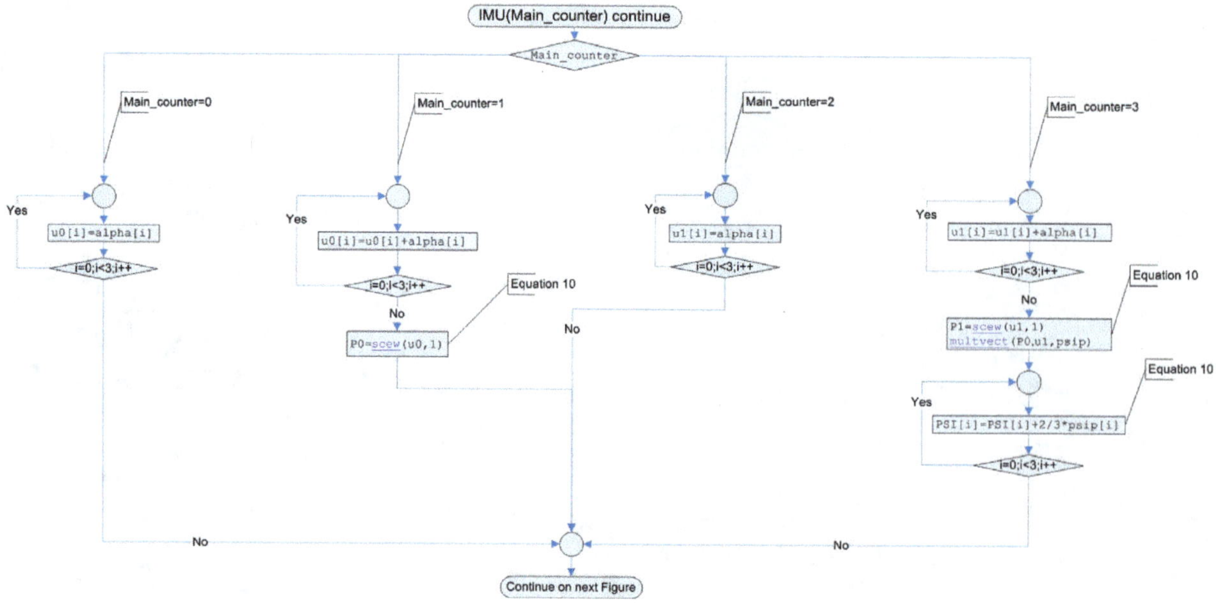

Figure 5. IMU procedure flowchart—part two.

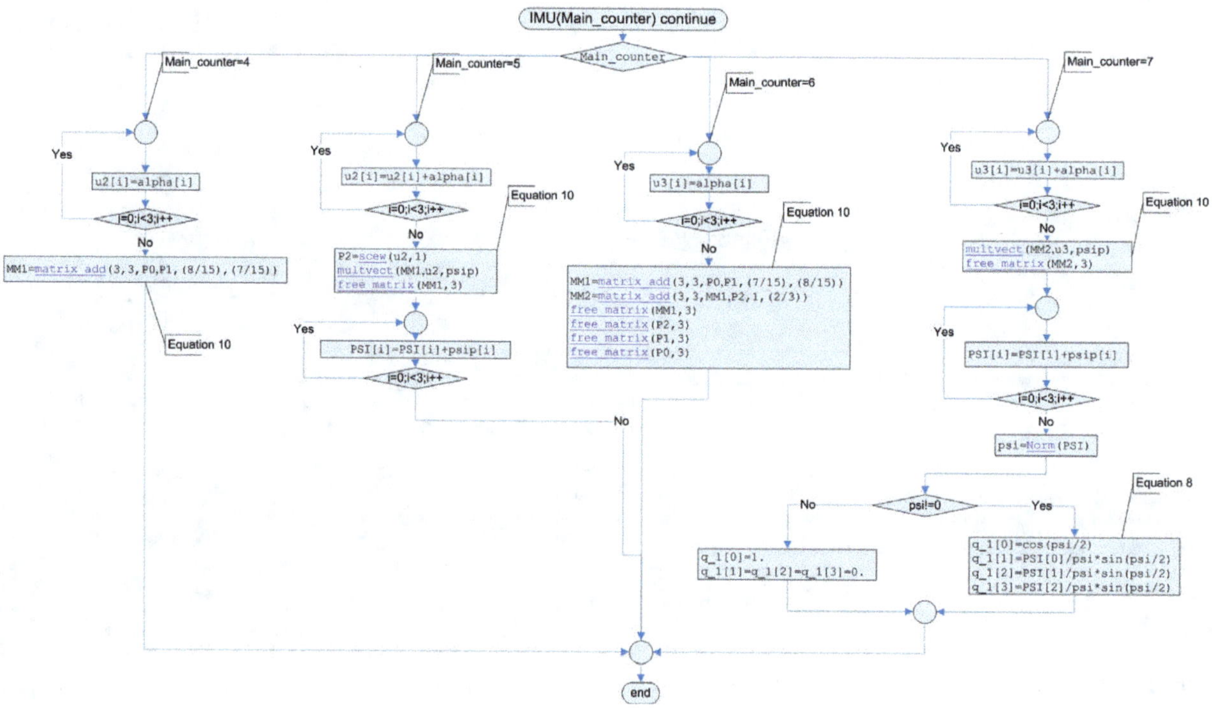

Figure 6. IMU procedure flowchart—part three.

Similar to the IMU algorithm, the navigation procedure is also divided into several steps shown in **Figure 7**.

Availability of output data calculated by all three SINS algorithms is presented in **Table 1**.

From **Table 1**, it can be seen that forward Euler algorithm provides all SINS output values in every step unlike regular and divided SINS which will provide outputs eight times slower.

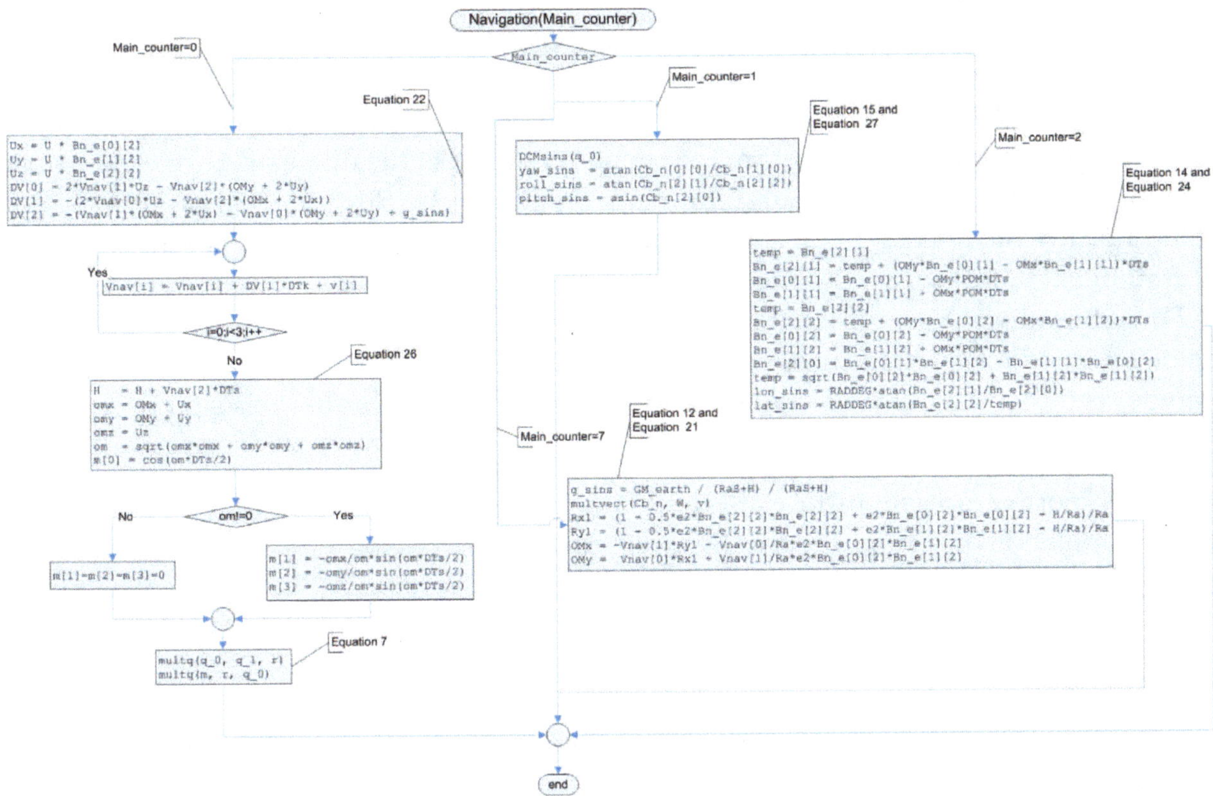

Figure 7. Navigation procedure flowchart.

Algorithm step	0	1	2	3	4	5	6	7
Forward Euler method	V_{xyz}, $\mathbf{B}^n_{ECEF}$, H, lat, lon, $\mathbf{C}^n_b$, ϕ, θ, ψ	V_{xyz}, $\mathbf{B}^n_{ECEF}$, H, lat, lon, $\mathbf{C}^n_b$, ϕ, θ, ψ	V_{xyz}, $\mathbf{B}^n_{ECEF}$, H, lat, lon, $\mathbf{C}^n_b$, ϕ, θ, ψ	V_{xyz}, $\mathbf{B}^n_{ECEF}$, H, lat, lon, $\mathbf{C}^n_b$, ϕ, θ, ψ	V_{xyz}, $\mathbf{B}^n_{ECEF}$, H, lat, lon, $\mathbf{C}^n_b$, ϕ, θ, ψ	V_{xyz}, $\mathbf{B}^n_{ECEF}$, H, lat, lon, $\mathbf{C}^n_b$, ϕ, θ, ψ	V_{xyz}, $\mathbf{B}^n_{ECEF}$, H, lat, lon, $\mathbf{C}^n_b$, ϕ, θ, ψ	V_{xyz}, $\mathbf{B}^n_{ECEF}$, H, lat, lon, $\mathbf{C}^n_b$, ϕ, θ, ψ
Regular SINS	—	—	—	—	—	—	—	V_{xyz}, $\mathbf{B}^n_{ECEF}$, H, lat, lon, $\mathbf{C}^n_b$, ϕ, θ, ψ
Divided SINS	V_x, V_y, V_z, H	$\mathbf{C}^n_b$, ϕ, θ, ψ	$\mathbf{B}^n_{ECEF}$, lat, lon	—	—	—	—	—

Table 1. Comparison of available data in each step for different SINS algorithms.

Generally, guidance and autopilot algorithms do not require inputs with such high frequency, and both regular and divided SINS will usually satisfy requirements; see [8]. On the other hand, if we compare the regular and the divided SINS algorithm, we can see that in the case of the divided SINS algorithm, the entire mission algorithm can be optimized in these eight steps.

The regular and the divided SINS algorithms are based on the same numerical integration, and the results of those two algorithms are equal in time. On the other side, we can compare quaternion stability of forward Euler integration and regular SINS algorithm in time. Quaternion norm which needs to be equal to one for quaternion of rotation is sensitive to the integration step for forward Euler integration. Both algorithms were implemented in MATLAB Simulink. Norm of quaternion is presented in **Figure 8** for the same integration step of 2 ms and for the same input data of gyroscopes presented in **Figure 9**. From **Figure 8** it can be seen that the quaternion norm will be affected whenever there is significant movement of the object.

Quaternion norm error will further affect all outputs of SINS algorithm, and that will lead to error accumulation over time. **Figure 10** represents angle errors for the same simulation.

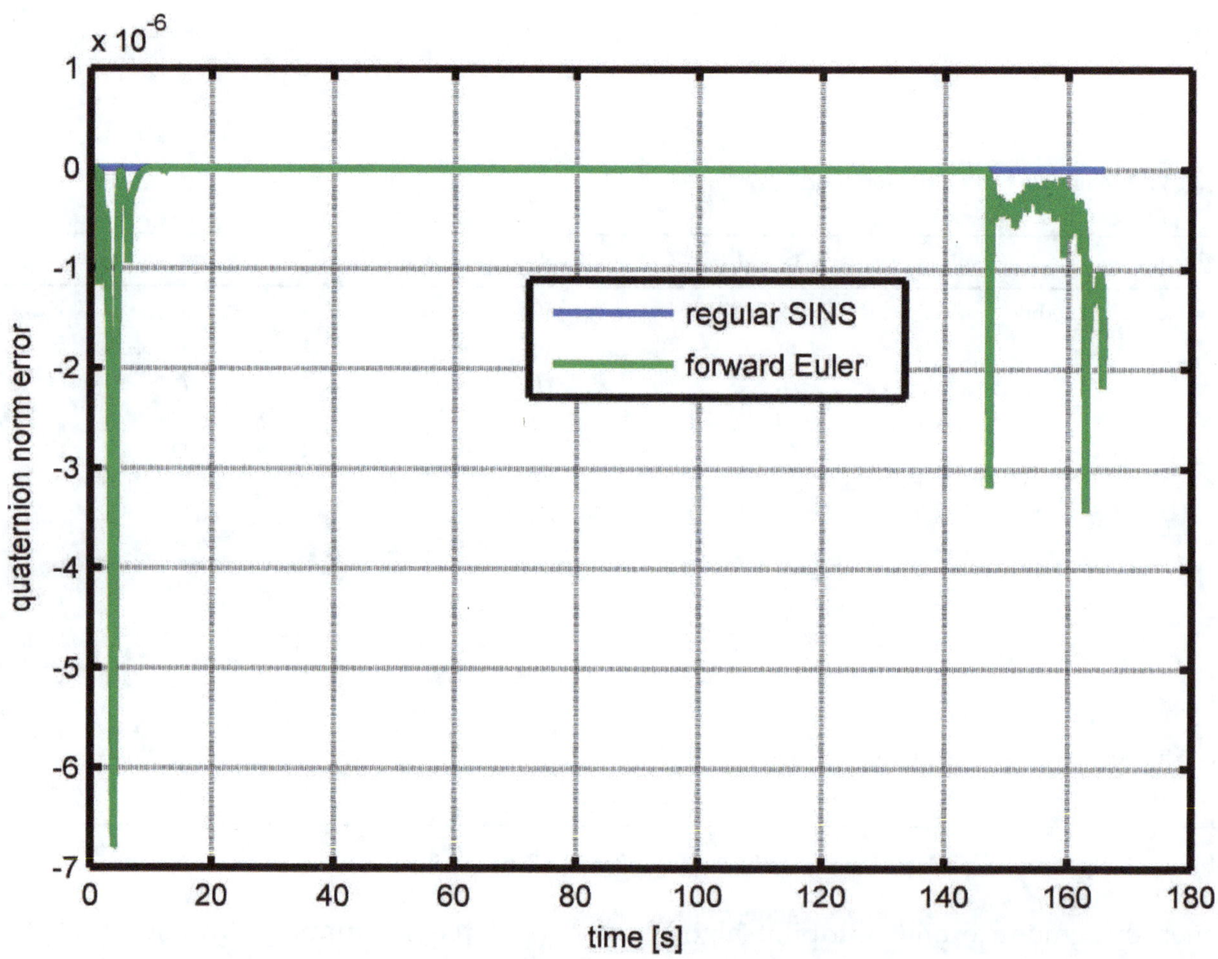

Figure 8. Quaternion norm error comparison.

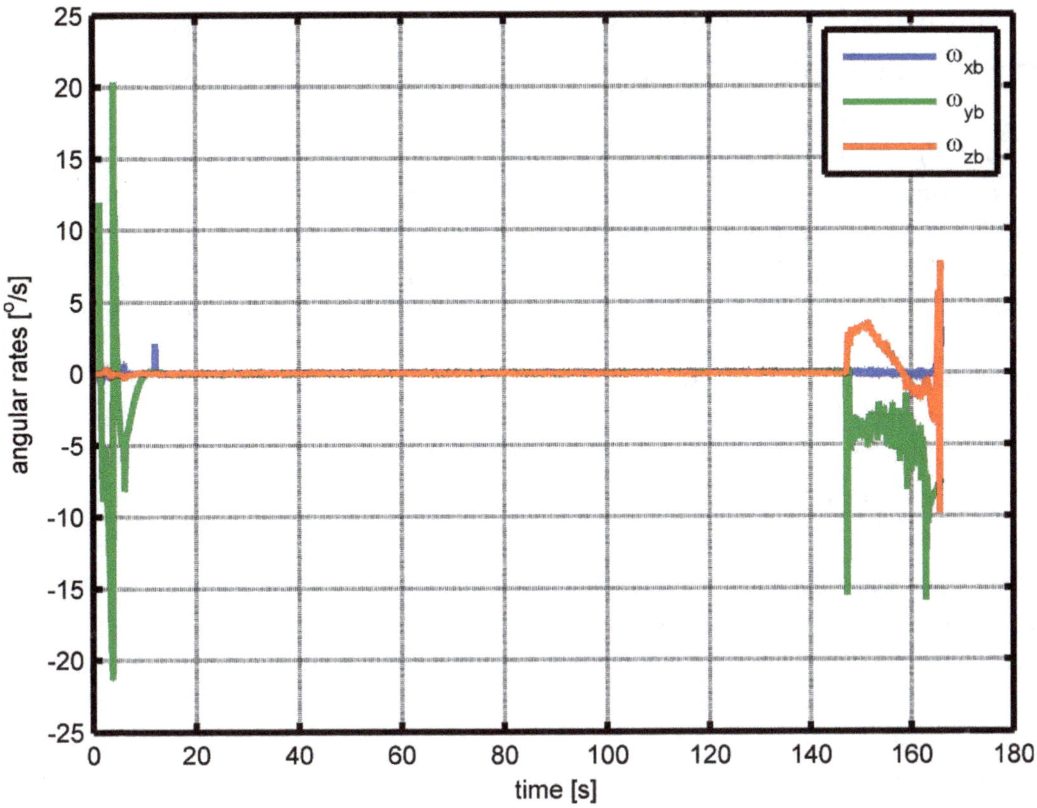

Figure 9. Input data from gyroscopes used for simulation.

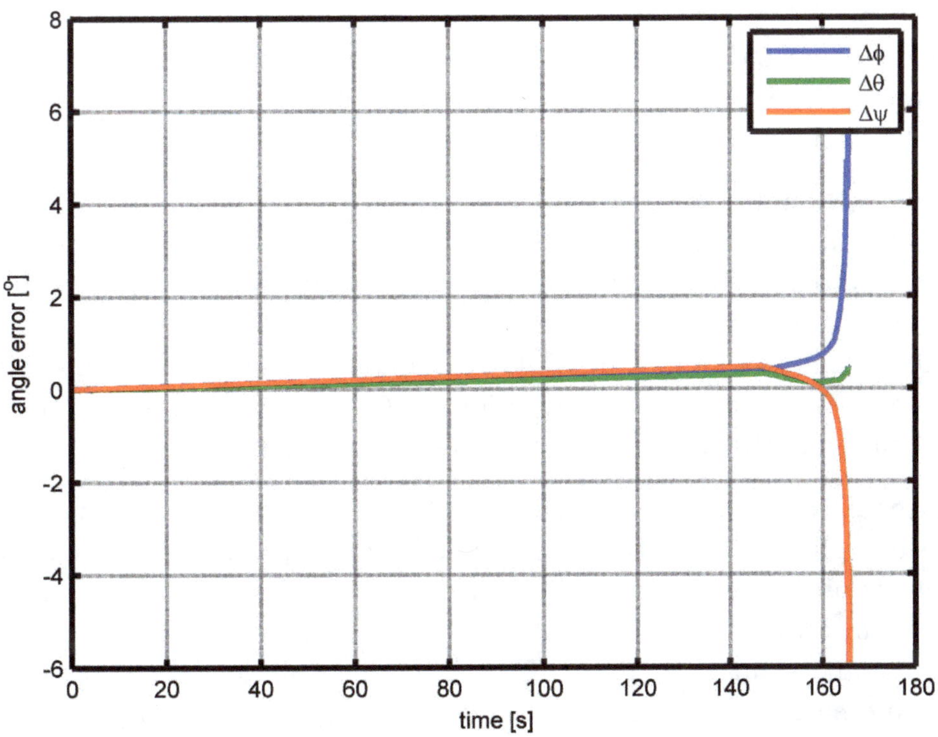

Figure 10. Angle error accumulation in time.

4. Time consumption and processor load comparison of the regular SINS, the divided SINS and the forward Euler algorithms

Forward Euler algorithm, regular SINS algorithm and SINS algorithm divided into eight different steps presented here were compared in terms of processor load and time it takes for all necessary calculations to complete. PC with Intel Core 2 Duo P8600 processor and 4 GB of RAM was used as a testbed for comparison of the three mentioned algorithms. Ubuntu 16.04 LTS operating system in real-time mode was used for time measurements and result generation.

Instead of using real sensors to feed the data to the algorithm, the data were read from the files that contained recorded sensor outputs from INS tests previously performed. All the data were memory mapped to avoid any loss of time due to IO operations, thus making the algorithm exclusively CPU bound. Real-time interval timer set to 2 ms was used as the time frame generator for the INS algorithm in order to mimic real-life operation. Every 2 ms, an interrupt would occur causing the next piece of data to be fed to the algorithm, and the next step of the algorithm would be performed. In the case of the regular SINS algorithm, the entire quaternion calculation will be performed in every eighth step. In the case of the divided SINS algorithm, a piece of that calculation will be calculated in all of those seven middle steps as well as in the final eighth step, thus optimizing processor load and dividing calculation time across all steps in the algorithm evenly.

Statistics that are compared after the completion of the two SINS algorithms are the total time spent in every eight steps of the algorithm, average amount of time spent in every step and average processor load in each of the steps of the algorithm. Total time spent in every step of the algorithm depends on the number of steps and as such is not important as a performance measure. Average time and average processor load in each step of the algorithm are used for performance comparison. In Linux, there are three distinct time measures of process execution. Those are wall clock time, user time and system time.

Wall clock time is the amount of calendar time that elapsed from starting the process or the stopwatch until moment "now". Thus, wall clock time includes the time the process has spent waiting for its turn on the CPU besides the time it actually spent running on the CPU. User time is the time the process spent executing on the CPU in user mode, while system time is the time the process spent executing on the CPU in system or kernel mode. User and system time measure the actual time the process spent using the CPU, and total amount of time spent on the CPU is calculated as the sum of these two time measurements.

All of these considered, wall clock represents the time that would be measured using a stopwatch. Although wall clock time heavily relies on the operating system load, on the scheduling policy used by the operating system and on the number of cores the CPU has, it can be used as a measure of time since all versions of the algorithm are subjected to the same conditions during the testing procedure. Even though the wall clock time is measured, it is not actually used in time comparison of the two mentioned algorithms. Instead, user and system time are used for comparison, because they rely only on the performance of the CPU, and the actual time it takes for calculations in the algorithm is the sum of these two times.

All times mentioned are given in microseconds. Total amount of time spent in the step of the algorithm is calculated as the sum of user and system time. Average processor load is calculated as

$$PL = \frac{\overline{u} + \overline{s}}{t} \tag{28}$$

where $\overline{u}$ is the average user time, $\overline{s}$ is the average system time, t is the time sample duration.

The results obtained after time measurements of the regular SINS algorithm are presented in **Table 2**.

The results obtained after time measurements of the divided SINS algorithm are presented in **Table 3**.

Results presented in the tables are not comparable to the execution times on faster or slower processors. Even though exact times are not comparable when a switch to a different CPU is made, their ratio will still hold. Relative time gain and processor load gain of the divided SINS over the regular SINS are presented in **Table 4**.

For the sake of completeness, forward Euler version of the SINS algorithm that performs all calculations in each timer interrupt was also taken into consideration. Every 2 ms both quaternions are calculated as well as navigation parameters. Basically, this approach has no notable steps, so the previous method of time measurement is not applicable here. Instead, the average time necessary for the calculation of both quaternions and navigation parameters is taken as the performance measure.

On average, it takes 6.16023 μs for the forward Euler algorithm to perform all calculations. This translates to average processor load of 0.00308 which is significantly worse compared to

Algorithm step	0	1	2	3	4	5	6	7
Elapsed time (μs)	2.07395	2.17831	3.11257	2.71898	2.04108	2.00986	2.06820	4.82662
Processor load	0.00104	0.00109	0.00156	0.00136	0.00102	0.00100	0.00103	0.00241

Table 2. Measured time of the regular SINS.

Algorithm step	0	1	2	3	4	5	6	7
Elapsed time (μs)	2.08052	2.19571	3.09868	3.07395	2.07148	2.10682	2.27033	2.35067
Processor load	0.00104	0.00109	0.00154	0.00153	0.00103	0.00105	0.00113	0.00115

Table 3. Measured time of divided SINS.

Algorithm step	0	1	2	3	4	5	6	7
Time (%)	+0.32	+0.80	−0.45	+13.05	+1.49	+4.82	+9.77	−51.30
Processor load (%)	+0.00	+0.00	−1.3	+12.5	+0.98	+5.00	+9.71	−52.28

Table 4. Regular and divided SINS time and processor load ratio.

Algorithm step	0	1	2	3	4	5	6	7
RSINS T (%)	−66.33	−64.64	−49.47	−55.86	−66.87	−67.37	−66.43	−21.65
RSINS PL (%)	−66.23	−64.61	−49.35	−55.84	−66.88	−67.35	−66.56	−21.75
DSINS T (%)	−66.23	−64.36	−49.70	−50.01	−66.37	−65.80	−63.14	−61.84
DSINS PL (%)	−66.23	−64.61	−50.00	−50.32	−66.56	−65.91	−63.31	−62.66

Table 5. Regular and divided SINS processor load and time gains over the forward Euler algorithm.

the regular SINS algorithm and even more so compared to the divided SINS algorithm. Processor load and time spent calculating in each step of the regular and the divided SINS algorithms vary from step to step, whereas the time it takes for the forward Euler algorithm to do its calculations can be considered as constant. Relative time gain (T) and processor load gain (PL) of the divided SINS (DSINS) and the regular SINS (RSINS) algorithm over the forward Euler algorithm are presented in **Table 5**.

5. Conclusion

In this chapter, navigation algorithm based on strapdown inertial navigation system algorithm optimized for coding in eight steps is presented. This algorithm proved to be a good option in situations where time and processor speed are limiting factors. Average time necessary for the regular SINS algorithm to complete all the steps and perform one full calculation is 21.02957 μs, whereas the divided SINS algorithm needs 19.24816 μs to perform the same operation, which scales to 8.47% improvement in time consumption. Even more important than time consumption improvement is the processor load in each timer interval, which is more uniformly distributed across all the steps in the divided SINS algorithm. Uniformly distributed processor load allows for easier design and development of multithreaded applications, as well as more free resources for the control computer to gather information about its surroundings and to issue commands to other devices in the control chain accordingly.

Also this algorithm proved to be mathematically more stable in term of quaternion norm, which mean that there is less error in angle computation and cumulatively in trajectory calculation.

Author details

Ivana Todić[1]* and Vladimir Kuzmanović[2]

*Address all correspondence to: itodic@mas.bg.ac.rs

1 Faculty of Mechanical Engineering, University of Belgrade, Belgrade, Republic of Serbia

2 Faculty of Mathematics, University of Belgrade, Belgrade, Republic of Serbia

References

[1] Titterton DH, Weston JL. Strapdown Inertial Navigation Technology. IEE Radar, Sonar, Navigation and Avionic Series. Vol. 17. 2nd ed. 2004. 576 p. ISBN: 0-86341-358-7

[2] Salychev O. Inertial Systems in Navigation and Geophysics. 1st ed. Moscow: Bauman MSTU Press; 1998. 352 p. ISBN: 5-7038-1346-8

[3] NIMA TR8350.2: Department of Defense World Geodetic System 1984, Its Definition and Relationship with Local Geodetic Systems. 3rd ed

[4] AIAA R-004: "Atmospheric and Space Flight Vehicle Coordinate Systems". American Institute of Aeronautics and Astronautics (AIAA); 1992. 69 p

[5] Salychev O. Applied Inertial Navigation: Problems and Solutions. Moscow: Bauman MSTU Press; 2004. 306 p. ISBN: 5-7038-2395-1

[6] Savage PG. Strapdown inertial navigation integration algorithm design Part1: Attitude algorithms. Journal of Guidance, Control, and Dynamics. 1998;**21**(1):19-28

[7] Savage PG. Strapdown inertial navigation integration algorithm design part 2: Velocity and position algorithms. Journal of Guidance, Control, and dynamics. 1998;**21**(2):208-221

[8] Siouris GM. Missile Guidance and Control Systems. 1st ed. New York: Springer; 2004. 666 p. ISBN: 0-387-00726-1

Permissions

All chapters in this book were first published in SF, by InTech Open; hereby published with permission under the Creative Commons Attribution License or equivalent. Every chapter published in this book has been scrutinized by our experts. Their significance has been extensively debated. The topics covered herein carry significant findings which will fuel the growth of the discipline. They may even be implemented as practical applications or may be referred to as a beginning point for another development.

The contributors of this book come from diverse backgrounds, making this book a truly international effort. This book will bring forth new frontiers with its revolutionizing research information and detailed analysis of the nascent developments around the world.

We would like to thank all the contributing authors for lending their expertise to make the book truly unique. They have played a crucial role in the development of this book. Without their invaluable contributions this book wouldn't have been possible. They have made vital efforts to compile up to date information on the varied aspects of this subject to make this book a valuable addition to the collection of many professionals and students.

This book was conceptualized with the vision of imparting up-to-date information and advanced data in this field. To ensure the same, a matchless editorial board was set up. Every individual on the board went through rigorous rounds of assessment to prove their worth. After which they invested a large part of their time researching and compiling the most relevant data for our readers.

The editorial board has been involved in producing this book since its inception. They have spent rigorous hours researching and exploring the diverse topics which have resulted in the successful publishing of this book. They have passed on their knowledge of decades through this book. To expedite this challenging task, the publisher supported the team at every step. A small team of assistant editors was also appointed to further simplify the editing procedure and attain best results for the readers.

Apart from the editorial board, the designing team has also invested a significant amount of their time in understanding the subject and creating the most relevant covers. They scrutinized every image to scout for the most suitable representation of the subject and create an appropriate cover for the book.

The publishing team has been an ardent support to the editorial, designing and production team. Their endless efforts to recruit the best for this project, has resulted in the accomplishment of this book. They are a veteran in the field of academics and their pool of knowledge is as vast as their experience in printing. Their expertise and guidance has proved useful at every step. Their uncompromising quality standards have made this book an exceptional effort. Their encouragement from time to time has been an inspiration for everyone.

The publisher and the editorial board hope that this book will prove to be a valuable piece of knowledge for researchers, students, practitioners and scholars across the globe.

List of Contributors

Daniel Condurache
Technical University of Iasi, Iaşi, Romania

Yongjie Ding, Hong Li and Daren Yu
School of Energy Science and Engineering, Harbin Institute of Technology, Harbin, People's Republic of China

Liqiu Wei
Academy of Fundamental an Interdisciplinary Science, Harbin Institute of Technology, Harbin, People's Republic of China

Zhengshi Yu, Pingyuan Cui, Rui Xu and Shengying Zhu
School of Aerospace Engineering, Beijing Institute of Technology, Beijing, China
Key Laboratory of Autonomous Navigation and Control for Deep Space Exploration, Ministry of Industry and Information Technology, Beijing, China

Barbara Bermudez-Reyes
Faculty of Mechanical and Electrical Engineering, Autonomous University of Nuevo Leon, Nuevo León, México

Frederic Trillaud and Lauro Santiago-Cruz
Institute of Engineering, National Autonomous University of Mexico, México City, México

Fernando Velazquez-Villegas
Center of Advanced Engineering, Faculty of Engineering, National Autonomous University of Mexico, México City, México

Jonathan Remba-Uribe
Remtronic Telecommunications, Guanajuato, México

Ana M. Arizmendi-Morquecho
Center of Research in Advanced Materials-Monterrey, Nuevo León, México

Alberto Caballero-Ruíz and Leopoldo Ruiz-Huerta
National Laboratory of Additive Manufacturing, 3D digitalization and Computed Tomography, Center of Applied Sciences and Technological Development, National Autonomous University of Mexico, México City, México

Mario A. Mendoza-Barcenas and Rafael Prieto-Melendez
Aerospace Development Center, National Polytechnic Institute, México City, México

Paula S. Morgan
Jet Propulsion Laboratory, California Institute of Technology, Pasadena, California, United States of America

Marco Gómez Jenkins
Space Systems Laboratory, Costa Rica Institute of Technology, Cartago, Costa Rica

Jose Antonio Castro Nieto
Ad Astra Rocket Company, Costa Rica

Kobi Gozlan, Kfir Cohen and Boaz Ben-Moshe
K&CG lab, Ariel University, Israel

Yuval Reuveni
K&CG lab, Ariel University, Israel
Department of Management, Bar-Ilan University, Israel
Department of Physics, Ariel University, Israel
Eastern R&D Center, Ariel, Israel

Eyal Berliner
K&CG lab, Ariel University, Israel
Department of Management, Bar-Ilan University, Israel
School of Sustainability, Interdisciplinary Center (IDC) Herzliya, Israel

George Dekoulis
Aeronautical and Space Engineering Department, Aerospace Engineering Institute (AEI), Nicosia, Cyprus

Alexander P. Yefremov
Institute of Gravitation and Cosmology, RUDN University, Moscow, Russia

Innocent Okoloko
Department of Electrical Engineering, Universidad de Ingenieria y Tecnologia, Lima, Peru

Roy Stevenson Soler Chisabas
National Institute for Space Research, INPE, São Paulo, Brazil
Integration and Testing Laboratory, LIT, São Paulo, Brazil
Systems Concurrent Engineering Laboratory, LSIS, São Paulo, Brazil

Geilson Loureiro and Carlos de Oliveira Lino
National Institute for Space Research, INPE, São Paulo, Brazil

Ivana Todić
Faculty of Mechanical Engineering, University of Belgrade, Belgrade, Republic of Serbia

Vladimir Kuzmanović
Faculty of Mathematics, University of Belgrade, Belgrade, Republic of Serbia

Index

www.ingramcontent.com/pod-product-compliance
Lightning Source LLC
LaVergne TN
LVHW081030210926
848485LV00004B/989